Undergraduate Texts in Mathematics

(continued after index)

Tom M. Apostol

Introduction to Analytic Number Theory

With 24 Illustrations

Tom M. Apostol
Department of Mathematics
California Institute of Technology
Pasadena, California 91125
U.S.A.

Mathematics Subject Classification (2000): 11-01, 11AXX

Library of Congress Cataloging-in-Publication Data
Apostol, Tom M.
Introduction to analytic number theory.
(Undergraduate texts in mathematics)
"Evolved from a course (Mathematics 160) offered at the California Institute of Technology during the last 25 years."
Bibliography : p. 329.
Includes index.
1. Numbers, Theory of. 2. Arithmetic functions.
3. Numbers, Prime. I. Title.
QA241.A6 512'73 75-37697

ISBN 978-0-387-90163-3 Printed on acid-free paper.

9

springeronline.com

Preface

This is the first volume of a two-volume textbook[1] which evolved from a course (Mathematics 160) offered at the California Institute of Technology during the last 25 years. It provides an introduction to analytic number theory suitable for undergraduates with some background in advanced calculus, but with no previous knowledge of number theory. Actually, a great deal of the book requires no calculus at all and could profitably be studied by sophisticated high school students.

Number theory is such a vast and rich field that a one-year course cannot do justice to all its parts. The choice of topics included here is intended to provide some variety and some depth. Problems which have fascinated generations of professional and amateur mathematicians are discussed together with some of the techniques for solving them.

One of the goals of this course has been to nurture the intrinsic interest that many young mathematics students seem to have in number theory and to open some doors for them to the current periodical literature. It has been gratifying to note that many of the students who have taken this course during the past 25 years have become professional mathematicians, and some have made notable contributions of their own to number theory. To all of them this book is dedicated.

[1] The second volume is scheduled to appear in the Springer-Verlag Series Graduate Texts in Mathematics under the title **Modular Functions and Dirichlet Series in Number Theory.**

Contents

Chapter 3

Averages of Arithmetical Functions

Chapter 4

Some Elementary Theorems on the Distribution of Prime Numbers

Chapter 5

Congruences

Chapter 6

Finite Abelian Groups and Their Characters

Chapter 7

Dirichlet's Theorem on Primes in Arithmetic Progressions

Chapter 8

Periodic Arithmetical Functions and Gauss Sums

Chapter 9
Quadratic Residues and the Quadratic Reciprocity Law

Chapter 10
Primitive Roots

Chapter 11
Dirichlet Series and Euler Products

Chapter 14
Partitions

Historical Introduction

The theory of numbers is that branch of mathematics which deals with properties of the whole numbers,

$$1, 2, 3, 4, 5, \ldots$$

also called the *counting numbers*, or *positive integers*.

The positive integers are undoubtedly man's first mathematical creation. It is hardly possible to imagine human beings without the ability to count, at least within a limited range. Historical record shows that as early as 3500 BC the ancient Sumerians kept a calendar, so they must have developed some form of arithmetic.

By 2500 BC the Sumerians had developed a number system using 60 as a base. This was passed on to the Babylonians, who became highly skilled calculators. Babylonian clay tablets containing elaborate mathematical tables have been found, dating back to 2000 BC.

When ancient civilizations reached a level which provided leisure time to ponder about things, some people began to speculate about the nature and properties of numbers. This curiosity developed into a sort of number-mysticism or numerology, and even today numbers such as 3, 7, 11, and 13 are considered omens of good or bad luck.

Numbers were used for keeping records and for commercial transactions for over 2000 years before anyone thought of studying numbers themselves in a systematic way. The first scientific approach to the study of integers, that is, the true origin of the theory of numbers, is generally attributed to the Greeks. Around 600 BC Pythagoras and his disciples made rather thorough studies of the integers. They were the first to classify integers in various ways:

Even numbers: 2, 4, 6, 8, 10, 12, 14, 16, ...
Odd numbers: 1, 3, 5, 7, 9, 11, 13, 15, ...
Prime numbers: 2, 3, 5, 7, 11, 13, 17, 19, 23, 29, 31, 37, 41, 43, 47, 53, 59, 61, 67, 71, 73, 79, 83, 89, 97, ...
Composite numbers: 4, 6, 8, 9, 10, 12, 14, 15, 16, 18, 20, ...

A *prime number* is a number greater than 1 whose only divisors are 1 and the number itself. Numbers that are not prime are called *composite*, except that the number 1 is considered neither prime nor composite.

The Pythagoreans also linked numbers with geometry. They introduced the idea of *polygonal numbers*: triangular numbers, square numbers, pentagonal numbers, etc. The reason for this geometrical nomenclature is clear when the numbers are represented by dots arranged in the form of triangles, squares, pentagons, etc., as shown in Figure I.1.

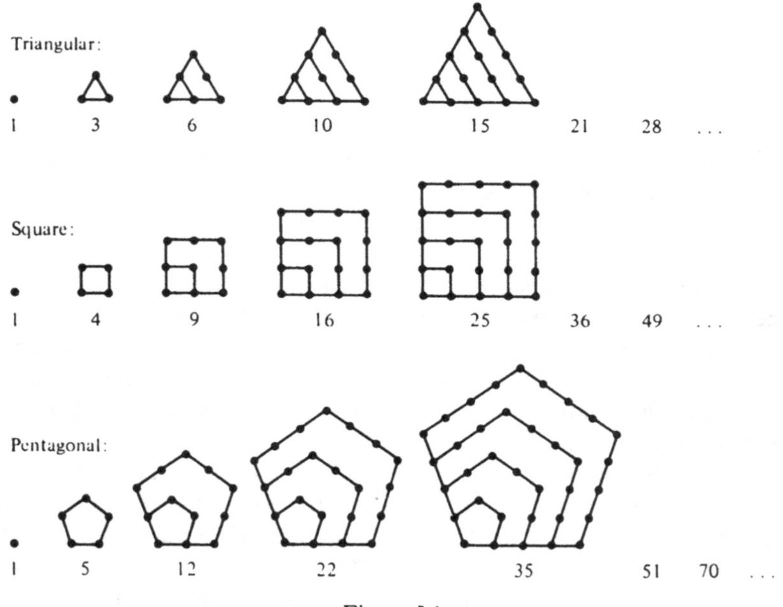

Figure I.1

Another link with geometry came from the famous Theorem of Pythagoras which states that in any right triangle the square of the length of the hypotenuse is the sum of the squares of the lengths of the two legs (see Figure I.2). The Pythagoreans were interested in right triangles whose sides are integers, as in Figure I.3. Such triangles are now called *Pythagorean triangles*. The corresponding triple of numbers (x, y, z) representing the lengths of the sides is called a *Pythagorean triple*.

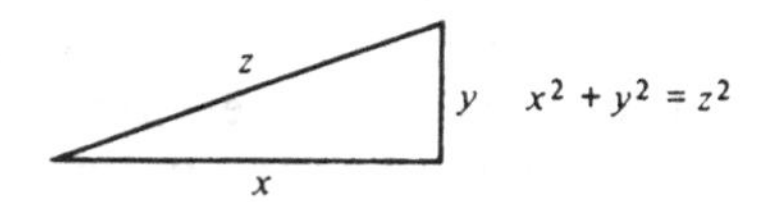

Figure I.2

A Babylonian tablet has been found, dating from about 1700 BC, which contains an extensive list of Pythagorean triples, some of the numbers being quite large. The Pythagoreans were the first to give a method for determining infinitely many triples. In modern notation it can be described as follows: Let n be any odd number greater than 1, and let

$$x = n, \qquad y = \tfrac{1}{2}(n^2 - 1), \qquad z = \tfrac{1}{2}(n^2 + 1).$$

The resulting triple (x, y, z) will always be a Pythagorean triple with $z = y + 1$. Here are some examples:

x	3	5	7	9	11	13	15	17	19
y	4	12	24	40	60	84	112	144	180
z	5	13	25	41	61	85	113	145	181

There are other Pythagorean triples besides these; for example:

x	8	12	16	20
y	15	35	63	99
z	17	37	65	101

In these examples we have $z = y + 2$. Plato (430–349 BC) found a method for determining all these triples; in modern notation they are given by the formulas

$$x = 4n, \qquad y = 4n^2 - 1, \qquad z = 4n^2 + 1.$$

Around 300 BC an important event occurred in the history of mathematics. The appearance of Euclid's *Elements*, a collection of 13 books, transformed mathematics from numerology into a deductive science. Euclid was the first to present mathematical facts along with rigorous proofs of these facts.

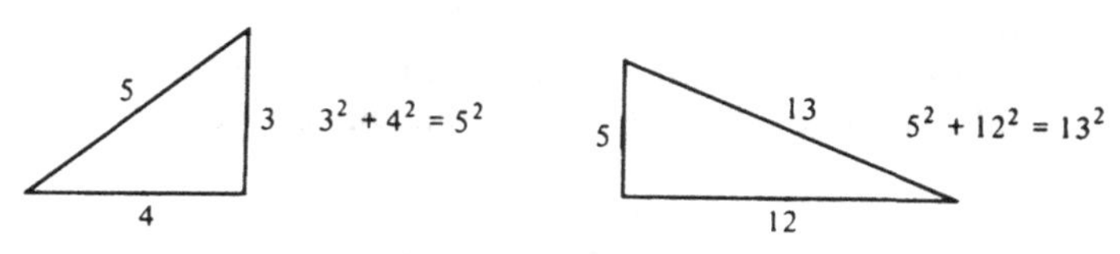

Figure I.3

Three of the thirteen books were devoted to the theory of numbers (Books VII, IX, and X). In Book IX Euclid proved that there are infinitely many primes. His proof is still taught in the classroom today. In Book X he gave a method for obtaining all Pythagorean triples although he gave no proof that his method did, indeed, give them all. The method can be summarized by the formulas

$$x = t(a^2 - b^2), \qquad y = 2tab, \qquad z = t(a^2 + b^2),$$

where t, a, and b, are arbitrary positive integers such that $a > b$, a and b have no prime factors in common, and one of a or b is odd, the other even.

Euclid also made an important contribution to another problem posed by the Pythagoreans—that of finding all perfect numbers. The number 6 was called a perfect number because $6 = 1 + 2 + 3$, the sum of all its proper divisors (that is, the sum of all divisors less than 6). Another example of a perfect number is 28 because $28 = 1 + 2 + 4 + 7 + 14$, and 1, 2, 4, 7, and 14 are the divisors of 28 less than 28. The Greeks referred to the proper divisors of a number as its "parts." They called 6 and 28 perfect numbers because in each case the number is equal to the sum of all its parts.

In Book IX, Euclid found all *even* perfect numbers. He proved that an even number is perfect if it has the form

$$2^{p-1}(2^p - 1),$$

where both p and $2^p - 1$ are primes.

Two thousand years later, Euler proved the converse of Euclid's theorem. That is, every even perfect number must be of Euclid's type. For example, for 6 and 28 we have

$$6 = 2^{2-1}(2^2 - 1) = 2 \cdot 3 \qquad \text{and } 28 = 2^{3-1}(2^3 - 1) = 4 \cdot 7.$$

The first five even perfect numbers are

$$6, 28, 496, 8128 \qquad \text{and } 33{,}550{,}336.$$

Perfect numbers are very rare indeed. At the present time (1983) only 29 perfect numbers are known. They correspond to the following values of p in Euclid's formula:

2, 3, 5, 7, 13, 17, 19, 31, 61, 89, 107, 127, 521, 607, 1279, 2203, 2281, 3217, 4253, 4423, 9689, 9941, 11,213, 19,937, 21,701, 23,209, 44,497, 86,243, 132,049

Numbers of the form $2^p - 1$, where p is prime, are now called *Mersenne numbers* and are denoted by M_p in honor of Mersenne, who studied them in 1644. It is known that M_p is prime for the 29 primes listed above and composite for all values of $p < 44{,}497$. For the following primes,

$$p = 137, 139, 149, 199, 227, 257$$

although M_p is composite, no prime factor of M_p is known.

No *odd* perfect numbers are known; it is not even known if any exist. But if any do exist they must be very large; in fact, greater than 10^{50} (see Hagis [29]).

We turn now to a brief description of the history of the theory of numbers since Euclid's time.

After Euclid in 300 BC no significant advances were made in number theory until about AD 250 when another Greek mathematician, Diophantus of Alexandria, published 13 books, six of which have been preserved. This was the first Greek work to make systematic use of algebraic symbols. Although his algebraic notation seems awkward by present-day standards, Diophantus was able to solve certain algebraic equations involving two or three unknowns. Many of his problems originated from number theory and it was natural for him to seek *integer* solutions of equations. Equations to be solved with integer values of the unknowns are now called *Diophantine equations*, and the study of such equations is known as *Diophantine analysis*. The equation $x^2 + y^2 = z^2$ for Pythagorean triples is an example of a Diophantine equation.

After Diophantus, not much progress was made in the theory of numbers until the seventeenth century, although there is some evidence that the subject began to flourish in the Far East—especially in India—in the period between AD 500 and AD 1200.

In the seventeenth century the subject was revived in Western Europe, largely through the efforts of a remarkable French mathematician, Pierre de Fermat (1601–1665), who is generally acknowledged to be the father of modern number theory. Fermat derived much of his inspiration from the works of Diophantus. He was the first to discover really deep properties of the integers. For example, Fermat proved the following surprising theorems:

Every integer is either a triangular number or a sum of 2 *or* 3 *triangular numbers; every integer is either a square or a sum of* 2, 3, *or* 4 *squares; every integer is either a pentagonal number or the sum of* 2, 3, 4, *or* 5 *pentagonal numbers*, and so on.

Fermat also discovered that every prime number of the form $4n + 1$ such as 5, 13, 17, 29, 37, 41, etc., is a sum of two squares. For example,

$$5 = 1^2 + 2^2, \quad 13 = 2^2 + 3^2, \quad 17 = 1^2 + 4^2, \quad 29 = 2^2 + 5^2,$$
$$37 = 1^2 + 6^2, \quad 41 = 4^2 + 5^2.$$

Shortly after Fermat's time, the names of Euler (1707–1783), Lagrange (1736–1813), Legendre (1752–1833), Gauss (1777–1855), and Dirichlet (1805–1859) became prominent in the further development of the subject. The first textbook in number theory was published by Legendre in 1798. Three years later Gauss published *Disquisitiones Arithmeticae*, a book which transformed the subject into a systematic and beautiful science. Although he made a wealth of contributions to other branches of mathematics, as well as to other sciences, Gauss himself considered his book on number theory to be his greatest work.

In the last hundred years or so since Gauss's time there has been an intensive development of the subject in many different directions. It would be impossible to give in a few pages a fair cross-section of the types of problems that are studied in the theory of numbers. The field is vast and some parts require a profound knowledge of higher mathematics. Nevertheless, there are many problems in number theory which are very easy to state. Some of these deal with prime numbers, and we devote the rest of this introduction to such problems.

The primes less than 100 have been listed above. A table listing all primes less than 10 million was published in 1914 by an American mathematician, D. N. Lehmer [43]. There are exactly 664,579 primes less than 10 million, or about $6\frac{1}{2}\%$. More recently D. H. Lehmer (the son of D. N. Lehmer) calculated the total number of primes less than 10 billion; there are exactly 455,052,511 such primes, or about $4\frac{1}{2}\%$, although not all these primes are known individually (see Lehmer [41]).

A close examination of a table of primes reveals that they are distributed in a very irregular fashion. The tables show long gaps between primes. For example, the prime 370,261 is followed by 111 composite numbers. There are no primes between 20,831,323 and 20,831,533. It is easy to prove that arbitrarily large gaps between prime numbers must eventually occur.

On the other hand, the tables indicate that consecutive primes, such as 3 and 5, or 101 and 103, keep recurring. Such pairs of primes which differ only by 2 are known as *twin primes*. There are over 1000 such pairs below 100,000 and over 8000 below 1,000,000. The largest pair known to date (see Williams and Zarnke [76]) is $76 \cdot 3^{139} - 1$ and $76 \cdot 3^{139} + 1$. Many mathematicians think there are infinitely many such pairs, but no one has been able to prove this as yet.

One of the reasons for this irregularity in distribution of primes is that no simple formula exists for producing all the primes. Some formulas do yield many primes. For example, the expression

$$x^2 - x + 41$$

gives a prime for $x = 0, 1, 2, \ldots, 40$, whereas

$$x^2 - 79x + 1601$$

gives a prime for $x = 0, 1, 2, \ldots, 79$. However, no such simple formula can give a prime for all x, even if cubes and higher powers are used. In fact, in 1752 Goldbach proved that no polynomial in x with integer coefficients can be prime for all x, or even for all sufficiently large x.

Some polynomials represent infinitely many primes. For example, as x runs through the integers $0, 1, 2, 3, \ldots$, the linear polynomial

$$2x + 1$$

gives all the odd numbers hence infinitely many primes. Also, each of the polynomials

$$4x + 1 \qquad \text{and} \qquad 4x + 3$$

represents infinitely many primes. In a famous memoir [15] published in 1837, Dirichlet proved that, if a and b are positive integers with no prime factor in common, the polynomial

$$ax + b$$

gives infinitely many primes as x runs through all the positive integers. This result is now known as Dirichlet's theorem on the existence of primes in a given arithmetical progression.

To prove this theorem, Dirichlet went outside the realm of integers and introduced tools of analysis such as limits and continuity. By so doing he laid the foundations for a new branch of mathematics called *analytic number theory*, in which ideas and methods of real and complex analysis are brought to bear on problems about the integers.

It is not known if there is any quadratic polynomial $ax^2 + bx + c$ with $a \neq 0$ which represents infinitely many primes. However, Dirichlet [16] used his powerful analytic methods to prove that, if a, $2b$, and c have no prime factor in common, the quadratic polynomial in two variables

$$ax^2 + 2bxy + cy^2$$

represents infinitely many primes as x and y run through the positive integers.

Fermat thought that the formula $2^{2^n} + 1$ would always give a prime for $n = 0, 1, 2, \ldots$ These numbers are called *Fermat numbers* and are denoted by F_n. The first five are

$$F_0 = 3, \qquad F_1 = 5, \qquad F_2 = 17, \qquad F_3 = 257 \qquad \text{and } F_4 = 65{,}537,$$

and they are all primes. However, in 1732 Euler found that F_5 is composite; in fact,

$$F_5 = 2^{32} + 1 = (641)(6{,}700{,}417).$$

These numbers are also of interest in plane geometry. Gauss proved that if F_n is a prime, say $F_n = p$, then a regular polygon of p sides can be constructed with straightedge and compass.

Beyond F_5, *no further Fermat primes have been found.* In fact, for $5 \leq n \leq 19$ each Fermat number F_n is composite. Also, F_n is known to be composite for the following further isolated values of n:

$n =$ 21, 23, 25, 26, 27, 29, 30, 32, 36, 38, 39, 42, 52, 55, 58, 62, 63, 66, 71, 73, 77, 81, 91, 93, 99, 117, 125, 144, 147, 150, 201, 207, 215, 226, 228, 250, 255, 267, 268, 284, 287, 298, 316, 329, 416, 452, 544, 556, 692, 744, 1551, 1945, 2023, 2456, 3310, 4724, and 6537.

(See Robinson [59] and Wrathall [77]. More recent work is described in Gostin and McLaughlin, *Math. Comp.* 38 (1982), 645–649.)

It was mentioned earlier that there is no simple formula that gives all the primes. In this connection, we should mention a result discovered recently by Davis, Matijasevič, Putnam and Robinson. They have shown how to construct a polynomial $P(x_1, \ldots, x_k)$, all of whose positive values are primes for nonnegative integer values of $x_1, \ldots, x_k$ and for which the positive values run through all the primes but the negative values are composite. (See Jones, Sata, Wada and Wiens, *Amer. Math. Monthly* 83 (1976), 449–65 for references.)

The foregoing results illustrate the irregularity of the distribution of the prime numbers. However, by examining large blocks of primes one finds that their average distribution seems to be quite regular. Although there is no end to the primes, they become more widely spaced, on the average, as we go further and further in the table. The question of the diminishing frequency of primes was the subject of much speculation in the early nineteenth century. To study this distribution, we consider a function, denoted by $\pi(x)$, which counts the number of primes $\leq x$. Thus,

$$\pi(x) = \text{the number of primes } p \text{ satisfying } 2 \leq p \leq x.$$

Here is a brief table of this function and its comparison with $x/\log x$, where $\log x$ is the natural logarithm of x.

x	$\pi(x)$	$x/\log x$	$\pi(x)\Big/\dfrac{x}{\log x}$
10	4	4.3	0.93
10^2	25	21.7	1.15
10^3	168	144.8	1.16
10^4	1,229	1,086	1.13
10^5	9,592	8,686	1.10
10^6	78,498	72,382	1.08
10^7	664,579	620,420	1.07
10^8	5,761,455	5,428,681	1.06
10^9	50,847,534	48,254,942	1.05
10^{10}	455,052,511	434,294,482	1.048

By examining a table like this for $x \leq 10^6$, Gauss [24] and Legendre [40] proposed independently that for large x the ratio

$$\pi(x)\Big/\frac{x}{\log x}$$

was nearly 1 and they conjectured that this ratio would approach 1 as x approaches ∞. Both Gauss and Legendre attempted to prove this statement but did not succeed. The problem of deciding the truth or falsehood of this conjecture attracted the attention of eminent mathematicians for nearly 100 years.

In 1851 the Russian mathematician Chebyshev [9] made an important step forward by proving that *if* the ratio did tend to a limit, then this limit must be 1. However he was unable to prove that the ratio *does* tend to a limit.

In 1859 Riemann [58] attacked the problem with analytic methods, using a formula discovered by Euler in 1737 which relates the prime numbers to the function

$$\zeta(s) = \sum_{n=1}^{\infty} \frac{1}{n^s}$$

for real $s > 1$. Riemann considered complex values of s and outlined an ingenious method for connecting the distribution of primes to properties of the function $\zeta(s)$. The mathematics needed to justify all the details of his method had not been fully developed and Riemann was unable to completely settle the problem before his death in 1866.

Thirty years later the necessary analytic tools were at hand and in 1896 J. Hadamard [28] and C. J. de la Vallée Poussin [71] independently and almost simultaneously succeeded in proving that

$$\lim_{x \to \infty} \frac{\pi(x)\log x}{x} = 1.$$

This remarkable result is called the *prime number theorem*, and its proof was one of the crowning achievements of analytic number theory.

In 1949, two contemporary mathematicians, Atle Selberg [62] and Paul Erdös [19] caused a sensation in the mathematical world when they discovered an elementary proof of the prime number theorem. Their proof, though very intricate, makes no use of $\zeta(s)$ nor of complex function theory and in principle is accessible to anyone familiar with elementary calculus.

One of the most famous problems concerning prime numbers is the so-called *Goldbach conjecture*. In 1742, Goldbach [26] wrote to Euler suggesting that every even number ≥ 4 is a sum of two primes. For example

$$4 = 2 + 2, \qquad 6 = 3 + 3, \qquad 8 = 3 + 5,$$
$$10 = 3 + 7 = 5 + 5, \qquad 12 = 5 + 7.$$

This conjecture is undecided to this day, although in recent years some progress has been made to indicate that it is probably true. Now why do mathematicians think it is *probably* true if they haven't been able to prove it? First of all, the conjecture has been verified by actual computation for all even numbers less than 33×10^6. It has been found that every even number greater than 6 and less than 33×10^6 is, in fact, not only the sum of two odd primes but the sum of two *distinct* odd primes (see Shen [66]). But in number theory verification of a few thousand cases is not enough evidence to convince mathematicians that something is *probably* true. For example, all the odd primes fall into two categories, those of the form $4n + 1$ and those of the form $4n + 3$. Let $\pi_1(x)$ denote all the primes $\leq x$ that are of the form $4n + 1$, and let $\pi_3(x)$ denote the number that are of the form $4n + 3$. It is known that there are infinitely many primes of both types. By computation it was found that $\pi_1(x) \leq \pi_3(x)$ for all $x < 26{,}861$. But in 1957, J. Leech [39] found that for $x = 26{,}861$ we have $\pi_1(x) = 1473$ and $\pi_3(x) = 1472$, so the inequality was reversed. In 1914, Littlewood [49] proved that this inequality reverses back and forth infinitely often. That is, there are infinitely many x for which $\pi_1(x) < \pi_3(x)$ and also infinitely many x for which $\pi_3(x) < \pi_1(x)$. Conjectures about prime numbers can be erroneous even if they are verified by computation in thousands of cases.

Therefore, the fact that Goldbach's conjecture has been verified for all even numbers less than 33×10^6 is only a tiny bit of evidence in its favor.

Another way that mathematicians collect evidence about the truth of a particular conjecture is by proving other theorems which are somewhat similar to the conjecture. For example, in 1930 the Russian mathematician Schnirelmann [61] proved that there is a number M such that every number n from some point on is a sum of M or fewer primes:

$$n = p_1 + p_2 + \cdots + p_M \qquad \text{(for sufficiently large } n\text{)}.$$

If we knew that M were equal to 2 for all even n, this would prove Goldbach's conjecture for all sufficiently large n. In 1956 the Chinese mathematician Yin Wen-Lin [78] proved that $M \leq 18$. That is, every number n from some point on is a sum of 18 or fewer primes. Schnirelmann's result is considered a giant step toward a proof of Goldbach's conjecture. It was the first real progress made on this problem in nearly 200 years.

A much closer approach to a solution of Goldbach's problem was made in 1937 by another Russian mathematician, I. M. Vinogradov [73], who proved that from some point on every *odd* number is the sum of *three* primes:

$$n = p_1 + p_2 + p_3 \qquad (n \text{ odd}, n \text{ sufficiently large}).$$

In fact, this is true for all odd n greater than $3^{3^{15}}$ (see Borodzkin [5]). To date, this is the strongest piece of evidence in favor of Goldbach's conjecture. Because every large even number is equal to an odd number plus 3, Vinogradov's result implies that $M \leq 4$ in Schnirelmann's theorem. Thus, Vinogradov's theorem supersedes the work of Schnirelmann and Yin Wen-Lin. But Vinogradov's proof uses the most powerful tools of analytic number theory, while the methods of the other authors are essentially elementary.

Another piece of evidence in favor of Goldbach's conjecture was found in 1948 by the Hungarian mathematician Rényi [57] who proved that there is a number M such that every sufficiently large even number n can be written as a prime plus another number which has no more than M distinct prime factors:

$$n = p + A$$

where A has no more than M distinct prime factors (n even, n sufficiently large). If we knew that $M = 1$ then Goldbach's conjecture would be true for all sufficiently large n. In 1965 A. A. Buhštab [6] and A. I. Vinogradov [72] proved that $M \leq 3$, and in 1966 Chen Jing-run [10] proved that $M \leq 2$.

We conclude this introduction with a brief mention of some outstanding unsolved problems concerning prime numbers.

1. (Goldbach's problem). Is there an even number >2 which is not the sum of two primes?
2. Is there an even number >2 which is not the difference of two primes?
3. Are there infinitely many twin primes?
4. Are there infinitely many Mersenne primes, that is, primes of the form $2^p - 1$ where p is prime?
5. Are there infinitely many composite Mersenne numbers?
6. Are there infinitely many Fermat primes, that is, primes of the form $2^{2^n} + 1$?
7. Are there infinitely many composite Fermat numbers?
8. Are there infinitely many primes of the form $x^2 + 1$, where x is an integer? (It is known that there are infinitely many of the form $x^2 + y^2$, and of the form $x^2 + y^2 + 1$, and of the form $x^2 + y^2 + z^2 + 1$).
9. Are there infinitely many primes of the form $x^2 + k$, (k given)?
10. Does there always exist at least one prime between n^2 and $(n + 1)^2$ for every integer $n \geq 1$?
11. Does there always exist at least one prime between n^2 and $n^2 + n$ for every integer $n > 1$?
12. Are there infinitely many primes whose digits (in base 10) are all ones? (Here are two examples: 11 and 11,111,111,111,111,111,111,111.)

The professional mathematician is attracted to number theory because of the way all the weapons of modern mathematics can be brought to bear on its problems. As a matter of fact, many important branches of mathematics had their origin in number theory. For example, the early attempts to prove the prime number theorem stimulated the development of the theory of functions of a complex variable, especially the theory of entire functions. Attempts to prove that the Diophantine equation $x^n + y^n = z^n$ has no nontrivial solution if $n \geq 3$ (Fermat's conjecture) led to the development of algebraic number theory, one of the most active areas of modern mathematical research. The conjecture itself seems unimportant compared to the vast amount of valuable mathematics that was created by those working on it. (A. Wiles announced a proof of Fermat's conjecture in 1994.) Another example is the theory of partitions which has been an important factor in the development of combinatorial analysis and in the study of modular functions.

There are hundreds of unsolved problems in number theory. New problems arise more rapidly than the old ones are solved, and many of the old ones have remained unsolved for centuries. As the mathematician Sierpinski once said, "...the progress of our knowledge of numbers is advanced not only by what we already know about them, but also by realizing what we yet do not know about them."

Note. Every serious student of number theory should become acquainted with Dickson's three-volume *History of the Theory of Numbers* [13], and LeVeque's six-volume *Reviews in Number Theory* [45]. Dickson's *History* gives an encyclopedic account of the entire literature of number theory up until 1918. LeVeque's volumes reproduce all the reviews in Volumes 1–44 of *Mathematical Reviews* (1940–1972) which bear directly on questions commonly regarded as part of number theory. These two valuable collections provide a history of virtually all important discoveries in number theory from antiquity until 1972.

The Fundamental Theorem of Arithmetic 1

1.1 Introduction

This chapter introduces basic concepts of elementary number theory such as divisibility, greatest common divisor, and prime and composite numbers. The principal results are Theorem 1.2, which establishes the existence of the greatest common divisor of any two integers, and Theorem 1.10 (the fundamental theorem of arithmetic), which shows that every integer greater than 1 can be represented as a product of prime factors in only one way (apart from the order of the factors). Many of the proofs make use of the following property of integers.

The principle of induction *If Q is a set of integers such that*

(a) $1 \in Q$,
(b) $n \in Q$ *implies* $n + 1 \in Q$,

then

(c) *all integers* ≥ 1 *belong to* Q.

There are, of course, alternate formulations of this principle. For example, in statement (a), the integer 1 can be replaced by any integer k, provided that the inequality ≥ 1 is replaced by $\geq k$ in (c). Also, (b) can be replaced by the statement $1, 2, 3, \ldots, n \in Q$ implies $(n + 1) \in Q$.

We assume that the reader is familiar with this principle and its use in proving theorems by induction. We also assume familiarity with the following principle, which is logically equivalent to the principle of induction.

The well-ordering principle *If A is a nonempty set of positive integers, then A contains a smallest member.*

Again, this principle has equivalent formulations. For example, "positive integers" can be replaced by "integers $\geq k$ for some k."

1.2 Divisibility

Notation In this chapter, small latin letters a, b, c, d, n, etc., denote integers; they can be positive, negative, or zero.

Definition of divisibility We say d divides n and we write $d \mid n$ whenever $n = cd$ for some c. We also say that n is a multiple of d, that d is a divisor of n, or that d is a factor of n. If d does not divide n we write $d \nmid n$.

Divisibility establishes a relation between any two integers with the following elementary properties whose proofs we leave as exercises for the reader. (Unless otherwise indicated, the letters a, b, d, m, n in Theorem 1.1 represent arbitrary integers.)

Theorem 1.1 *Divisibility has the following properties:*

(a) $n \mid n$ (*reflexive property*)
(b) $d \mid n$ *and* $n \mid m$ *implies* $d \mid m$ (*transitive property*)
(c) $d \mid n$ *and* $d \mid m$ *implies* $d \mid (an + bm)$ (*linearity property*)
(d) $d \mid n$ *implies* $ad \mid an$ (*multiplication property*)
(e) $ad \mid an$ *and* $a \neq 0$ *implies* $d \mid n$ (*cancellation law*)
(f) $1 \mid n$ (*1 divides every integer*)
(g) $n \mid 0$ (*every integer divides zero*)
(h) $0 \mid n$ *implies* $n = 0$ (*zero divides only zero*)
(i) $d \mid n$ *and* $n \neq 0$ *implies* $|d| \leq |n|$ (*comparison property*)
(j) $d \mid n$ *and* $n \mid d$ *implies* $|d| = |n|$
(k) $d \mid n$ *and* $d \neq 0$ *implies* $(n/d) \mid n$.

Note. If $d \mid n$ then n/d is called the divisor conjugate to d.

1.3 Greatest common divisor

If d divides two integers a and b, then d is called a *common divisor* of a and b. Thus, 1 is a common divisor of every pair of integers a and b. We prove now that every pair of integers a and b has a common divisor which can be expressed as a linear combination of a and b.

Theorem 1.2 *Given any two integers a and b, there is a common divisor d of a and b of the form*

$$d = ax + by,$$

where x and y are integers. Moreover, every common divisor of a and b divides this d.

PROOF. First we assume that $a \geq 0$ and $b \geq 0$. We use induction on n, where $n = a + b$. If $n = 0$ then $a = b = 0$ and we can take $d = 0$ with $x = y = 0$. Assume, then, that the theorem has been proved for $0, 1, 2, \ldots, n - 1$. By symmetry, we can assume $a \geq b$. If $b = 0$ take $d = a$, $x = 1$, $y = 0$. If $b \geq 1$ apply the theorem to $a - b$ and b. Since $(a - b) + b = a = n - b \leq n - 1$, the induction assumption is applicable and there is a common divisor d of $a - b$ and b of the form $d = (a - b)x + by$. This d also divides $(a - b) + b = a$ so d is a common divisor of a and b and we have $d = ax + (y - x)b$, a linear combination of a and b. To complete the proof we need to show that every common divisor divides d. But a common divisor divides a and b and hence, by linearity, divides d.

If $a < 0$ or $b < 0$ (or both), we can apply the result just proved to $|a|$ and $|b|$. Then there is a common divisor d of $|a|$ and $|b|$ of the form

$$d = |a|x + |b|y.$$

If $a < 0$, $|a|x = -ax = a(-x)$. Similarly, if $b < 0$, $|b|y = b(-y)$. Hence d is again a linear combination of a and b. □

Theorem 1.3 *Given integers a and b, there is one and only one number d with the following properties:*

(a) $d \geq 0$	*(d is nonnegative)*
(b) $d \mid a$ *and* $d \mid b$	*(d is a common divisor of a and b)*
(c) $e \mid a$ *and* $e \mid b$ *implies* $e \mid d$	*(every common divisor divides d).*

PROOF. By Theorem 1.2 there is at least one d satisfying conditions (b) and (c). Also, $-d$ satisfies these conditions. But if d' satisfies (b) and (c), then $d \mid d'$ and $d' \mid d$, so $|d| = |d'|$. Hence there is exactly one $d \geq 0$ satisfying (b) and (c). □

Note. In Theorem 1.3, $d = 0$ if, and only if, $a = b = 0$. Otherwise $d \geq 1$.

Definition The number d of Theorem 1.3 is called the greatest common divisor (gcd) of a and b and is denoted by (a, b) or by aDb. If $(a, b) = 1$ then a and b are said to be relatively prime.

The notation aDb arises from interpreting the gcd as an operation performed on a and b. However, the most common notation in use is (a, b) and this is the one we shall adopt, although in the next theorem we also use the notation aDb to emphasize the algebraic properties of the operation D.

Theorem 1.4 *The gcd has the following properties:*

(a) $(a, b) = (b, a)$
$aDb = bDa$ (*commutative law*)
(b) $(a, (b, c)) = ((a, b), c)$
$aD(bDc) = (aDb)Dc$ (*associative law*)
(c) $(ac, bc) = |c|(a, b)$
$(ca)D(cb) = |c|(aDb)$ (*distributive law*)
(d) $(a, 1) = (1, a) = 1,\quad (a, 0) = (0, a) = |a|.$
$aD1 = 1Da = 1,\quad aD0 = 0Da = |a|.$

PROOF. We prove only (c). Proofs of the other statements are left as exercises for the reader.

Let $d = (a, b)$ and let $e = (ac, bc)$. We wish to prove that $e = |c|d$. Write $d = ax + by$. Then we have

$$cd = acx + bcy. \tag{1}$$

Therefore $cd|e$ because cd divides both ac and bc. Also, Equation (1) shows that $e|cd$ because $e|ac$ and $e|bc$. Hence $|e| = |cd|$, or $e = |c|d$. □

Theorem 1.5 Euclid's lemma. *If $a|bc$ and if $(a, b) = 1$, then $a|c$.*

PROOF. Since $(a, b) = 1$ we can write $1 = ax + by$. Therefore $c = acx + bcy$. But $a|acx$ and $a|bcy$, so $a|c$. □

1.4 Prime numbers

Definition An integer n is called prime if $n > 1$ and if the only positive divisors of n are 1 and n. If $n > 1$ and if n is not prime, then n is called composite.

EXAMPLES The prime numbers less than 100 are 2, 3, 5, 7, 11, 13, 17, 19, 23, 29, 31, 37, 41, 43, 47, 53, 59, 61, 67, 71, 73, 79, 83, 89, and 97.

Notation Prime numbers are usually denoted by p, p', p_i, q, q', q_i.

Theorem 1.6 *Every integer $n > 1$ is either a prime number or a product of prime numbers.*

PROOF. We use induction on n. The theorem is clearly true for $n = 2$. Assume it is true for every integer $<n$. Then if n is not prime it has a positive divisor $d \neq 1, d \neq n$. Hence $n = cd$, where $c \neq n$. But both c and d are $<n$ and >1 so each of c, d is a product of prime numbers, hence so is n. □

Theorem 1.7 Euclid. *There are infinitely many prime numbers.*

EUCLID'S PROOF. Suppose there are only a finite number, say $p_1, p_2, \ldots, p_n$. Let $N = 1 + p_1 p_2 \cdots p_n$. Now $N > 1$ so either N is prime or N is a product of primes. Of course N is not prime since it exceeds each p_i. Moreover, no p_i divides N (if $p_i | N$ then p_i divides the difference $N - p_1 p_2 \cdots p_n = 1$). This contradicts Theorem 1.6. □

Theorem 1.8 *If a prime p does not divide a, then* $(p, a) = 1$.

PROOF. Let $d = (p, a)$. Then $d|p$ so $d = 1$ or $d = p$. But $d|a$ so $d \neq p$ because $p \nmid a$. Hence $d = 1$. □

Theorem 1.9 *If a prime p divides ab, then $p|a$ or $p|b$. More generally, if a prime p divides a product $a_1 \cdots a_n$, then p divides at least one of the factors.*

PROOF. Assume $p|ab$ and that $p \nmid a$. We shall prove that $p|b$. By Theorem 1.8, $(p, a) = 1$ so, by Euclid's lemma, $p|b$.

To prove the more general statement we use induction on n, the number of factors. Details are left to the reader. □

1.5 The fundamental theorem of arithmetic

Theorem 1.10 Fundamental theorem of arithmetic. *Every integer $n > 1$ can be represented as a product of prime factors in only one way, apart from the order of the factors.*

PROOF. We use induction on n. The theorem is true for $n = 2$. Assume, then, that it is true for all integers greater than 1 and less than n. We shall prove it is also true for n. If n is prime there is nothing more to prove. Assume, then, that n is composite and that n has two factorizations, say

$$n = p_1 p_2 \cdots p_s = q_1 q_2 \cdots q_t. \tag{2}$$

We wish to show that $s = t$ and that each p equals some q. Since p_1 divides the product $q_1 q_2 \cdots q_t$ it must divide at least one factor. Relabel $q_1, q_2, \ldots, q_t$ so that $p_1 | q_1$. Then $p_1 = q_1$ since both p_1 and q_1 are primes. In (2) we may cancel p_1 on both sides to obtain

$$n/p_1 = p_2 \cdots p_s = q_2 \cdots q_t.$$

If $s > 1$ or $t > 1$ then $1 < n/p_1 < n$. The induction hypothesis tells us that the two factorizations of n/p_1 must be identical, apart from the order of the factors. Therefore $s = t$ and the factorizations in (2) are also identical, apart from order. This completes the proof. □

Note. In the factorization of an integer n, a particular prime p may occur more than once. If the *distinct* prime factors of n are $p_1, \ldots, p_r$ and if p_i occurs as a factor a_i times, we can write

$$n = p_1^{a_1} \cdots p_r^{a_r}$$

or, more briefly,

$$n = \prod_{i=1}^{r} p_i^{a_i}.$$

This is called the factorization of n into prime powers. We can also express 1 in this form by taking each exponent a_i to be 0.

Theorem 1.11 *If $n = \prod_{i=1}^{r} p_i^{a_i}$, the set of positive divisors of n is the set of numbers of the form $\prod_{i=1}^{r} p_i^{c_i}$, where $0 \le c_i \le a_i$ for $i = 1, 2, \ldots, r$.*

PROOF. Exercise.

Note. If we label the primes in increasing order, thus:

$$p_1 = 2, \qquad p_2 = 3, \qquad p_3 = 5, \ldots, \qquad p_n = \text{the } n\text{th prime},$$

every positive integer n (including 1) can be expressed in the form

$$n = \prod_{i=1}^{\infty} p_i^{a_i}$$

where now each exponent $a_i \ge 0$. The positive divisors of n are all numbers of the form

$$\prod_{i=1}^{\infty} p_i^{c_i}$$

where $0 \le c_i \le a_i$. The products are, of course, *finite.*

Theorem 1.12 *If two positive integers a and b have the factorizations*

$$a = \prod_{i=1}^{\infty} p_i^{a_i}, \qquad b = \prod_{i=1}^{\infty} p_i^{b_i},$$

then their gcd *has the factorization*

$$(a, b) = \prod_{i=1}^{\infty} p_i^{c_i}$$

where each $c_i =$ min $\{a_i, b_i\}$, the smaller of a_i and b_i.

PROOF. Let $d = \prod_{i=1}^{\infty} p_i^{c_i}$. Since $c_i \le a_i$ and $c_i \le b_i$ we have $d|a$ and $d|b$ so d is a common divisor of a and b. Let e be any common divisor of a and b, and write $e = \prod_{i=1}^{\infty} p_i^{e_i}$. Then $e_i \le a_i$ and $e_i \le b_i$ so $e_i \le c_i$. Hence $e|d$, so d is the gcd of a and b. □

1.6 The series of reciprocals of the primes

Theorem 1.13 *The infinite series $\sum_{n=1}^{\infty} 1/p_n$ diverges.*

PROOF. The following short proof of this theorem is due to Clarkson [11]. We assume the series converges and obtain a contradiction. If the series converges there is an integer k such that

$$\sum_{m=k+1}^{\infty} \frac{1}{p_m} < \frac{1}{2}.$$

Let $Q = p_1 \cdots p_k$, and consider the numbers $1 + nQ$ for $n = 1, 2, \ldots$ None of these is divisible by any of the primes $p_1, \ldots, p_k$. Therefore, all the prime factors of $1 + nQ$ occur among the primes $p_{k+1}, p_{k+2}, \ldots$ Therefore for each $r \geq 1$ we have

$$\sum_{n=1}^{r} \frac{1}{1 + nQ} \leq \sum_{t=1}^{\infty} \left(\sum_{m=k+1}^{\infty} \frac{1}{p_m} \right)^t,$$

since the sum on the right includes among its terms all the terms on the left. But the right-hand side of this inequality is dominated by the convergent geometric series

$$\sum_{t=1}^{\infty} \left(\frac{1}{2}\right)^t.$$

Therefore the series $\sum_{n=1}^{\infty} 1/(1 + nQ)$ has bounded partial sums and hence converges. But this is a contradiction because the integral test or the limit comparison test shows that this series diverges.

Note. The divergence of the series $\sum 1/p_n$ was first proved in 1737 by Euler [20] who noted that it implies Euclid's theorem on the existence of infinitely many primes.

In a later chapter we shall obtain an asymptotic formula which shows that the partial sums $\sum_{k=1}^{n} 1/p_k$ tend to infinity like $\log(\log n)$.

1.7 The Euclidean algorithm

Theorem 1.12 provides a practical method for computing the gcd (a, b) when the prime-power factorizations of a and b are known. However, considerable calculation may be required to obtain these prime-power factorizations and it is desirable to have an alternative procedure that requires less computation. There is a useful process, known as Euclid's algorithm, which does not require the factorizations of a and b. This process is based on successive divisions and makes use of the following theorem.

Theorem 1.14 The division algorithm. *Given integers a and b with $b > 0$, there exists a unique pair of integers q and r such that*

$$a = bq + r, \quad \textit{with } 0 \leq r < b.$$

Moreover, $r = 0$ if, and only if, $b \mid a$.

Note. We say that q is the *quotient* and r the *remainder* obtained when b is divided into a.

Proof. Let S be the set of nonnegative integers given by

$$S = \{y : y = a - bx, x \text{ is an integer}, y \geq 0\}.$$

This is a nonempty set of nonnegative integers so it has a smallest member, say $a - bq$. Let $r = a - bq$. Then $a = bq + r$ and $r \geq 0$. Now we show that $r < b$. Assume $r \geq b$. Then $0 \leq r - b < r$. But $r - b \in S$ since $r - b = a - b(q + 1)$. Hence $r - b$ is a member of S smaller than its smallest member, r. This contradiction shows that $r < b$. The pair q, r is unique, for if there were another such pair, say q', r', then $bq + r = bq' + r'$ so $b(q - q') = r' - r$. Hence $b \mid (r' - r)$. If $r' - r \neq 0$ this implies $b \leq |r - r'|$, a contradiction. Therefore $r' = r$ and $q' = q$. Finally, it is clear that $r = 0$ if, and only if, $b \mid a$. □

Note. Although Theorem 1.14 is an *existence* theorem, its proof actually gives us a method for computing the quotient q and the remainder r. We subtract from a (or add to a) enough multiples of b until it is clear that we have obtained the smallest nonnegative number of the form $a - bx$.

Theorem 1.15 The Euclidean algorithm. *Given positive integers a and b, where $b \nmid a$. Let $r_0 = a$, $r_1 = b$, and apply the division algorithm repeatedly to obtain a set of remainders $r_2, r_3, \ldots, r_n, r_{n+1}$ defined successively by the relations*

$$\begin{aligned} r_0 &= r_1 q_1 + r_2, & 0 &< r_2 < r_1, \\ r_1 &= r_2 q_2 + r_3, & 0 &< r_3 < r_2, \\ &\vdots \\ r_{n-2} &= r_{n-1} q_{n-1} + r_n, & 0 &< r_n < r_{n-1}, \\ r_{n-1} &= r_n q_n + r_{n+1}, & r_{n+1} &= 0. \end{aligned}$$

Then r_n, the last nonzero remainder in this process, is (a, b), the gcd *of a and b.*

Proof. There is a stage at which $r_{n+1} = 0$ because the r_i are decreasing and nonnegative. The last relation, $r_{n-1} = r_n q_n$ shows that $r_n \mid r_{n-1}$. The next to last shows that $r_n \mid r_{n-2}$. By induction we see that r_n divides each r_i. In particular $r_n \mid r_1 = b$ and $r_n \mid r_0 = a$, so r_n is a common divisor of a and b. Now let d be any common divisor of a and b. The definition of r_2 shows that $d \mid r_2$. The next relation shows that $d \mid r_3$. By induction, d divides each r_i so $d \mid r_n$. Hence r_n is the required gcd. □

1.8 The greatest common divisor of more than two numbers

The greatest common divisor of three integers a, b, c is denoted by (a, b, c) and is defined by the relation

$$(a, b, c) = (a, (b, c)).$$

By Theorem 1.4(b) we have $(a, (b, c)) = ((a, b), c)$ so the gcd depends only on a, b, c and not on the order in which they are written.

Similarly, the gcd of n integers $a_1, \ldots, a_n$ is defined inductively by the relation

$$(a_1, \ldots, a_n) = (a_1, (a_2, \ldots, a_n)).$$

Again, this number is independent of the order in which the a_i appear.

If $d = (a_1, \ldots, a_n)$ it is easy to verify that d divides each of the a_i and that every common divisor divides d. Moreover, d is a linear combination of the a_i. That is, there exist integers $x_1, \ldots, x_n$ such that

$$(a_1, \ldots, a_n) = a_1x_1 + \cdots + a_nx_n.$$

If $d = 1$ the numbers are said to be *relatively prime*. For example, 2, 3, and 10 are relatively prime.

If $(a_i, a_j) = 1$ whenever $i \neq j$ the numbers $a_1, \ldots, a_n$ are said to be *relatively prime in pairs*. If $a_1, \ldots, a_n$ are relatively prime in pairs then $(a_1, \ldots, a_n) = 1$. However, the example (2, 3, 10) shows that the converse is not necessarily true.

Exercises for Chapter 1

In these exercises lower case latin letters $a, b, c, \ldots, x, y, z$ represent integers.

Prove each of the statements in Exercises 1 through 6.

1. If $(a, b) = 1$ and if $c \mid a$ and $d \mid b$, then $(c, d) = 1$.

2. If $(a, b) = (a, c) = 1$, then $(a, bc) = 1$.

3. If $(a, b) = 1$, then $(a^n, b^k) = 1$ for all $n \geq 1, k \geq 1$.

4. If $(a, b) = 1$, then $(a + b, a - b)$ is either 1 or 2.

5. If $(a, b) = 1$, then $(a + b, a^2 - ab + b^2)$ is either 1 or 3.

6. If $(a, b) = 1$ and if $d \mid (a + b)$, then $(a, d) = (b, d) = 1$.

7. A rational number a/b with $(a, b) = 1$ is called a *reduced fraction*. If the sum of two reduced fractions is an integer, say $(a/b) + (c/d) = n$, prove that $|b| = |d|$.

8. An integer is called *squarefree* if it is not divisible by the square of any prime. Prove that for every $n \geq 1$ there exist uniquely determined $a > 0$ and $b > 0$ such that $n = a^2b$, where b is squarefree.

9. For each of the following statements, either give a proof or exhibit a counter example.
(a) If $b^2 \mid n$ and $a^2 \mid n$ and $a^2 \leq b^2$, then $a \mid b$.
(b) If b^2 is the largest square divisor of n, then $a^2 \mid n$ implies $a \mid b$.

10. Given x and y, let $m = ax + by$, $n = cx + dy$, where $ad - bc = \pm 1$. Prove that $(m, n) = (x, y)$.

11. Prove that $n^4 + 4$ is composite if $n > 1$.

In Exercises 12, 13, and 14, a, b, c, m, n denote *positive* integers.

12. For each of the following statements either give a proof or exhibit a counter example.

(a) If $a^n | b^n$ then $a | b$.
(b) If $n^n | m^m$ then $n | m$.
(c) If $a^n | 2b^n$ and $n > 1$, then $a | b$.

13. If $(a, b) = 1$ and $(a/b)^m = n$, prove that $b = 1$.
(b) If n is not the mth power of a positive integer, prove that $n^{1/m}$ is irrational.

14. If $(a, b) = 1$ and $ab = c^n$, prove that $a = x^n$ and $b = y^n$ for some x and y. [*Hint*: Consider $d = (a, c)$.]

15. Prove that every $n \geq 12$ is the sum of two composite numbers.

16. Prove that if $2^n - 1$ is prime, then n is prime.

17. Prove that if $2^n + 1$ is prime, then n is a power of 2.

18. If $m \neq n$ compute the gcd $(a^{2^m} + 1, a^{2^n} + 1)$ in terms of a. [*Hint*: Let $A_n = a^{2^n} + 1$ and show that $A_n | (A_m - 2)$ if $m > n$.]

19. The *Fibonacci sequence* 1, 1, 2, 3, 5, 8, 13, 21, 34, ... is defined by the recursion formula $a_{n+1} = a_n + a_{n-1}$, with $a_1 = a_2 = 1$. Prove that $(a_n, a_{n+1}) = 1$ for each n.

20. Let $d = (826, 1890)$. Use the Euclidean algorithm to compute d, then express d as a linear combination of 826 and 1890.

21. The least common multiple (lcm) of two integers a and b is denoted by $[a, b]$ or by aMb, and is defined as follows:

$$[a, b] = |ab|/(a, b) \quad \text{if } a \neq 0 \text{ and } b \neq 0,$$
$$[a, b] = 0 \quad \text{if } a = 0 \text{ or } b = 0.$$

Prove that the lcm has the following properties:

(a) If $a = \prod_{i=1}^{\infty} p_i^{a_i}$ and $b = \prod_{i=1}^{\infty} p_i^{b_i}$ then $[a, b] = \prod_{i=1}^{\infty} p_i^{c_i}$, where $c_i = \max\{a_i, b_i\}$.
(b) $(aDb)Mc = (aMc)D(bMc)$.
(c) $(aMb)Dc = (aDc)M(bDc)$.

(D and M are distributive with respect to each other)

22. Prove that $(a, b) = (a + b, [a, b])$.

23. The sum of two positive integers is 5264 and their least common multiple is 200,340. Determine the two integers.

24. Prove the following multiplicative property of the gcd:

$$(ah, bk) = (a, b)(h, k)\left(\frac{a}{(a, b)}, \frac{k}{(h, k)}\right)\left(\frac{b}{(a, b)}, \frac{h}{(h, k)}\right).$$

In particular this shows that $(ah, bk) = (a, k)(b, h)$ whenever $(a, b) = (h, k) = 1$.

Prove each of the statements in Exercises 25 through 28. All integers are positive.

25. If $(a, b) = 1$ there exist $x > 0$ and $y > 0$ such that $ax - by = 1$.

26. If $(a, b) = 1$ and $x^a = y^b$ then $x = n^b$ and $y = n^a$ for some n. [*Hint:* Use Exercises 25 and 13.]

27. (a) If $(a, b) = 1$ then for every $n > ab$ there exist positive x and y such that $n = ax + by$.
(b) If $(a, b) = 1$ there are no positive x and y such that $ab = ax + by$.

28. If $a > 1$ then $(a^m - 1, a^n - 1) = a^{(m,n)} - 1$.

29. Given $n > 0$, let S be a set whose elements are positive integers $\leq 2n$ such that if a and b are in S and $a \neq b$ then $a \nmid b$. What is the maximum number of integers that S can contain? [*Hint:* S can contain at most one of the integers $1, 2, 2^2, 2^3, \ldots$, at most one of $3, 3 \cdot 2, 3 \cdot 2^2, \ldots$, etc.]

30. If $n > 1$ prove that the sum

$$\sum_{k=1}^{n} \frac{1}{k}$$

is not an integer.

2 Arithmetical Functions and Dirichlet Multiplication

2.1 Introduction

Number theory, like many other branches of mathematics, is often concerned with sequences of real or complex numbers. In number theory such sequences are called *arithmetical functions.*

Definition A real- or complex-valued function defined on the positive integers is called an arithmetical function or a number-theoretic function.

This chapter introduces several arithmetical functions which play an important role in the study of divisibility properties of integers and the distribution of primes. The chapter also discusses Dirichlet multiplication, a concept which helps clarify interrelationships between various arithmetical functions.

We begin with two important examples, the *Möbius function* $\mu(n)$ and the *Euler totient function* $\varphi(n)$.

2.2 The Möbius function $\mu(n)$

Definition The Möbius function μ is defined as follows:

$$\mu(1) = 1;$$

If $n > 1$, write $n = p_1^{a_1} \cdots p_k^{a_k}$. Then

$$\mu(n) = (-1)^k \text{ if } a_1 = a_2 = \cdots = a_k = 1,$$
$$\mu(n) = 0 \text{ otherwise.}$$

Note that $\mu(n) = 0$ if and only if n has a square factor > 1.

Here is a short table of values of $\mu(n)$:

n:	1	2	3	4	5	6	7	8	9	10
$\mu(n)$:	1	-1	-1	0	-1	1	-1	0	0	1

The Möbius function arises in many different places in number theory. One of its fundamental properties is a remarkably simple formula for the divisor sum $\sum_{d|n} \mu(d)$, extended over the positive divisors of n. In this formula, $[x]$ denotes the greatest integer $\le x$.

Theorem 2.1 *If $n \ge 1$ we have*

$$\sum_{d|n} \mu(d) = \left[\frac{1}{n}\right] = \begin{cases} 1 & \text{if } n = 1, \\ 0 & \text{if } n > 1. \end{cases}$$

PROOF. The formula is clearly true if $n = 1$. Assume, then, that $n > 1$ and write $n = p_1^{a_1} \cdots p_k^{a_k}$. In the sum $\sum_{d|n} \mu(d)$ the only nonzero terms come from $d = 1$ and from those divisors of n which are products of distinct primes. Thus

$$\begin{aligned} \sum_{d|n} \mu(d) &= \mu(1) + \mu(p_1) + \cdots + \mu(p_k) + \mu(p_1p_2) + \cdots + \mu(p_{k-1}p_k) \\ &\quad + \cdots + \mu(p_1p_2\cdots p_k) \\ &= 1 + \binom{k}{1}(-1) + \binom{k}{2}(-1)^2 + \cdots + \binom{k}{k}(-1)^k = (1-1)^k = 0. \quad \square \end{aligned}$$

2.3 The Euler totient function $\varphi(n)$

Definition If $n \ge 1$ the Euler totient $\varphi(n)$ is defined to be the number of positive integers not exceeding n which are relatively prime to n; thus,

$$\varphi(n) = \sum_{k=1}^{n}{}' \, 1, \tag{1}$$

where the $'$ indicates that the sum is extended over those k relatively prime to n.

Here is a short table of values of $\varphi(n)$:

n:	1	2	3	4	5	6	7	8	9	10
$\varphi(n)$:	1	1	2	2	4	2	6	4	6	4

As in the case of $\mu(n)$ there is a simple formula for the divisor sum $\sum_{d|n} \varphi(d)$.

Theorem 2.2 *If* $n \geq 1$ *we have*

$$\sum_{d|n} \varphi(d) = n.$$

PROOF. Let S denote the set $\{1, 2, \ldots, n\}$. We distribute the integers of S into disjoint sets as follows. For each divisor d of n, let

$$A(d) = \{k : (k, n) = d, 1 \leq k \leq n\}.$$

That is, $A(d)$ contains those elements of S which have the gcd d with n. The sets $A(d)$ form a disjoint collection whose union is S. Therefore if $f(d)$ denotes the number of integers in $A(d)$ we have

$$\sum_{d|n} f(d) = n. \tag{2}$$

But $(k, n) = d$ if and only if $(k/d, n/d) = 1$, and $0 < k \leq n$ if and only if $0 < k/d \leq n/d$. Therefore, if we let $q = k/d$, there is a one-to-one correspondence between the elements in $A(d)$ and those integers q satisfying $0 < q \leq n/d$, $(q, n/d) = 1$. The number of such q is $\varphi(n/d)$. Hence $f(d) = \varphi(n/d)$ and (2) becomes

$$\sum_{d|n} \varphi(n/d) = n.$$

But this is equivalent to the statement $\sum_{d|n} \varphi(d) = n$ because when d runs through all divisors of n so does n/d. This completes the proof. □

2.4 A relation connecting φ and μ

The Euler totient is related to the Möbius function through the following formula:

Theorem 2.3 *If* $n \geq 1$ *we have*

$$\varphi(n) = \sum_{d|n} \mu(d) \frac{n}{d}.$$

PROOF. The sum (1) defining $\varphi(n)$ can be rewritten in the form

$$\varphi(n) = \sum_{k=1}^{n} \left[\frac{1}{(n, k)} \right],$$

where now k runs through all integers $\leq n$. Now we use Theorem 2.1 with n replaced by (n, k) to obtain

$$\varphi(n) = \sum_{k=1}^{n} \sum_{d|(n,k)} \mu(d) = \sum_{k=1}^{n} \sum_{\substack{d|n \\ d|k}} \mu(d).$$

For a fixed divisor d of n we must sum over all those k in the range $1 \le k \le n$ which are multiples of d. If we write $k = qd$ then $1 \le k \le n$ if and only if $1 \le q \le n/d$. Hence the last sum for $\varphi(n)$ can be written as

$$\varphi(n) = \sum_{d|n} \sum_{q=1}^{n/d} \mu(d) = \sum_{d|n} \mu(d) \sum_{q=1}^{n/d} 1 = \sum_{d|n} \mu(d) \frac{n}{d}.$$

This proves the theorem. □

2.5 A product formula for $\varphi(n)$

The sum for $\varphi(n)$ in Theorem 2.3 can also be expressed as a product extended over the distinct prime divisors of n.

Theorem 2.4 *For $n \ge 1$ we have*

$$\varphi(n) = n \prod_{p|n} \left(1 - \frac{1}{p}\right). \tag{3}$$

PROOF. For $n = 1$ the product is empty since there are no primes which divide 1. In this case it is understood that the product is to be assigned the value 1.

Suppose, then, that $n > 1$ and let $p_1, \ldots, p_r$ be the distinct prime divisors of n. The product can be written as

$$\prod_{p|n} \left(1 - \frac{1}{p}\right) = \prod_{i=1}^{r} \left(1 - \frac{1}{p_i}\right) \tag{4}$$

$$= 1 - \sum \frac{1}{p_i} + \sum \frac{1}{p_i p_j} - \sum \frac{1}{p_i p_j p_k} + \cdots + \frac{(-1)^r}{p_1 p_2 \cdots p_r}.$$

On the right, in a term such as $\sum 1/p_i p_j p_k$ it is understood that we consider all possible products $p_i p_j p_k$ of distinct prime factors of n taken three at a time. Note that each term on the right of (4) is of the form $\pm 1/d$ where d is a divisor of n which is either 1 or a product of distinct primes. The numerator ± 1 is exactly $\mu(d)$. Since $\mu(d) = 0$ if d is divisible by the square of any p_i we see that the sum in (4) is exactly the same as

$$\sum_{d|n} \frac{\mu(d)}{d}.$$

This proves the theorem. □

Many properties of $\varphi(n)$ can be easily deduced from this product formula. Some of these are listed in the next theorem.

Theorem 2.5 *Euler's totient has the following properties:*

(a) $\varphi(p^\alpha) = p^\alpha - p^{\alpha-1}$ *for prime* p *and* $\alpha \geq 1$.
(b) $\varphi(mn) = \varphi(m)\varphi(n)(d/\varphi(d))$, *where* $d = (m, n)$.
(c) $\varphi(mn) = \varphi(m)\varphi(n)$ *if* $(m, n) = 1$.
(d) $a \mid b$ *implies* $\varphi(a) \mid \varphi(b)$.
(e) $\varphi(n)$ *is even for* $n \geq 3$. *Moreover, if* n *has* r *distinct odd prime factors, then* $2^r \mid \varphi(n)$.

PROOF. Part (a) follows at once by taking $n = p^\alpha$ in (3). To prove part (b) we write

$$\frac{\varphi(n)}{n} = \prod_{p|n}\left(1 - \frac{1}{p}\right).$$

Next we note that each prime divisor of mn is either a prime divisor of m or of n, and those primes which divide both m and n also divide (m, n). Hence

$$\frac{\varphi(mn)}{mn} = \prod_{p|mn}\left(1 - \frac{1}{p}\right) = \frac{\displaystyle\prod_{p|m}\left(1 - \frac{1}{p}\right)\prod_{p|n}\left(1 - \frac{1}{p}\right)}{\displaystyle\prod_{p|(m,n)}\left(1 - \frac{1}{p}\right)} = \frac{\dfrac{\varphi(m)}{m}\dfrac{\varphi(n)}{n}}{\dfrac{\varphi(d)}{d}},$$

for which we get (b). Part (c) is a special case of (b).

Next we deduce (d) from (b). Since $a \mid b$ we have $b = ac$ where $1 \leq c \leq b$. If $c = b$ then $a = 1$ and part (d) is trivially satisfied. Therefore, assume $c < b$. From (b) we have

$$\varphi(b) = \varphi(ac) = \varphi(a)\varphi(c)\frac{d}{\varphi(d)} = d\varphi(a)\frac{\varphi(c)}{\varphi(d)}, \tag{5}$$

where $d = (a, c)$. Now the result follows by induction on b. For $b = 1$ it holds trivially. Suppose, then, that (d) holds for all integers $<b$. Then it holds for c so $\varphi(d) \mid \varphi(c)$ since $d \mid c$. Hence the right member of (5) is a multiple of $\varphi(a)$ which means $\varphi(a) \mid \varphi(b)$. This proves (d).

Now we prove (e). If $n = 2^\alpha$, $\alpha \geq 2$, part (a) shows that $\varphi(n)$ is even. If n has at least one odd prime factor we write

$$\varphi(n) = n\prod_{p|n}\frac{p-1}{p} = \frac{n}{\prod_{p|n} p}\prod_{p|n}(p - 1) = c(n)\prod_{p|n}(p - 1),$$

where $c(n)$ is an integer. The product multiplying $c(n)$ is even so $\varphi(n)$ is even. Moreover, each odd prime p contributes a factor 2 to this product, so $2^r \mid \varphi(n)$ if n has r distinct odd prime factors. □

2.6 The Dirichlet product of arithmetical functions

In Theorem 2.3 we proved that

$$\varphi(n) = \sum_{d|n} \mu(d) \frac{n}{d}.$$

The sum on the right is of a type that occurs frequently in number theory. These sums have the form

$$\sum_{d|n} f(d)g\left(\frac{n}{d}\right)$$

where f and g are arithmetical functions, and it is worthwhile to study some properties which these sums have in common. We shall find later that sums of this type arise naturally in the theory of Dirichlet series. It is fruitful to treat these sums as a new kind of multiplication of arithmetical functions, a point of view introduced by E. T. Bell [4] in 1915.

Definition If f and g are two arithmetical functions we define their Dirichlet product (or Dirichlet convolution) to be the arithmetical function h defined by the equation

$$h(n) = \sum_{d|n} f(d)g\left(\frac{n}{d}\right).$$

Notation We write $f * g$ for h and $(f * g)(n)$ for $h(n)$. The symbol N will be used for the arithmetical function for which $N(n) = n$ for all n. In this notation, Theorem 2.3 can be stated in the form

$$\varphi = \mu * N.$$

The next theorem describes algebraic properties of Dirichlet multiplication.

Theorem 2.6 *Dirichlet multiplication is commutative and associative. That is, for any arithmetical functions f, g, k we have*

$$f * g = g * f \qquad \textit{(commutative law)}$$

$$(f * g) * k = f * (g * k) \qquad \textit{(associative law)}.$$

PROOF. First we note that the definition of $f * g$ can also be expressed as follows:

$$(f * g)(n) = \sum_{a \cdot b = n} f(a)g(b),$$

where a and b vary over all positive integers whose product is n. This makes the commutative property self-evident.

To prove the associative property we let $A = g * k$ and consider $f * A = f * (g * k)$. We have

$$(f * A)(n) = \sum_{a \cdot d = n} f(a)A(d) = \sum_{a \cdot d = n} f(a) \sum_{b \cdot c = d} g(b)k(c)$$

$$= \sum_{a \cdot b \cdot c = n} f(a)g(b)k(c).$$

In the same way, if we let $B = f * g$ and consider $B * k$ we are led to the same formula for $(B * k)(n)$. Hence $f * A = B * k$ which means that Dirichlet multiplication is associative. □

We now introduce an identity element for this multiplication.

Definition The arithmetical function I given by

$$I(n) = \left[\frac{1}{n}\right] = \begin{cases} 1 & \text{if } n = 1, \\ 0 & \text{if } n > 1, \end{cases}$$

is called the identity function.

Theorem 2.7 *For all f we have $I * f = f * I = f$.*

PROOF. We have

$$(f * I)(n) = \sum_{d \mid n} f(d) I\left(\frac{n}{d}\right) = \sum_{d \mid n} f(d)\left[\frac{d}{n}\right] = f(n)$$

since $[d/n] = 0$ if $d < n$. □

2.7 Dirichlet inverses and the Möbius inversion formula

Theorem 2.8 *If f is an arithmetical function with $f(1) \neq 0$ there is a unique arithmetical function f^{-1}, called the Dirichlet inverse of f, such that*

$$f * f^{-1} = f^{-1} * f = I.$$

Moreover, f^{-1} is given by the recursion formulas

$$f^{-1}(1) = \frac{1}{f(1)}, \qquad f^{-1}(n) = \frac{-1}{f(1)} \sum_{\substack{d \mid n \\ d < n}} f\left(\frac{n}{d}\right) f^{-1}(d) \quad \text{for } n > 1.$$

PROOF. Given f, we shall show that the equation $(f * f^{-1})(n) = I(n)$ has a unique solution for the function values $f^{-1}(n)$. For $n = 1$ we have to solve the equation

$$(f * f^{-1})(1) = I(1)$$

which reduces to

$$f(1)f^{-1}(1) = 1.$$

Since $f(1) \neq 0$ there is one and only one solution, namely $f^{-1}(1) = 1/f(1)$. Assume now that the function values $f^{-1}(k)$ have been uniquely determined for all $k < n$. Then we have to solve the equation $(f * f^{-1})(n) = I(n)$, or

$$\sum_{d|n} f\left(\frac{n}{d}\right)f^{-1}(d) = 0.$$

This can be written as

$$f(1)f^{-1}(n) + \sum_{\substack{d|n \\ d<n}} f\left(\frac{n}{d}\right)f^{-1}(d) = 0.$$

If the values $f^{-1}(d)$ are known for all divisors $d < n$, there is a uniquely determined value for $f^{-1}(n)$, namely,

$$f^{-1}(n) = \frac{-1}{f(1)} \sum_{\substack{d|n \\ d<n}} f\left(\frac{n}{d}\right)f^{-1}(d),$$

since $f(1) \neq 0$. This establishes the existence and uniqueness of f^{-1} by induction. □

Note. We have $(f * g)(1) = f(1)g(1)$. Hence, if $f(1) \neq 0$ and $g(1) \neq 0$ then $(f * g)(1) \neq 0$. This fact, along with Theorems 2.6, 2.7, and 2.8, tells us that, in the language of group theory, the set of all arithmetical functions f with $f(1) \neq 0$ forms an abelian group with respect to the operation $*$, the identity element being the function I. The reader can easily verify that

$$(f * g)^{-1} = f^{-1} * g^{-1} \quad \text{if } f(1) \neq 0 \text{ and } g(1) \neq 0.$$

Definition We define the unit function u to be the arithmetical function such that $u(n) = 1$ for all n.

Theorem 2.1 states that $\sum_{d|n} \mu(d) = I(n)$. In the notation of Dirichlet multiplication this becomes

$$\mu * u = I.$$

Thus u and μ are Dirichlet inverses of each other:

$$u = \mu^{-1} \qquad \text{and} \qquad \mu = u^{-1}.$$

This simple property of the Möbius function, along with the associative property of Dirichlet multiplication, enables us to give a simple proof of the next theorem.

Theorem 2.9 Möbius inversion formula. *The equation*

$$f(n) = \sum_{d|n} g(d) \tag{6}$$

implies

$$g(n) = \sum_{d|n} f(d)\mu\left(\frac{n}{d}\right). \tag{7}$$

Conversely, (7) *implies* (6).

PROOF. Equation (6) states that $f = g * u$. Multiplication by μ gives $f * \mu = (g * u) * \mu = g * (u * \mu) = g * I = g$, which is (7). Conversely, multiplication of $f * \mu = g$ by u gives (6). □

The Möbius inversion formula has already been illustrated by the pair of formulas in Theorems 2.2 and 2.3:

$$n = \sum_{d|n} \varphi(d), \qquad \varphi(n) = \sum_{d|n} d\mu\left(\frac{n}{d}\right).$$

2.8 The Mangoldt function $\Lambda(n)$

We introduce next Mangoldt's function Λ which plays a central role in the distribution of primes.

Definition For every integer $n \geq 1$ we define

$$\Lambda(n) = \begin{cases} \log p & \text{if } n = p^m \text{ for some prime } p \text{ and some } m \geq 1, \\ 0 & \text{otherwise.} \end{cases}$$

Here is a short table of values of $\Lambda(n)$:

n:	1	2	3	4	5	6	7	8	9	10
$\Lambda(n)$:	0	log 2	log 3	log 2	log 5	0	log 7	log 2	log 3	0

The proof of the next theorem shows how this function arises naturally from the fundamental theorem of arithmetic.

Theorem 2.10 *If* $n \geq 1$ *we have*

$$\log n = \sum_{d|n} \Lambda(d). \tag{8}$$

PROOF. The theorem is true if $n = 1$ since both members are 0. Therefore, assume that $n > 1$ and write

$$n = \prod_{k=1}^{r} p_k^{a_k}.$$

Taking logarithms we have

$$\log n = \sum_{k=1}^{r} a_k \log p_k.$$

Now consider the sum on the right of (8). The only nonzero terms in the sum come from those divisors d of the form $p_k{}^m$ for $m = 1, 2, \ldots, a_k$ and $k = 1, 2, \ldots, r$. Hence

$$\sum_{d|n} \Lambda(d) = \sum_{k=1}^{r} \sum_{m=1}^{a_k} \Lambda(p_k{}^m) = \sum_{k=1}^{r} \sum_{m=1}^{a_k} \log p_k = \sum_{k=1}^{r} a_k \log p_k = \log n,$$

which proves (8). □

Now we use Möbius inversion to express $\Lambda(n)$ in terms of the logarithm.

Theorem 2.11 *If* $n \geq 1$ *we have*

$$\Lambda(n) = \sum_{d|n} \mu(d) \log \frac{n}{d} = -\sum_{d|n} \mu(d) \log d.$$

PROOF. Inverting (8) by the Möbius inversion formula we obtain

$$\Lambda(n) = \sum_{d|n} \mu(d) \log \frac{n}{d} = \log n \sum_{d|n} \mu(d) - \sum_{d|n} \mu(d) \log d$$

$$= I(n) \log n - \sum_{d|n} \mu(d) \log d.$$

Since $I(n)\log n = 0$ for all n the proof is complete. □

2.9 Multiplicative functions

We have already noted that the set of all arithmetical functions f with $f(1) \neq 0$ forms an abelian group under Dirichlet multiplication. In this section we discuss an important subgroup of this group, the so-called *multiplicative* functions.

Definition An arithmetical function f is called multiplicative if f is not identically zero and if

$$f(mn) = f(m)f(n) \quad \text{whenever } (m, n) = 1.$$

A multiplicative function f is called completely multiplicative if we also have

$$f(mn) = f(m)f(n) \quad \text{for all } m, n.$$

EXAMPLE 1 Let $f_\alpha(n) = n^\alpha$, where α is a fixed real or complex number. This function is completely multiplicative. In particular, the unit function $u = f_0$

is completely multiplicative. We denote the function f_α by N^α and call it the power function.

EXAMPLE 2 The identity function $I(n) = [1/n]$ is completely multiplicative.

EXAMPLE 3 The Möbius function is multiplicative but not completely multiplicative. This is easily seen from the definition of $\mu(n)$. Consider two relatively prime integers m and n. If either m or n has a prime-square factor then so does mn, and both $\mu(mn)$ and $\mu(m)\mu(n)$ are zero. If neither has a square factor write $m = p_1 \cdots p_s$ and $n = q_1 \cdots q_t$ where the p_i and q_i are distinct primes. Then $\mu(m) = (-1)^s$, $\mu(n) = (-1)^t$ and $\mu(mn) = (-1)^{s+t} = \mu(m)\mu(n)$. This shows that μ is multiplicative. It is not completely multiplicative since $\mu(4) = 0$ but $\mu(2)\mu(2) = 1$.

EXAMPLE 4 The Euler totient $\varphi(n)$ is multiplicative. This is part (c) of Theorem 2.5. It is not completely multiplicative since $\varphi(4) = 2$ whereas $\varphi(2)\varphi(2) = 1$.

EXAMPLE 5 The ordinary product fg of two arithmetical functions f and g is defined by the usual formula

$$(fg)(n) = f(n)g(n).$$

Similarly, the quotient f/g is defined by the formula

$$\left(\frac{f}{g}\right)(n) = \frac{f(n)}{g(n)} \qquad \text{whenever } g(n) \neq 0.$$

If f and g are multiplicative, so are fg and f/g. If f and g are completely multiplicative, so are fg and f/g.

We now derive some properties common to all multiplicative functions.

Theorem 2.12 *If f is multiplicative then $f(1) = 1$.*

PROOF. We have $f(n) = f(1)f(n)$ since $(n, 1) = 1$ for all n. Since f is not identically zero we have $f(n) \neq 0$ for some n, so $f(1) = 1$. □

Note. Since $\Lambda(1) = 0$, the Mangoldt function is not multiplicative.

Theorem 2.13 *Given f with $f(1) = 1$. Then:*

(a) *f is multiplicative if, and only if,*

$$f(p_1^{a_1} \cdots p_r^{a_r}) = f(p_1^{a_1}) \cdots f(p_r^{a_r})$$

for all primes p_i and all integers $a_i \geq 1$.

(b) *If f is multiplicative, then f is completely multiplicative if, and only if,*

$$f(p^a) = f(p)^a$$

for all primes p and all integers $a \geq 1$.

PROOF. The proof follows easily from the definitions and is left as an exercise for the reader.

2.10 Multiplicative functions and Dirichlet multiplication

Theorem 2.14 *If f and g are multiplicative, so is their Dirichlet product $f * g$.*

PROOF. Let $h = f * g$ and choose relatively prime integers m and n. Then

$$h(mn) = \sum_{c \mid mn} f(c) g\left(\frac{mn}{c}\right).$$

Now every divisor c of mn can be expressed in the form $c = ab$ where $a \mid m$ and $b \mid n$. Moreover, $(a, b) = 1$, $(m/a, n/b) = 1$, and there is a one-to-one correspondence between the set of products ab and the divisors c of mn. Hence

$$h(mn) = \sum_{\substack{a \mid m \\ b \mid n}} f(ab) g\left(\frac{mn}{ab}\right) = \sum_{\substack{a \mid m \\ b \mid n}} f(a) f(b) g\left(\frac{m}{a}\right) g\left(\frac{n}{b}\right)$$

$$= \sum_{a \mid m} f(a) g\left(\frac{m}{a}\right) \sum_{b \mid n} f(b) g\left(\frac{n}{b}\right) = h(m)h(n).$$

This completes the proof. □

Warning. The Dirichlet product of two completely multiplicative functions need not be completely multiplicative.

A slight modification of the foregoing proof enables us to prove:

Theorem 2.15 *If both g and $f * g$ are multiplicative, then f is also multiplicative.*

PROOF. We shall assume that f is not multiplicative and deduce that $f * g$ is also not multiplicative. Let $h = f * g$. Since f is not multiplicative there exist positive integers m and n with $(m, n) = 1$ such that

$$f(mn) \neq f(m)f(n).$$

We choose such a pair m and n for which the product mn is as small as possible.

If $mn = 1$ then $f(1) \neq f(1)f(1)$ so $f(1) \neq 1$. Since $h(1) = f(1)g(1) = f(1) \neq 1$, this shows that h is not multiplicative.

If $mn > 1$, then we have $f(ab) = f(a)f(b)$ for all positive integers a and b with $(a, b) = 1$ and $ab < mn$. Now we argue as in the proof of Theorem 2.14, except that in the sum defining $h(mn)$ we separate the term corresponding to $a = m$, $b = n$. We then have

$$h(mn) = \sum_{\substack{a|m \\ b|n \\ ab < mn}} f(ab)g\left(\frac{mn}{ab}\right) + f(mn)g(1) = \sum_{\substack{a|m \\ b|n \\ ab < mn}} f(a)f(b)g\left(\frac{m}{a}\right)g\left(\frac{n}{b}\right) + f(mn)$$

$$= \sum_{a|m} f(a)g\left(\frac{m}{a}\right) \sum_{b|n} f(b)g\left(\frac{n}{b}\right) - f(m)f(n) + f(mn)$$

$$= h(m)h(n) - f(m)f(n) + f(mn).$$

Since $f(mn) \neq f(m)f(n)$ this shows that $h(mn) \neq h(m)h(n)$ so h is not multiplicative. This contradiction completes the proof. □

Theorem 2.16 *If g is multiplicative, so is g^{-1}, its Dirichlet inverse.*

PROOF. This follows at once from Theorem 2.15 since both g and $g * g^{-1} = I$ are multiplicative. (See Exercise 2.34 for an alternate proof.) □

Note. Theorems 2.14 and 2.16 together show that the set of multiplicative functions is a subgroup of the group of all arithmetical functions f with $f(1) \neq 0$.

2.11 The inverse of a completely multiplicative function

The Dirichlet inverse of a completely multiplicative function is especially easy to determine.

Theorem 2.17 *Let f be multiplicative. Then f is completely multiplicative if, and only if,*

$$f^{-1}(n) = \mu(n)f(n) \quad \text{for all } n \geq 1.$$

PROOF. Let $g(n) = \mu(n)f(n)$. If f is completely multiplicative we have

$$(g * f)(n) = \sum_{d|n} \mu(d)f(d)f\left(\frac{n}{d}\right) = f(n)\sum_{d|n} \mu(d) = f(n)I(n) = I(n)$$

since $f(1) = 1$ and $I(n) = 0$ for $n > 1$. Hence $g = f^{-1}$.

Conversely, assume $f^{-1}(n) = \mu(n)f(n)$. To show that f is completely multiplicative it suffices to prove that $f(p^a) = f(p)^a$ for prime powers. The equation $f^{-1}(n) = \mu(n)f(n)$ implies that

$$\sum_{d|n} \mu(d)f(d)f\left(\frac{n}{d}\right) = 0 \qquad \text{for all } n > 1.$$

Hence, taking $n = p^a$ we have

$$\mu(1)f(1)f(p^a) + \mu(p)f(p)f(p^{a-1}) = 0,$$

from which we find $f(p^a) = f(p)f(p^{a-1})$. This implies $f(p^a) = f(p)^a$, so f is completely multiplicative. □

EXAMPLE The inverse of Euler's φ function. Since $\varphi = \mu * N$ we have $\varphi^{-1} = \mu^{-1} * N^{-1}$. But $N^{-1} = \mu N$ since N is completely multiplicative, so

$$\varphi^{-1} = \mu^{-1} * \mu N = u * \mu N.$$

Thus

$$\varphi^{-1}(n) = \sum_{d|n} d\mu(d).$$

The next theorem shows that

$$\varphi^{-1}(n) = \prod_{p|n} (1 - p).$$

Theorem 2.18 *If f is multiplicative we have*

$$\sum_{d|n} \mu(d)f(d) = \prod_{p|n} (1 - f(p)).$$

PROOF. Let

$$g(n) = \sum_{d|n} \mu(d)f(d).$$

Then g is multiplicative, so to determine $g(n)$ it suffices to compute $g(p^a)$. But

$$g(p^a) = \sum_{d|p^a} \mu(d)f(d) = \mu(1)f(1) + \mu(p)f(p) = 1 - f(p).$$

Hence

$$g(n) = \prod_{p|n} g(p^a) = \prod_{p|n} (1 - f(p)).$$ □

2.12 Liouville's function $\lambda(n)$

An important example of a completely multiplicative function is Liouville's function λ, which is defined as follows.

Definition We define $\lambda(1) = 1$, and if $n = p_1^{a_1} \cdots p_k^{a_k}$ we define

$$\lambda(n) = (-1)^{a_1 + \cdots + a_k}.$$

The definition shows at once that λ is completely multiplicative. The next theorem describes the divisor sum of λ.

Theorem 2.19 *For every $n \geq 1$ we have*

$$\sum_{d|n} \lambda(d) = \begin{cases} 1 & \text{if } n \text{ is a square,} \\ 0 & \text{otherwise.} \end{cases}$$

Also, $\lambda^{-1}(n) = |\mu(n)|$ for all n.

PROOF. Let $g(n) = \sum_{d|n} \lambda(d)$. Then g is multiplicative, so to determine $g(n)$ we need only compute $g(p^a)$ for prime powers. We have

$$g(p^a) = \sum_{d|p^a} \lambda(d) = 1 + \lambda(p) + \lambda(p^2) + \cdots + \lambda(p^a)$$

$$= 1 - 1 + 1 - \cdots + (-1)^a = \begin{cases} 0 & \text{if } a \text{ is odd,} \\ 1 & \text{if } a \text{ is even.} \end{cases}$$

Hence if $n = \prod_{i=1}^k p_i^{a_i}$ we have $g(n) = \prod_{i=1}^k g(p_i^{a_i})$. If any exponent a_i is odd then $g(p_i^{a_i}) = 0$ so $g(n) = 0$. If all the exponents a_i are even then $g(p_i^{a_i}) = 1$ for all i and $g(n) = 1$. This shows that $g(n) = 1$ if n is a square, and $g(n) = 0$ otherwise. Also, $\lambda^{-1}(n) = \mu(n)\lambda(n) = \mu^2(n) = |\mu(n)|$. □

2.13 The divisor functions $\sigma_\alpha(n)$

Definition For real or complex α and any integer $n \geq 1$ we define

$$\sigma_\alpha(n) = \sum_{d|n} d^\alpha,$$

the sum of the αth powers of the divisors of n.

The functions σ_α so defined are called *divisor functions*. They are multiplicative because $\sigma_\alpha = u * N^\alpha$, the Dirichlet product of two multiplicative functions

When $\alpha = 0$, $\sigma_0(n)$ is the *number* of divisors of n; this is often denoted by $d(n)$.

When $\alpha = 1$, $\sigma_1(n)$ is the *sum* of the divisors of n; this is often denoted by $\sigma(n)$.

Since σ_α is multiplicative we have

$$\sigma_\alpha(p_1^{a_1} \cdots p_k^{a_k}) = \sigma_\alpha(p_1^{a_1}) \cdots \sigma_\alpha(p_k^{a_k}).$$

To compute $\sigma_\alpha(p^a)$ we note that the divisors of a prime power p^a are

$$1, p, p^2, \ldots, p^a,$$

hence

$$\sigma_\alpha(p^a) = 1^\alpha + p^\alpha + p^{2\alpha} + \cdots + p^{a\alpha} = \frac{p^{\alpha(a+1)} - 1}{p^\alpha - 1} \quad \text{if } \alpha \neq 0$$

$$= a + 1 \qquad \text{if } \alpha = 0.$$

The Dirichlet inverse of σ_α can also be expressed as a linear combination of the αth powers of the divisors of n.

Theorem 2.20 *For $n \geq 1$ we have*

$$\sigma_\alpha^{-1}(n) = \sum_{d|n} d^\alpha \mu(d)\mu\left(\frac{n}{d}\right).$$

Proof. Since $\sigma_\alpha = N^\alpha * u$ and N^α is completely multiplicative we have

$$\sigma_\alpha^{-1} = (\mu N^\alpha) * u^{-1} = (\mu N^\alpha) * \mu. \qquad \square$$

2.14 Generalized convolutions

Throughout this section F denotes a real or complex-valued function defined on the positive real axis $(0, +\infty)$ such that $F(x) = 0$ for $0 < x < 1$. Sums of the type

$$\sum_{n \leq x} \alpha(n) F\left(\frac{x}{n}\right)$$

arise frequently in number theory. Here α is any arithmetical function. The sum defines a new function G on $(0, +\infty)$ which also vanishes for $0 < x < 1$. We denote this function G by $\alpha \circ F$. Thus,

$$(\alpha \circ F)(x) = \sum_{n \leq x} \alpha(n) F\left(\frac{x}{n}\right).$$

If $F(x) = 0$ for all nonintegral x, the restriction of F to the integers is an arithmetical function and we find that

$$(\alpha \circ F)(m) = (\alpha * F)(m)$$

for all integers $m \geq 1$, so the operation $\circ$ can be regarded as a generalization of the Dirichlet convolution $*$.

The operation $\circ$ is, in general, neither commutative nor associative. However, the following theorem serves as a useful substitute for the associative law.

Theorem 2.21 Associative property relating $\circ$ and $*$. *For any arithmetical functions α and β we have*

$$\alpha \circ (\beta \circ F) = (\alpha * \beta) \circ F. \tag{9}$$

PROOF. For $x > 0$ we have

$$\begin{aligned}\{\alpha \circ (\beta \circ F)\}(x) &= \sum_{n \le x} \alpha(n) \sum_{m \le x/n} \beta(m) F\left(\frac{x}{mn}\right) = \sum_{mn \le x} \alpha(n)\beta(m) F\left(\frac{x}{mn}\right) \\ &= \sum_{k \le x} \left(\sum_{n|k} \alpha(n)\beta\left(\frac{k}{n}\right)\right) F\left(\frac{x}{k}\right) = \sum_{k \le x} (\alpha * \beta)(k) F\left(\frac{x}{k}\right) \\ &= \{(\alpha * \beta) \circ F\}(x).\end{aligned}$$

This completes the proof. □

Next we note that the identity function $I(n) = [1/n]$ for Dirichlet convolution is also a left identity for the operation $\circ$. That is, we have

$$(I \circ F)(x) = \sum_{n \le x} \left[\frac{1}{n}\right] F\left(\frac{x}{n}\right) = F(x).$$

Now we use this fact along with the associative property to prove the following inversion formula.

Theorem 2.22 Generalized inversion formula. *If α has a Dirichlet inverse α^{-1}, then the equation*

$$G(x) = \sum_{n \le x} \alpha(n) F\left(\frac{x}{n}\right) \tag{10}$$

implies

$$F(x) = \sum_{n \le x} \alpha^{-1}(n) G\left(\frac{x}{n}\right). \tag{11}$$

Conversely, (11) *implies* (10).

PROOF. If $G = \alpha \circ F$ then

$$\alpha^{-1} \circ G = \alpha^{-1} \circ (\alpha \circ F) = (\alpha^{-1} * \alpha) \circ F = I \circ F = F.$$

Thus (10) implies (11). The converse is similarly proved. □

The following special case is of particular importance.

Theorem 2.23 Generalized Möbius inversion formula. *If α is completely multiplicative we have*

$$G(x) = \sum_{n \le x} \alpha(n) F\left(\frac{x}{n}\right) \quad \textit{if, and only if,}\ F(x) = \sum_{n \le x} \mu(n)\alpha(n) G\left(\frac{x}{n}\right).$$

PROOF. In this case $\alpha^{-1}(n) = \mu(n)\alpha(n)$. □

2.15 Formal power series

In calculus an infinite series of the form

$$\sum_{n=0}^{\infty} a(n)x^n = a(0) + a(1)x + a(2)x^2 + \cdots + a(n)x^n + \cdots \tag{12}$$

is called a power series in x. Both x and the coefficients $a(n)$ are real or complex numbers. To each power series there corresponds a radius of convergence $r \geq 0$ such that the series converges absolutely if $|x| < r$ and diverges if $|x| > r$. (The radius r can be $+\infty$.)

In this section we consider power series from a different point of view. We call them *formal* power series to distinguish them from the ordinary power series of calculus. In the theory of formal power series x is never assigned a numerical value, and questions of convergence or divergence are not of interest.

The object of interest is the sequence of coefficients

$$(a(0), a(1), \ldots, a(n), \ldots). \tag{13}$$

All that we do with formal power series could also be done by treating the sequence of coefficients as though it were an infinite-dimensional vector with components $a(0)$, $a(1)$, ... But for our purposes it is more convenient to display the terms as coefficients of a power series as in (12) rather than as components of a vector as in (13). The symbol x^n is simply a device for locating the position of the nth coefficient $a(n)$. The coefficient $a(0)$ is called the *constant coefficient* of the series.

We operate on formal power series algebraically as though they were convergent power series. If $A(x)$ and $B(x)$ are two formal power series, say

$$A(x) = \sum_{n=0}^{\infty} a(n)x^n \quad \text{and } B(x) = \sum_{n=0}^{\infty} b(n)x^n,$$

we define:

Equality: $A(x) = B(x)$ means that $a(n) = b(n)$ for all $n \geq 0$.
Sum: $A(x) + B(x) = \sum_{n=0}^{\infty} (a(n) + b(n))x^n$.
Product: $A(x)B(x) = \sum_{n=0}^{\infty} c(n)x^n$, where

$$c(n) = \sum_{k=0}^{n} a(k)b(n-k). \tag{14}$$

The sequence $\{c(n)\}$ determined by (14) is called the *Cauchy product* of the sequences $\{a(n)\}$ and $\{b(n)\}$.

The reader can easily verify that these two operations satisfy the commutative and associative laws, and that multiplication is distributive with respect to addition. In the language of modern algebra, formal power series form a *ring*. This ring has a *zero element* for addition which we denote by 0,

$$0 = \sum_{n=0}^{\infty} a(n)x^n, \quad \text{where } a(n) = 0 \text{ for all } n \geq 0,$$

and an *identity element* for multiplication which we denote by 1,

$$1 = \sum_{n=0}^{\infty} a(n)x^n, \quad \text{where } a(0) = 1 \text{ and } a(n) = 0 \text{ for } n \geq 1.$$

A formal power series is called a *formal polynomial* if all its coefficients are 0 from some point on.

For each formal power series $A(x) = \sum_{n=0}^{\infty} a(n)x^n$ with constant coefficient $a(0) \neq 0$ there is a uniquely determined formal power series $B(x) = \sum_{n=0}^{\infty} b(n)x^n$ such that $A(x)B(x) = 1$. Its coefficients can be determined by solving the infinite system of equations

$$\begin{aligned} &a(0)b(0) = 1 \\ &a(0)b(1) + a(1)b(0) = 0, \\ &a(0)b(2) + a(1)b(1) + a(2)b(0) = 0, \\ &\qquad\vdots \end{aligned}$$

in succession for $b(0), b(1), b(2), \ldots$ The series $B(x)$ is called the *inverse* of $A(x)$ and is denoted by $A(x)^{-1}$ or by $1/A(x)$.

The special series

$$A(x) = 1 + \sum_{n=1}^{\infty} a^n x^n$$

is called a *geometric series*. Here a is an arbitrary real or complex number. Its inverse is the formal polynomial

$$B(x) = 1 - ax.$$

In other words, we have

$$\frac{1}{1 - ax} = 1 + \sum_{n=1}^{\infty} a^n x^n.$$

2.16 The Bell series of an arithmetical function

E. T. Bell used formal power series to study properties of multiplicative arithmetical functions.

Definition Given an arithmetical function f and a prime p, we denote by

$f_p(x)$ the formal power series

$$f_p(x) = \sum_{n=0}^{\infty} f(p^n)x^n$$

and call this the Bell series of f modulo p.

Bell series are especially useful when f is multiplicative.

Theorem 2.24 Uniqueness theorem. *Let f and g be multiplicative functions. Then $f = g$ if, and only if,*

$$f_p(x) = g_p(x) \quad \textit{for all primes } p.$$

PROOF. If $f = g$ then $f(p^n) = g(p^n)$ for all p and all $n \geq 0$, so $f_p(x) = g_p(x)$. Conversely, if $f_p(x) = g_p(x)$ for all p then $f(p^n) = g(p^n)$ for all $n \geq 0$. Since f and g are multiplicative and agree at all prime powers they agree at all the positive integers, so $f = g$. □

It is easy to determine the Bell series for some of the multiplicative functions introduced earlier in this chapter.

EXAMPLE 1 Möbius function μ. Since $\mu(p) = -1$ and $\mu(p^n) = 0$ for $n \geq 2$ we have

$$\mu_p(x) = 1 - x.$$

EXAMPLE 2 Euler's totient φ. Since $\varphi(p^n) = p^n - p^{n-1}$ for $n \geq 1$ we have

$$\varphi_p(x) = 1 + \sum_{n=1}^{\infty} (p^n - p^{n-1})x^n = \sum_{n=0}^{\infty} p^n x^n - x \sum_{n=0}^{\infty} p^n x^n$$

$$= (1 - x) \sum_{n=0}^{\infty} p^n x^n = \frac{1 - x}{1 - px}.$$

EXAMPLE 3 Completely multiplicative functions. If f is completely multiplicative then $f(p^n) = f(p)^n$ for all $n \geq 0$ so the Bell series $f_p(x)$ is a geometric series,

$$f_p(x) = \sum_{n=0}^{\infty} f(p)^n x^n = \frac{1}{1 - f(p)x}.$$

In particular we have the following Bell series for the identity function I, the unit function u, the power function N^α, and Liouville's function λ:

$$I_p(x) = 1.$$

$$u_p(x) = \sum_{n=0}^{\infty} x^n = \frac{1}{1-x}.$$

$$N_p^\alpha(x) = 1 + \sum_{n=1}^{\infty} p^{\alpha n} x^n = \frac{1}{1 - p^\alpha x}.$$

$$\lambda_p(x) = \sum_{n=0}^{\infty} (-1)^n x^n = \frac{1}{1+x}.$$

2.17 Bell series and Dirichlet multiplication

The next theorem relates multiplication of Bell series to Dirichlet multiplication.

Theorem 2.25 *For any two arithmetical functions f and g let $h = f * g$. Then for every prime p we have*

$$h_p(x) = f_p(x) g_p(x).$$

PROOF. Since the divisors of p^n are $1, p, p^2, \ldots, p^n$ we have

$$h(p^n) = \sum_{d \mid p^n} f(d) g\left(\frac{p^n}{d}\right) = \sum_{k=0}^{n} f(p^k) g(p^{n-k}).$$

This completes the proof because the last sum is the Cauchy product of the sequences $\{f(p^n)\}$ and $\{g(p^n)\}$. □

EXAMPLE 1 Since $\mu^2(n) = \lambda^{-1}(n)$ the Bell series of μ^2 modulo p is

$$\mu_p^2(x) = \frac{1}{\lambda_p(x)} = 1 + x.$$

EXAMPLE 2 Since $\sigma_\alpha = N^\alpha * u$ the Bell series of σ_α modulo p is

$$(\sigma_\alpha)_p(x) = N_p^\alpha(x) u_p(x) = \frac{1}{1 - p^\alpha x} \cdot \frac{1}{1 - x} = \frac{1}{1 - \sigma_\alpha(p) x + p^\alpha x^2}.$$

EXAMPLE 3 This example illustrates how Bell series can be used to discover identities involving arithmetical functions. Let

$$f(n) = 2^{\nu(n)},$$

where $\nu(1) = 0$ and $\nu(n) = k$ if $n = p_1^{a_1} \cdots p_k^{a_k}$. Then f is multiplicative and its Bell series modulo p is

$$f_p(x) = 1 + \sum_{n=1}^{\infty} 2^{\nu(p^n)}x^n = 1 + \sum_{n=1}^{\infty} 2x^n = 1 + \frac{2x}{1-x} = \frac{1+x}{1-x}.$$

Hence

$$f_p(x) = \mu_p^2(x)u_p(x)$$

which implies $f = \mu^2 * u$, or

$$2^{\nu(n)} = \sum_{d|n} \mu^2(d).$$

2.18 Derivatives of arithmetical functions

Definition For any arithmetical function f we define its derivative f' to be the arithmetical function given by the equation

$$f'(n) = f(n)\log n \quad \text{for } n \geq 1.$$

EXAMPLES Since $I(n)\log n = 0$ for all n we have $I' = 0$. Since $u(n) = 1$ for all n we have $u'(n) = \log n$. Hence, the formula $\sum_{d|n} \Lambda(d) = \log n$ can be written as

(15) $$\Lambda * u = u'.$$

This concept of derivative shares many of the properties of the ordinary derivative discussed in elementary calculus. For example, the usual rules for differentiating sums and products also hold if the products are Dirichlet products.

Theorem 2.26 *If f and g are arithmetical functions we have:*

(a) $(f + g)' = f' + g'$.
(b) $(f * g)' = f' * g + f * g'$.
(c) $(f^{-1})' = -f' * (f * f)^{-1}$, *provided that* $f(1) \neq 0$.

PROOF. The proof of (a) is immediate. Of course, it is understood that $f + g$ is the function for which $(f + g)(n) = f(n) + g(n)$ for all n.

To prove (b) we use the identity $\log n = \log d + \log(n/d)$ to write

$$\begin{aligned}(f * g)'(n) &= \sum_{d|n} f(d)g\left(\frac{n}{d}\right)\log n \\ &= \sum_{d|n} f(d)\log d\, g\left(\frac{n}{d}\right) + \sum_{d|n} f(d)g\left(\frac{n}{d}\right)\log\left(\frac{n}{d}\right) \\ &= (f' * g)(n) + (f * g')(n).\end{aligned}$$

To prove (c) we apply part (b) to the formula $I' = 0$, remembering that $I = f * f^{-1}$. This gives us

$$0 = (f * f^{-1})' = f' * f^{-1} + f * (f^{-1})'$$

so

$$f * (f^{-1})' = -f' * f^{-1}.$$

Multiplication by f^{-1} now gives us

$$(f^{-1})' = -(f' * f^{-1}) * f^{-1} = -f' * (f^{-1} * f^{-1}).$$

But $f^{-1} * f^{-1} = (f * f)^{-1}$ so (c) is proved. □

2.19 The Selberg identity

Using the concept of derivative we can quickly derive a formula of Selberg which is sometimes used as the starting point of an elementary proof of the prime number theorem.

Theorem 2.27 The Selberg identity. *For $n \geq 1$ we have*

$$\Lambda(n)\log n + \sum_{d|n} \Lambda(d)\Lambda\left(\frac{n}{d}\right) = \sum_{d|n} \mu(d)\log^2 \frac{n}{d}.$$

PROOF. Equation (15) states that $\Lambda * u = u'$. Differentiation of this equation gives us

$$\Lambda' * u + \Lambda * u' = u''$$

or, since $u' = \Lambda * u$,

$$\Lambda' * u + \Lambda * (\Lambda * u) = u''.$$

Now we multiply both sides by $\mu = u^{-1}$ to obtain

$$\Lambda' + \Lambda * \Lambda = u'' * \mu.$$

This is the required identity. □

Exercises for Chapter 2

1. Find all integers n such that

   (a) $\varphi(n) = n/2$, (b) $\varphi(n) = \varphi(2n)$, (c) $\varphi(n) = 12$.

2. For each of the following statements either give a proof or exhibit a counter example.

   (a) If $(m, n) = 1$ then $(\varphi(m), \varphi(n)) = 1$.
   (b) If n is composite, then $(n, \varphi(n)) > 1$.
   (c) If the same primes divide m and n, then $n\varphi(m) = m\varphi(n)$.

3. Prove that

$$\frac{n}{\varphi(n)} = \sum_{d|n} \frac{\mu^2(d)}{\varphi(d)}.$$

4. Prove that $\varphi(n) > n/6$ for all n with at most 8 distinct prime factors.

5. Define $\nu(1) = 0$, and for $n > 1$ let $\nu(n)$ be the number of distinct prime factors of n. Let $f = \mu * \nu$ and prove that $f(n)$ is either 0 or 1.

6. Prove that

$$\sum_{d^2|n} \mu(d) = \mu^2(n)$$

and, more generally,

$$\sum_{d^k|n} \mu(d) = \begin{cases} 0 & \text{if } m^k | n \text{ for some } m > 1, \\ 1 & \text{otherwise.} \end{cases}$$

The last sum is extended over all positive divisors d of n whose kth power also divide n.

7. Let $\mu(p, d)$ denote the value of the Möbius function at the gcd of p and d. Prove that for every prime p we have

$$\sum_{d|n} \mu(d)\mu(p, d) = \begin{cases} 1 & \text{if } n = 1, \\ 2 & \text{if } n = p^a, a \geq 1, \\ 0 & \text{otherwise.} \end{cases}$$

8. Prove that

$$\sum_{d|n} \mu(d)\log^m d = 0$$

if $m \geq 1$ and n has more than m distinct prime factors. [*Hint:* Induction.]

9. If x is real, $x \geq 1$, let $\varphi(x, n)$ denote the number of positive integers $\leq x$ that are relatively prime to n. [Note that $\varphi(n, n) = \varphi(n)$.] Prove that

$$\varphi(x, n) = \sum_{d|n} \mu(d)\left[\frac{x}{d}\right] \quad \text{and} \quad \sum_{d|n} \varphi\left(\frac{x}{d}, \frac{n}{d}\right) = [x].$$

In Exercises 10, 11, and 12, $d(n)$ denotes the number of positive divisors of n.

10. Prove that $\prod_{t|n} t = n^{d(n)/2}$.

11. Prove that $d(n)$ is odd if, and only if, n is a square.

12. Prove that $\sum_{t|n} d(t)^3 = (\sum_{t|n} d(t))^2$.

13. *Product form of the Möbius inversion formula.* If $f(n) > 0$ for all n and if $a(n)$ is real, $a(1) \neq 0$, prove that

$$g(n) = \prod_{d|n} f(d)^{a(n/d)} \text{ if, and only if, } f(n) = \prod_{d|n} g(d)^{b(n/d)},$$

where $b = a^{-1}$, the Dirichlet inverse of a.

14. Let $f(x)$ be defined for all rational x in $0 \le x \le 1$ and let

$$F(n) = \sum_{k=1}^{n} f\left(\frac{k}{n}\right), \qquad F^*(n) = \sum_{\substack{k=1 \\ (k,n)=1}}^{n} f\left(\frac{k}{n}\right).$$

(a) Prove that $F^* = \mu * F$, the Dirichlet product of μ and F.
(b) Use (a) or some other means to prove that $\mu(n)$ is the sum of the primitive nth roots of unity:

$$\mu(n) = \sum_{\substack{k=1 \\ (k,n)=1}}^{n} e^{2\pi ik/n}.$$

15. Let $\varphi_k(n)$ denote the sum of the kth powers of the numbers $\le n$ and relatively prime to n. Note that $\varphi_0(n) = \varphi(n)$. Use Exercise 14 or some other means to prove that

$$\sum_{d|n} \frac{\varphi_k(d)}{d^k} = \frac{1^k + \cdots + n^k}{n^k}.$$

16. Invert the formula in Exercise 15 to obtain, for $n > 1$,

$$\varphi_1(n) = \frac{1}{2}n\varphi(n), \qquad \text{and} \qquad \varphi_2(n) = \frac{1}{3}n^2\varphi(n) + \frac{n}{6}\prod_{p|n}(1 - p).$$

Derive a corresponding formula for $\varphi_3(n)$.

17. Jordan's totient J_k is a generalization of Euler's totient defined by

$$J_k(n) = n^k \prod_{p|n} (1 - p^{-k}).$$

(a) Prove that

$$J_k(n) = \sum_{d|n} \mu(d)\left(\frac{n}{d}\right)^k \qquad \text{and} \qquad n^k = \sum_{d|n} J_k(d).$$

(b) Determine the Bell series for J_k.

18. Prove that every number of the form $2^{a-1}(2^a - 1)$ is perfect if $2^a - 1$ is prime.

19. Prove that if n is *even* and perfect then $n = 2^{a-1}(2^a - 1)$ for some $a \ge 2$. It is not known if any odd perfect numbers exist. It is known that there are no odd perfect numbers with less than 7 prime factors.

20. Let $P(n)$ be the product of the positive integers which are $\le n$ and relatively prime to n. Prove that

$$P(n) = n^{\varphi(n)} \prod_{d|n} \left(\frac{d!}{d^d}\right)^{\mu(n/d)}.$$

21. Let $f(n) = [\sqrt{n}] - [\sqrt{n-1}]$. Prove that f is multiplicative but not completely multiplicative.

22. Prove that

$$\sigma_1(n) = \sum_{d|n} \varphi(d)\sigma_0\left(\frac{n}{d}\right),$$

and derive a generalization involving $\sigma_\alpha(n)$. (More than one generalization is possible.)

23. Prove the following statement or exhibit a counter example. If f is multiplicative, then $F(n) = \prod_{d|n} f(d)$ is multiplicative.

24. Let $A(x)$ and $B(x)$ be formal power series. If the product $A(x)B(x)$ is the zero series, prove that at least one factor is zero. In other words, the ring of formal power series has no zero divisors.

25. Assume f is multiplicative. Prove that:

(a) $f^{-1}(n) = \mu(n)f(n)$ for every squarefree n.
(b) $f^{-1}(p^2) = f(p)^2 - f(p^2)$ for every prime p.

26. Assume f is multiplicative. Prove that f is completely multiplicative if, and only if, $f^{-1}(p^a) = 0$ for all primes p and all integers $a \geq 2$.

27. (a) If f is completely multiplicative, prove that

$$f \cdot (g * h) = (f \cdot g) * (f \cdot h)$$

for all arithmetical functions g and h, where $f \cdot g$ denotes the ordinary product, $(f \cdot g)(n) = f(n)g(n)$.

(b) If f is multiplicative and if the relation in (a) holds for $g = \mu$ and $h = \mu^{-1}$, prove that f is completely multiplicative.

28. (a) If f is completely multiplicative, prove that

$$(f \cdot g)^{-1} = f \cdot g^{-1}$$

for every arithmetical function g with $g(1) \neq 0$.

(b) If f is multiplicative and the relation in (a) holds for $g = \mu^{-1}$, prove that f is completely multiplicative.

29. Prove that there is a multiplicative arithmetical function g such that

$$\sum_{k=1}^{n} f((k, n)) = \sum_{d|n} f(d)g\left(\frac{n}{d}\right)$$

for every arithmetical function f. Here (k, n) is the gcd of n and k. Use this identity to prove that

$$\sum_{k=1}^{n} (k, n)\mu((k, n)) = \mu(n).$$

30. Let f be multiplicative and let g be any arithmetical function. Assume that

(a) $\qquad f(p^{n+1}) = f(p)f(p^n) - g(p)f(p^{n-1})$ for all primes p and all $n \geq 1$.

Prove that for each prime p the Bell series for f has the form

(b) $$f_p(x) = \frac{1}{1 - f(p)x + g(p)x^2}.$$

Conversely, prove that (b) implies (a).

31. (Continuation of Exercise 30.) If g is completely multiplicative prove that statement (a) of Exercise 30 implies

$$f(m)f(n) = \sum_{d|(m,n)} g(d)f\left(\frac{mn}{d^2}\right),$$

where the sum is extended over the positive divisors of the gcd (m, n). [*Hint*: Consider first the case $m = p^a$, $n = p^b$.]

32. Prove that

$$\sigma_\alpha(m)\sigma_\alpha(n) = \sum_{d|(m,n)} d^\alpha \sigma_\alpha\left(\frac{mn}{d^2}\right).$$

33. Prove that Liouville's function is given by the formula

$$\lambda(n) = \sum_{d^2|n} \mu\left(\frac{n}{d^2}\right).$$

34. This exercise describes an alternate proof of Theorem 2.16 which states that the Dirichlet inverse of a multiplicative function is multiplicative. Assume g is multiplicative and let $f = g^{-1}$.

(a) Prove that if p is prime then for $k \geq 1$ we have

$$f(p^k) = -\sum_{t=1}^{k} g(p^t)f(p^{k-t}).$$

(b) Let h be the uniquely determined multiplicative function which agrees with f at the prime powers. Show that $h * g$ agrees with the identity function I at the prime powers and deduce that $h * g = I$. This shows that $f = h$ so f is multiplicative.

35. If f and g are multiplicative and if a and b are positive integers with $a \geq b$, prove that the function h given by

$$h(n) = \sum_{d^a|n} f\left(\frac{n}{d^a}\right)g\left(\frac{n}{d^b}\right)$$

is also multiplicative. The sum is extended over those divisors d of n for which d^a divides n.

MÖBIUS FUNCTIONS OF ORDER k.

If $k \geq 1$ we define μ_k, the Möbius function of order k, as follows:

$$\mu_k(1) = 1,$$
$$\mu_k(n) = 0 \text{ if } p^{k+1}|n \quad \text{for some prime } p,$$
$$\mu_k(n) = (-1)^r \text{ if } n = p_1^{\,k} \cdots p_r^{\,k} \prod_{i>r} p_i^{\,a_i}, \qquad 0 \leq a_i < k,$$
$$\mu_k(n) = 1 \quad \text{otherwise.}$$

In other words, $\mu_k(n)$ vanishes if n is divisible by the $(k + 1)$st power of some prime; otherwise, $\mu_k(n)$ is 1 unless the prime factorization of n contains the

kth powers of exactly r distinct primes, in which case $\mu_k(n) = (-1)^r$. Note that $\mu_1 = \mu$, the usual Möbius function.

Prove the properties of the functions μ_k described in the following exercises.

36. If $k \geq 1$ then $\mu_k(n^k) = \mu(n)$.

37. Each function μ_k is multiplicative.

38. If $k \geq 2$ we have

$$\mu_k(n) = \sum_{d^k \mid n} \mu_{k-1}\left(\frac{n}{d^k}\right)\mu_{k-1}\left(\frac{n}{d}\right).$$

39. If $k \geq 1$ we have

$$|\mu_k(n)| = \sum_{d^{k+1} \mid n} \mu(d).$$

40. For each prime p the Bell series for μ_k is given by

$$(\mu_k)_p(x) = \frac{1 - 2x^k + x^{k+1}}{1 - x}.$$

3 Averages of Arithmetical Functions

3.1 Introduction

The last chapter discussed various identities satisfied by arithmetical functions such as $\mu(n)$, $\varphi(n)$, $\Lambda(n)$, and the divisor functions $\sigma_\alpha(n)$. We now inquire about the behavior of these and other arithmetical functions $f(n)$ for large values of n.

For example, consider $d(n)$, the number of divisors of n. This function takes on the value 2 infinitely often (when n is prime) and it also takes on arbitrarily large values when n has a large number of divisors. Thus the values of $d(n)$ fluctuate considerably as n increases.

Many arithmetical functions fluctuate in this manner and it is often difficult to determine their behavior for large n. Sometimes it is more fruitful to study the arithmetic mean

$$\tilde{f}(n) = \frac{1}{n}\sum_{k=1}^{n} f(k).$$

Averages smooth out fluctuations so it is reasonable to expect that the mean values $\tilde{f}(n)$ might behave more regularly than $f(n)$. This is indeed the case for the divisor function $d(n)$. We will prove later that the average $\tilde{d}(n)$ grows like $\log n$ for large n; more precisely,

$$\lim_{n\to\infty} \frac{\tilde{d}(n)}{\log n} = 1. \tag{1}$$

This is described by saying that the average order of $d(n)$ is $\log n$.

To study the average of an arbitrary function f we need a knowledge of its partial sums $\sum_{k=1}^{n} f(k)$. Sometimes it is convenient to replace the upper index n by an arbitrary positive real number x and to consider instead sums of the form

$$\sum_{k \le x} f(k).$$

Here it is understood that the index k varies from 1 to $[x]$, the greatest integer $\le x$. If $0 < x < 1$ the sum is empty and we assign it the value 0. Our goal is to determine the behavior of this sum as a function of x, especially for large x.

For the divisor function we will prove a result obtained by Dirichlet in 1849, which is stronger than (1), namely

$$\sum_{k \le x} d(k) = x \log x + (2C - 1)x + O(\sqrt{x}) \tag{2}$$

for all $x \ge 1$. Here C is Euler's constant, defined by the equation

$$C = \lim_{n \to \infty} \left(1 + \frac{1}{2} + \frac{1}{3} + \cdots + \frac{1}{n} - \log n \right). \tag{3}$$

The symbol $O(\sqrt{x})$ represents an unspecified function of x which grows no faster than some constant times $\sqrt{x}$. This is an example of the "big oh" notation which is defined as follows.

3.2 The big oh notation. Asymptotic equality of functions

Definition If $g(x) > 0$ for all $x \ge a$, we write

$$f(x) = O(g(x)) \qquad \text{(read: "}f(x)\text{ is big oh of }g(x)\text{")}$$

to mean that the quotient $f(x)/g(x)$ is bounded for $x \ge a$; that is, there exists a constant $M > 0$ such that

$$|f(x)| \le Mg(x) \quad \text{for all } x \ge a.$$

An equation of the form

$$f(x) = h(x) + O(g(x))$$

means that $f(x) - h(x) = O(g(x))$. We note that $f(t) = O(g(t))$ for $t \ge a$ implies $\int_a^x f(t)\,dt = O(\int_a^x g(t)\,dt)$ for $x \ge a$.

Definition If

$$\lim_{x \to \infty} \frac{f(x)}{g(x)} = 1$$

we say that $f(x)$ is asymptotic to $g(x)$ as $x \to \infty$, and we write

$$f(x) \sim g(x) \quad \text{as } x \to \infty.$$

For example, Equation (2) implies that

$$\sum_{k \le x} d(k) \sim x \log x \quad \text{as } x \to \infty.$$

In Equation (2) the term $x \log x$ is called the asymptotic value of the sum; the other two terms represent the error made by approximating the sum by its asymptotic value. If we denote this error by $E(x)$, then (2) states that

$$E(x) = (2C - 1)x + O(\sqrt{x}). \tag{4}$$

This could also be written $E(x) = O(x)$, an equation which is correct but which does not convey the more precise information in (4). Equation (4) tells us that the asymptotic value of $E(x)$ is $(2C - 1)x$.

3.3 Euler's summation formula

Sometimes the asymptotic value of a partial sum can be obtained by comparing it with an integral. A summation formula of Euler gives an exact expression for the error made in such an approximation. In this formula $[t]$ denotes the greatest integer $\le t$.

Theorem 3.1 Euler's summation formula. *If f has a continuous derivative f' on the interval $[y, x]$, where $0 < y < x$, then*

$$\sum_{y < n \le x} f(n) = \int_y^x f(t)\,dt + \int_y^x (t - [t])f'(t)\,dt + f(x)([x] - x) - f(y)([y] - y). \tag{5}$$

PROOF. Let $m = [y]$, $k = [x]$. For integers n and $n - 1$ in $[y, x]$ we have

$$\int_{n-1}^{n} [t]f'(t)\,dt = \int_{n-1}^{n} (n - 1)f'(t)\,dt = (n - 1)\{f(n) - f(n - 1)\}$$
$$= \{nf(n) - (n - 1)f(n - 1)\} - f(n).$$

Summing from $n = m + 2$ to $n = k$ we find the first sum telescopes, hence

$$\int_{m+1}^{k} [t]f'(t)\,dt = kf(k) - (m + 1)f(m + 1) - \sum_{n=m+2}^{k} f(n)$$
$$= kf(k) - mf(m + 1) - \sum_{y < n \le x} f(n).$$

Therefore

$$\sum_{y < n \le x} f(n) = -\int_{m+1}^{k} [t]f'(t)\,dt + kf(k) - mf(m + 1) \tag{6}$$
$$= -\int_y^x [t]f'(t)\,dt + kf(x) - mf(y).$$

Integration by parts gives us

$$\int_y^x f(t)\,dt = xf(x) - yf(y) - \int_y^x tf'(t)\,dt,$$

and when this is combined with (6) we obtain (5). □

3.4 Some elementary asymptotic formulas

The next theorem gives a number of asymptotic formulas which are easy consequences of Euler's summation formula. In part (a) the constant C is Euler's constant defined in (3). In part (b), $\zeta(s)$ denotes the Riemann zeta function which is defined by the equation

$$\zeta(s) = \sum_{n=1}^{\infty} \frac{1}{n^s} \quad \text{if } s > 1,$$

and by the equation

$$\zeta(s) = \lim_{x\to\infty} \left(\sum_{n\le x} \frac{1}{n^s} - \frac{x^{1-s}}{1-s} \right) \quad \text{if } 0 < s < 1.$$

Theorem 3.2 *If* $x \ge 1$ *we have*:

(a) $\displaystyle\sum_{n\le x} \frac{1}{n} = \log x + C + O\left(\frac{1}{x}\right).$

(b) $\displaystyle\sum_{n\le x} \frac{1}{n^s} = \frac{x^{1-s}}{1-s} + \zeta(s) + O(x^{-s})$ *if* $s > 0,\ s \ne 1.$

(c) $\displaystyle\sum_{n> x} \frac{1}{n^s} = O(x^{1-s})$ *if* $s > 1.$

(d) $\displaystyle\sum_{n\le x} n^\alpha = \frac{x^{\alpha+1}}{\alpha+1} + O(x^\alpha)$ *if* $\alpha \ge 0.$

Proof. For part (a) we take $f(t) = 1/t$ in Euler's summation formula to obtain

$$\begin{aligned}
\sum_{n\le x} \frac{1}{n} &= \int_1^x \frac{dt}{t} - \int_1^x \frac{t-[t]}{t^2}\,dt + 1 - \frac{x-[x]}{x} \\
&= \log x - \int_1^x \frac{t-[t]}{t^2}\,dt + 1 + O\left(\frac{1}{x}\right) \\
&= \log x + 1 - \int_1^\infty \frac{t-[t]}{t^2}\,dt + \int_x^\infty \frac{t-[t]}{t^2}\,dt + O\left(\frac{1}{x}\right).
\end{aligned}$$

The improper integral $\int_1^\infty (t - [t])t^{-2}\,dt$ exists since it is dominated by $\int_1^\infty t^{-2}\,dt$. Also,

$$0 \le \int_x^\infty \frac{t - [t]}{t^2}\,dt \le \int_x^\infty \frac{1}{t^2}\,dt = \frac{1}{x}$$

so the last equation becomes

$$\sum_{n \le x} \frac{1}{n} = \log x + 1 - \int_1^\infty \frac{t - [t]}{t^2}\,dt + O\left(\frac{1}{x}\right).$$

This proves (a) with

$$C = 1 - \int_1^\infty \frac{t - [t]}{t^2}\,dt.$$

Letting $x \to \infty$ in (a) we find that

$$\lim_{x \to \infty}\left(\sum_{n \le x} \frac{1}{n} - \log x\right) = 1 - \int_1^\infty \frac{t - [t]}{t^2}\,dt,$$

so C is also equal to Euler's constant.

To prove part (b) we use the same type of argument with $f(x) = x^{-s}$, where $s > 0$, $s \ne 1$. Euler's summation formula gives us

$$\begin{aligned}\sum_{n \le x} \frac{1}{n^s} &= \int_1^x \frac{dt}{t^s} - s\int_1^x \frac{t - [t]}{t^{s+1}}\,dt + 1 - \frac{x - [x]}{x^s}\\ &= \frac{x^{1-s}}{1-s} - \frac{1}{1-s} + 1 - s\int_1^\infty \frac{t - [t]}{t^{s+1}}\,dt + O(x^{-s}).\end{aligned}$$

Therefore

$$\sum_{n \le x} \frac{1}{n^s} = \frac{x^{1-s}}{1-s} + C(s) + O(x^{-s}), \tag{7}$$

where

$$C(s) = 1 - \frac{1}{1-s} - s\int_1^\infty \frac{t - [t]}{t^{s+1}}\,dt.$$

If $s > 1$, the left member of (7) approaches $\zeta(s)$ as $x \to \infty$ and the terms x^{1-s} and x^{-s} both approach 0. Hence $C(s) = \zeta(s)$ if $s > 1$. If $0 < s < 1$, $x^{-s} \to 0$ and (7) shows that

$$\lim_{x \to \infty}\left(\sum_{n \le x} \frac{1}{n^s} - \frac{x^{1-s}}{1-s}\right) = C(s).$$

Therefore $C(s)$ is also equal to $\zeta(s)$ if $0 < s < 1$. This proves (b).

To prove (c) we use (b) with $s > 1$ to obtain

$$\sum_{n>x} \frac{1}{n^s} = \zeta(s) - \sum_{n \le x} \frac{1}{n^s} = \frac{x^{1-s}}{s-1} + O(x^{-s}) = O(x^{1-s})$$

since $x^{-s} \le x^{1-s}$.

Finally, to prove (d) we use Euler's summation formula once more with $f(t) = t^\alpha$ to obtain

$$\begin{aligned}\sum_{n \le x} n^\alpha &= \int_1^x t^\alpha \, dt + \alpha \int_1^x t^{\alpha-1}(t - [t]) \, dt + 1 - (x - [x])x^\alpha \\ &= \frac{x^{\alpha+1}}{\alpha+1} - \frac{1}{\alpha+1} + O\left(\alpha \int_1^x t^{\alpha-1} \, dt\right) + O(x^\alpha) \\ &= \frac{x^{\alpha+1}}{\alpha+1} + O(x^\alpha).\end{aligned}$$

□

3.5 The average order of $d(n)$

In this section we derive Dirichlet's asymptotic formula for the partial sums of the divisor function $d(n)$.

Theorem 3.3 *For all $x \ge 1$ we have*

$$\sum_{n \le x} d(n) = x \log x + (2C - 1)x + O(\sqrt{x}), \tag{8}$$

where C is Euler's constant.

PROOF. Since $d(n) = \sum_{d|n} 1$ we have

$$\sum_{n \le x} d(n) = \sum_{n \le x} \sum_{d|n} 1.$$

This is a double sum extended over n and d. Since $d|n$ we can write $n = qd$ and extend the sum over all pairs of positive integers q, d with $qd \le x$. Thus,

$$\sum_{n \le x} d(n) = \sum_{\substack{q,d \\ qd \le x}} 1. \tag{9}$$

This can be interpreted as a sum extended over certain lattice points in the qd-plane, as suggested by Figure 3.1. (A lattice point is a point with integer coordinates.) The lattice points with $qd = n$ lie on a hyperbola, so the sum in (9) counts the number of lattice points which lie on the hyperbolas corresponding to $n = 1, 2, \ldots, [x]$. For each fixed $d \le x$ we can count first those

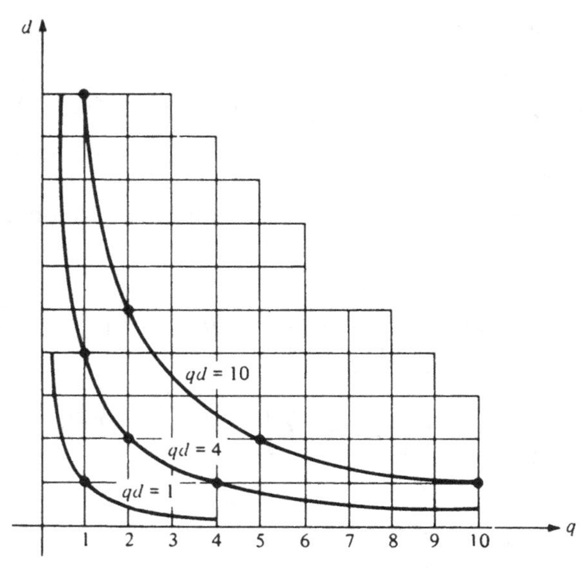

Figure 3.1

lattice points on the horizontal line segment $1 \leq q \leq x/d$, and then sum over all $d \leq x$. Thus (9) becomes

(10) $$\sum_{n \leq x} d(n) = \sum_{d \leq x} \sum_{q \leq x/d} 1.$$

Now we use part (d) of Theorem 3.2 with $\alpha = 0$ to obtain

$$\sum_{q \leq x/d} 1 = \frac{x}{d} + O(1).$$

Using this along with Theorem 3.2(a) we find

$$\sum_{n \leq x} d(n) = \sum_{d \leq x} \left\{\frac{x}{d} + O(1)\right\} = x \sum_{d \leq x} \frac{1}{d} + O(x)$$

$$= x\left\{\log x + C + O\left(\frac{1}{x}\right)\right\} + O(x) = x \log x + O(x).$$

This is a weak version of (8) which implies

$$\sum_{n \leq x} d(n) \sim x \log x \quad \text{as } x \to \infty$$

and gives $\log n$ as the average order of $d(n)$.

To prove the more precise formula (8) we return to the sum (9) which counts the number of lattice points in a hyperbolic region and take advantage of the symmetry of the region about the line $q = d$. The total number of lattice points in the region is equal to twice the number below the line $q = d$

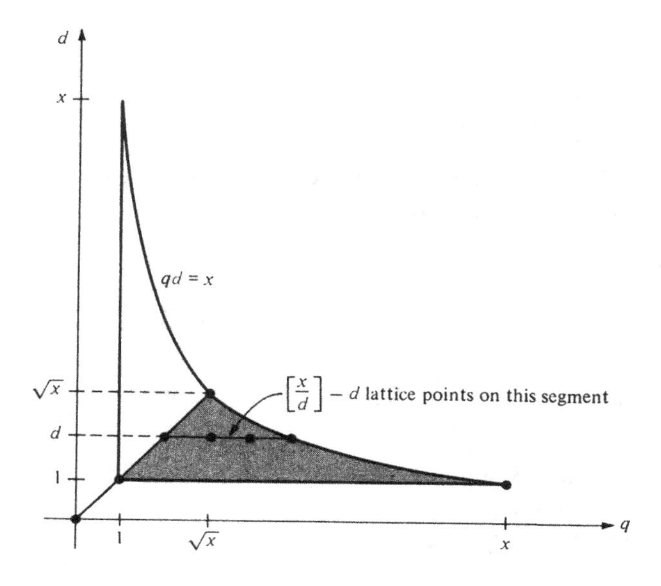

Figure 3.2

plus the number on the bisecting line segment. Referring to Figure 3.2 we see that

$$\sum_{n\le x} d(n) = 2\sum_{d\le\sqrt{x}}\left\{\left[\frac{x}{d}\right] - d\right\} + [\sqrt{x}].$$

Now we use the relation $[y] = y + O(1)$ and parts (a) and (d) of Theorem 3.2 to obtain

$$\begin{aligned}\sum_{n\le x} d(n) &= 2\sum_{d\le\sqrt{x}}\left\{\frac{x}{d} - d + O(1)\right\} + O(\sqrt{x})\\ &= 2x\sum_{d\le\sqrt{x}}\frac{1}{d} - 2\sum_{d\le\sqrt{x}} d + O(\sqrt{x})\\ &= 2x\left\{\log\sqrt{x} + C + O\left(\frac{1}{\sqrt{x}}\right)\right\} - 2\left\{\frac{x}{2} + O(\sqrt{x})\right\} + O(\sqrt{x})\\ &= x\log x + (2C - 1)x + O(\sqrt{x}).\end{aligned}$$

This completes the proof of Dirichlet's formula. □

Note. The error term $O(\sqrt{x})$ can be improved. In 1903 Voronoi proved that the error is $O(x^{1/3}\log x)$; in 1922 van der Corput improved this to $O(x^{33/100})$. The best estimate to date is $O(x^{(12/37)+\varepsilon})$ for every $\varepsilon > 0$, obtained by Kolesnik [35] in 1969. The determination of the infimum of all θ such that the error term is $O(x^\theta)$ is an unsolved problem known as Dirichlet's divisor problem. In 1915 Hardy and Landau showed that $\inf\theta \ge 1/4$.

3.6 The average order of the divisor functions $\sigma_\alpha(n)$

The case $\alpha = 0$ was considered in Theorem 3.3. Next we consider real $\alpha > 0$ and treat the case $\alpha = 1$ separately.

Theorem 3.4 *For all $x \geq 1$ we have*

$$\sum_{n\leq x} \sigma_1(n) = \frac{1}{2}\zeta(2)x^2 + O(x \log x). \tag{11}$$

Note. It can be shown that $\zeta(2) = \pi^2/6$. Therefore (11) shows that the average order of $\sigma_1(n)$ is $\pi^2 n/12$.

PROOF. The method is similar to that used to derive the weak version of Theorem 3.3. We have

$$\sum_{n\leq x} \sigma_1(n) = \sum_{n\leq x}\sum_{q|n} q = \sum_{\substack{q,d\\ qd\leq x}} q = \sum_{d\leq x}\sum_{q\leq x/d} q$$

$$= \sum_{d\leq x}\left\{\frac{1}{2}\left(\frac{x}{d}\right)^2 + O\left(\frac{x}{d}\right)\right\} = \frac{x^2}{2}\sum_{d\leq x}\frac{1}{d^2} + O\left(x\sum_{d\leq x}\frac{1}{d}\right)$$

$$= \frac{x^2}{2}\left\{-\frac{1}{x} + \zeta(2) + O\left(\frac{1}{x^2}\right)\right\} + O(x\log x) = \frac{1}{2}\zeta(2)x^2 + O(x\log x),$$

where we have used parts (a) and (b) of Theorem 3.2. □

Theorem 3.5 *If $x \geq 1$ and $\alpha > 0$, $\alpha \neq 1$, we have*

$$\sum_{n\leq x} \sigma_\alpha(n) = \frac{\zeta(\alpha+1)}{\alpha+1}x^{\alpha+1} + O(x^\beta),$$

where $\beta = \max\{1, \alpha\}$.

PROOF. This time we use parts (b) and (d) of Theorem 3.2 to obtain

$$\sum_{n\leq x} \sigma_\alpha(n) = \sum_{n\leq x}\sum_{q|n} q^\alpha = \sum_{d\leq x}\sum_{q\leq x/d} q^\alpha$$

$$= \sum_{d\leq x}\left\{\frac{1}{\alpha+1}\left(\frac{x}{d}\right)^{\alpha+1} + O\left(\frac{x^\alpha}{d^\alpha}\right)\right\} = \frac{x^{\alpha+1}}{\alpha+1}\sum_{d\leq x}\frac{1}{d^{\alpha+1}} + O\left(x^\alpha\sum_{d\leq x}\frac{1}{d^\alpha}\right)$$

$$= \frac{x^{\alpha+1}}{\alpha+1}\left\{\frac{x^{-\alpha}}{-\alpha} + \zeta(\alpha+1) + O(x^{-\alpha-1})\right\}$$

$$+ O\left(x^\alpha\left\{\frac{x^{1-\alpha}}{1-\alpha} + \zeta(\alpha) + O(x^{-\alpha})\right\}\right)$$

$$= \frac{\zeta(\alpha+1)}{\alpha+1}x^{\alpha+1} + O(x) + O(1) + O(x^\alpha) = \frac{\zeta(\alpha+1)}{\alpha+1}x^{\alpha+1} + O(x^\beta)$$

where $\beta = \max\{1, \alpha\}$. □

To find the average order of $\sigma_\alpha(n)$ for negative α we write $\alpha = -\beta$, where $\beta > 0$.

Theorem 3.6 *If* $\beta > 0$ *let* $\delta = \max\{0, 1 - \beta\}$. *Then if* $x > 1$ *we have*

$$\sum_{n \le x} \sigma_{-\beta}(n) = \zeta(\beta + 1)x + O(x^\delta) \quad \textit{if } \beta \ne 1,$$

$$= \zeta(2)x + O(\log x) \quad \textit{if } \beta = 1.$$

PROOF. We have

$$\sum_{n \le x} \sigma_{-\beta}(n) = \sum_{n \le x} \sum_{d|n} \frac{1}{d^\beta} = \sum_{d \le x} \frac{1}{d^\beta} \sum_{q \le x/d} 1$$

$$= \sum_{d \le x} \frac{1}{d^\beta} \left\{ \frac{x}{d} + O(1) \right\} = x \sum_{d \le x} \frac{1}{d^{\beta+1}} + O\left(\sum_{d \le x} \frac{1}{d^\beta} \right).$$

The last term is $O(\log x)$ if $\beta = 1$ and $O(x^\delta)$ if $\beta \ne 1$. Since

$$x \sum_{d \le x} \frac{1}{d^{\beta+1}} = \frac{x^{1-\beta}}{-\beta} + \zeta(\beta + 1)x + O(x^{-\beta}) = \zeta(\beta + 1)x + O(x^{1-\beta})$$

this completes the proof. □

3.7 The average order of $\varphi(n)$

The asymptotic formula for the partial sums of Euler's totient involves the sum of the series

$$\sum_{n=1}^{\infty} \frac{\mu(n)}{n^2}.$$

This series converges absolutely since it is dominated by $\sum_{n=1}^{\infty} n^{-2}$. In a later chapter we will prove that

$$\sum_{n=1}^{\infty} \frac{\mu(n)}{n^2} = \frac{1}{\zeta(2)} = \frac{6}{\pi^2}. \tag{12}$$

Assuming this result for the time being we have

$$\sum_{n \le x} \frac{\mu(n)}{n^2} = \sum_{n=1}^{\infty} \frac{\mu(n)}{n^2} - \sum_{n > x} \frac{\mu(n)}{n^2}$$

$$= \frac{6}{\pi^2} + O\left(\sum_{n > x} \frac{1}{n^2} \right) = \frac{6}{\pi^2} + O\left(\frac{1}{x} \right)$$

by part (c) of Theorem 3.2. We now use this to obtain the average order of $\varphi(n)$.

Theorem 3.7 *For $x > 1$ we have*

$$\sum_{n \le x} \varphi(n) = \frac{3}{\pi^2} x^2 + O(x \log x), \tag{13}$$

so the average order of $\varphi(n)$ is $3n/\pi^2$.

PROOF. The method is similar to that used for the divisor functions. We start with the relation

$$\varphi(n) = \sum_{d|n} \mu(d) \frac{n}{d}$$

and obtain

$$\sum_{n \le x} \varphi(n) = \sum_{n \le x} \sum_{d|n} \mu(d) \frac{n}{d} = \sum_{\substack{q, d \\ qd \le x}} \mu(d) q = \sum_{d \le x} \mu(d) \sum_{q \le x/d} q$$

$$= \sum_{d \le x} \mu(d) \left\{ \frac{1}{2} \left(\frac{x}{d} \right)^2 + O\left(\frac{x}{d} \right) \right\}$$

$$= \frac{1}{2} x^2 \sum_{d \le x} \frac{\mu(d)}{d^2} + O\left(x \sum_{d \le x} \frac{1}{d} \right)$$

$$= \frac{1}{2} x^2 \left\{ \frac{6}{\pi^2} + O\left(\frac{1}{x} \right) \right\} + O(x \log x) = \frac{3}{\pi^2} x^2 + O(x \log x). \qquad \square$$

3.8 An application to the distribution of lattice points visible from the origin

The asymptotic formula for the partial sums of $\varphi(n)$ has an interesting application to a theorem concerning the distribution of lattice points in the plane which are visible from the origin.

Definition Two lattice points P and Q are said to be mutually visible if the line segment which joins them contains no lattice points other than the endpoints P and Q.

Theorem 3.8 *Two lattice points (a, b) and (m, n) are mutually visible if, and only if, $a - m$ and $b - n$ are relatively prime.*

PROOF. It is clear that (a, b) and (m, n) are mutually visible if and only if $(a - m, b - n)$ is visible from the origin. Hence it suffices to prove the theorem when $(m, n) = (0, 0)$.

Assume (a, b) is visible from the origin, and let $d = (a, b)$. We wish to prove that $d = 1$. If $d > 1$ then $a = da'$, $b = db'$ and the lattice point (a', b') is on the line segment joining $(0, 0)$ to (a, b). This contradiction proves that $d = 1$.

Conversely, assume $(a, b) = 1$. If a lattice point (a', b') is on the line segment joining $(0, 0)$ to (a, b) we have

$$a' = ta, \qquad b' = tb, \quad \text{where } 0 < t < 1.$$

Hence t is rational, so $t = r/s$ where r, s are positive integers with $(r, s) = 1$. Thus

$$sa' = ar \qquad \text{and} \qquad sb' = br,$$

so $s \mid ar$, $s \mid br$. But $(s, r) = 1$ so $s \mid a$, $s \mid b$. Hence $s = 1$ since $(a, b) = 1$. This contradicts the inequality $0 < t < 1$. Therefore the lattice point (a, b) is visible from the origin. □

There are infinitely many lattice points visible from the origin and it is natural to ask how they are distributed in the plane.

Consider a large square region in the xy-plane defined by the inequalities

$$|x| \le r, \qquad |y| \le r.$$

Let $N(r)$ denote the number of lattice points in this square, and let $N'(r)$ denote the number which are visible from the origin. The quotient $N'(r)/N(r)$ measures the fraction of those lattice points in the square which are visible from the origin. The next theorem shows that this fraction tends to a limit as $r \to \infty$. We call this limit the *density* of the lattice points visible from the origin.

Theorem 3.9 *The set of lattice points visible from the origin has density* $6/\pi^2$.

PROOF. We shall prove that

$$\lim_{r \to \infty} \frac{N'(r)}{N(r)} = \frac{6}{\pi^2}.$$

The eight lattice points nearest the origin are all visible from the origin. (See Figure 3.3.) By symmetry, we see that $N'(r)$ is equal to 8, plus 8 times the number of visible points in the region

$$\{(x, y) : 2 \le x \le r, \qquad 1 \le y \le x\},$$

(the shaded region in Figure 3.3). This number is

$$N'(r) = 8 + 8 \sum_{2 \le n \le r} \sum_{\substack{1 \le m < n \\ (m, n) = 1}} 1 = 8 \sum_{1 \le n \le r} \varphi(n).$$

Using Theorem 3.7 we have

$$N'(r) = \frac{24}{\pi^2} r^2 + O(r \log r).$$

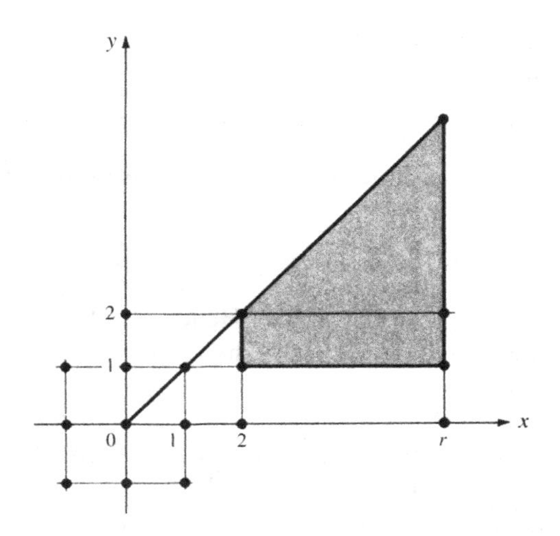

Figure 3.3

But the total number of lattice points in the square is

$$N(r) = (2[r] + 1)^2 = (2r + O(1))^2 = 4r^2 + O(r)$$

so

$$\frac{N'(r)}{N(r)} = \frac{\frac{24}{\pi^2}r^2 + O(r \log r)}{4r^2 + O(r)} = \frac{\frac{6}{\pi^2} + O\left(\frac{\log r}{r}\right)}{1 + O\left(\frac{1}{r}\right)}.$$

Hence as $r \to \infty$ we find $N'(r)/N(r) \to 6/\pi^2$. □

Note. The result of Theorem 3.9 is sometimes described by saying that a lattice point chosen at random has probability $6/\pi^2$ of being visible from the origin. Or, if two integers a and b are chosen at random, the probability that they are relatively prime is $6/\pi^2$.

3.9 The average order of $\mu(n)$ and of $\Lambda(n)$

The average orders of $\mu(n)$ and $\Lambda(n)$ are considerably more difficult to determine than those of $\varphi(n)$ and the divisor functions. It is known that $\mu(n)$ has average order 0 and that $\Lambda(n)$ has average order 1. That is,

$$\lim_{x\to\infty} \frac{1}{x} \sum_{n\leq x} \mu(n) = 0$$

and

$$\lim_{x\to\infty} \frac{1}{x} \sum_{n\leq x} \Lambda(n) = 1,$$

but the proofs are not simple. In the next chapter we will prove that both these results are equivalent to the prime number theorem,

$$\lim_{x\to\infty} \frac{\pi(x)\log x}{x} = 1,$$

where $\pi(x)$ is the number of primes $\le x$.

In this chapter we obtain some elementary identities involving $\mu(n)$ and $\Lambda(n)$ which will be used later in studying the distribution of primes. These will be derived from a general formula relating the partial sums of arbitrary arithmetical functions f and g with those of their Dirichlet product $f * g$.

3.10 The partial sums of a Dirichlet product

Theorem 3.10 *If $h = f * g$, let*

$$H(x) = \sum_{n\le x} h(n), \qquad F(x) = \sum_{n\le x} f(n), \qquad \text{and } G(x) = \sum_{n\le x} g(n).$$

Then we have

$$H(x) = \sum_{n\le x} f(n)G\left(\frac{x}{n}\right) = \sum_{n\le x} g(n)F\left(\frac{x}{n}\right). \tag{14}$$

Proof. We make use of the associative law (Theorem 2.21) which relates the operations $\circ$ and $*$. Let

$$U(x) = \begin{cases} 0 & \text{if } 0 < x < 1, \\ 1 & \text{if } x \ge 1. \end{cases}$$

Then $F = f \circ U$, $G = g \circ U$, and we have

$$f \circ G = f \circ (g \circ U) = (f * g) \circ U = H,$$
$$g \circ F = g \circ (f \circ U) = (g * f) \circ U = H.$$

This completes the proof. □

If $g(n) = 1$ for all n then $G(x) = [x]$, and (14) gives us the following corollary:

Theorem 3.11 *If $F(x) = \sum_{n\le x} f(n)$ we have*

$$\sum_{n\le x}\sum_{d|n} f(d) = \sum_{n\le x} f(n)\left[\frac{x}{n}\right] = \sum_{n\le x} F\left(\frac{x}{n}\right). \tag{15}$$

3.11 Applications to $\mu(n)$ and $\Lambda(n)$

Now we take $f(n) = \mu(n)$ and $\Lambda(n)$ in Theorem 3.11 to obtain the following identities which will be used later in studying the distribution of primes.

Theorem 3.12 *For $x \ge 1$ we have*

$$\sum_{n \le x} \mu(n)\left[\frac{x}{n}\right] = 1 \tag{16}$$

and

$$\sum_{n \le x} \Lambda(n)\left[\frac{x}{n}\right] = \log\,[x]!. \tag{17}$$

PROOF. From (15) we have

$$\sum_{n \le x} \mu(n)\left[\frac{x}{n}\right] = \sum_{n \le x}\sum_{d|n} \mu(d) = \sum_{n \le x}\left[\frac{1}{n}\right] = 1$$

and

$$\sum_{n \le x} \Lambda(n)\left[\frac{x}{n}\right] = \sum_{n \le x}\sum_{d|n} \Lambda(d) = \sum_{n \le x} \log n = \log[x]!. \qquad \square$$

Note. The sums in Theorem 3.12 can be regarded as weighted averages of the functions $\mu(n)$ and $\Lambda(n)$.

In Theorem 4.16 we will prove that the prime number theorem follows from the statement that the series

$$\sum_{n=1}^{\infty} \frac{\mu(n)}{n}$$

converges and has sum 0. Using (16) we can prove that this series has bounded partial sums.

Theorem 3.13 *For all $x \ge 1$ we have*

$$\left|\sum_{n \le x} \frac{\mu(n)}{n}\right| \le 1, \tag{18}$$

with equality holding only if $x < 2$.

PROOF. If $x < 2$ there is only one term in the sum, $\mu(1) = 1$. Now assume that $x \ge 2$. For each real y let $\{y\} = y - [y]$. Then

$$1 = \sum_{n \le x} \mu(n)\left[\frac{x}{n}\right] = \sum_{n \le x} \mu(n)\left(\frac{x}{n} - \left\{\frac{x}{n}\right\}\right) = x\sum_{n \le x} \frac{\mu(n)}{n} - \sum_{n \le x} \mu(n)\left\{\frac{x}{n}\right\}.$$

Since $0 \le \{y\} < 1$ this implies

$$x\left|\sum_{n\le x}\frac{\mu(n)}{n}\right| = \left|1 + \sum_{n\le x}\mu(n)\left\{\frac{x}{n}\right\}\right| \le 1 + \sum_{n\le x}\left\{\frac{x}{n}\right\}$$
$$= 1 + \{x\} + \sum_{2\le n\le x}\left\{\frac{x}{n}\right\} < 1 + \{x\} + [x] - 1 = x.$$

Dividing by x we obtain (18) with strict inequality. □

We turn next to identity (17) of Theorem 3.12,

$$\sum_{n\le x}\Lambda(n)\left[\frac{x}{n}\right] = \log[x]!, \tag{17}$$

and use it to determine the power of a prime which divides a factorial.

Theorem 3.14 Legendre's identity. *For every $x \ge 1$ we have*

$$[x]! = \prod_{p\le x} p^{\alpha(p)} \tag{19}$$

where the product is extended over all primes $\le x$, and

$$\alpha(p) = \sum_{m=1}^{\infty}\left[\frac{x}{p^m}\right]. \tag{20}$$

Note. The sum for $\alpha(p)$ is finite since $[x/p^m] = 0$ for $p > x$.

PROOF. Since $\Lambda(n) = 0$ unless n is a prime power, and $\Lambda(p^m) = \log p$, we have

$$\log[x]! = \sum_{n\le x}\Lambda(n)\left[\frac{x}{n}\right] = \sum_{p\le x}\sum_{m=1}^{\infty}\left[\frac{x}{p^m}\right]\log p = \sum_{p\le x}\alpha(p)\log p,$$

where $\alpha(p)$ is given by (20). The last sum is also the logarithm of the product in (19), so this completes the proof. □

Next we use Euler's summation formula to determine an asymptotic formula for $\log[x]!$.

Theorem 3.15 *If $x \ge 2$ we have*

$$\log[x]! = x\log x - x + O(\log x), \tag{21}$$

and hence

$$\sum_{n\le x}\Lambda(n)\left[\frac{x}{n}\right] = x\log x - x + O(\log x). \tag{22}$$

PROOF. Taking $f(t) = \log t$ in Euler's summation formula (Theorem 3.1) we obtain

$$\sum_{n \le x} \log n = \int_1^x \log t \, dt + \int_1^x \frac{t - [t]}{t} \, dt - (x - [x]) \log x$$

$$= x \log x - x + 1 + \int_1^x \frac{t - [t]}{t} \, dt + O(\log x).$$

This proves (21) since

$$\int_1^x \frac{t - [t]}{t} \, dt = O\left(\int_1^x \frac{1}{t} \, dt \right) = O(\log x),$$

and (22) follows from (17). □

The next theorem is a consequence of (22).

Theorem 3.16 *For $x \ge 2$ we have*

$$\sum_{p \le x} \left[\frac{x}{p} \right] \log p = x \log x + O(x), \tag{23}$$

where the sum is extended over all primes $\le x$.

PROOF. Since $\Lambda(n) = 0$ unless n is a prime power we have

$$\sum_{n \le x} \left[\frac{x}{n} \right] \Lambda(n) = \sum_{p} \sum_{\substack{m=1 \\ p^m \le x}}^{\infty} \left[\frac{x}{p^m} \right] \Lambda(p^m).$$

Now $p^m \le x$ implies $p \le x$. Also, $[x/p^m] = 0$ if $p > x$ so we can write the last sum as

$$\sum_{p \le x} \sum_{m=1}^{\infty} \left[\frac{x}{p^m} \right] \log p = \sum_{p \le x} \left[\frac{x}{p} \right] \log p + \sum_{p \le x} \sum_{m=2}^{\infty} \left[\frac{x}{p^m} \right] \log p.$$

Next we prove that the last sum is $O(x)$. We have

$$\sum_{p \le x} \log p \sum_{m=2}^{\infty} \left[\frac{x}{p^m} \right] \le \sum_{p \le x} \log p \sum_{m=2}^{\infty} \frac{x}{p^m} = x \sum_{p \le x} \log p \sum_{m=2}^{\infty} \left(\frac{1}{p} \right)^m$$

$$= x \sum_{p \le x} \log p \cdot \frac{1}{p^2} \cdot \frac{1}{1 - \dfrac{1}{p}} = x \sum_{p \le x} \frac{\log p}{p(p-1)}$$

$$\le x \sum_{n=2}^{\infty} \frac{\log n}{n(n-1)} = O(x).$$

Hence we have shown that

$$\sum_{n \le x} \left[\frac{x}{n}\right] \Lambda(n) = \sum_{p \le x} \left[\frac{x}{p}\right] \log p + O(x),$$

which, when used with (22), proves (23). □

Equation (23) will be used in the next chapter to derive an asymptotic formula for the partial sums of the divergent series $\sum (1/p)$.

3.12 Another identity for the partial sums of a Dirichlet product

We conclude this chapter with a more general version of Theorem 3.10 that will be used in Chapter 4 to study the partial sums of certain Dirichlet products.

As in Theorem 3.10 we write

$$F(x) = \sum_{n \le x} f(n), \qquad G(x) = \sum_{n \le x} g(n), \qquad \text{and } H(x) = \sum_{n \le x} (f * g)(n)$$

so that

$$H(x) = \sum_{n \le x} \sum_{d \mid n} f(d) g\left(\frac{n}{d}\right) = \sum_{\substack{q, d \\ qd \le x}} f(d) g(q).$$

Theorem 3.17 *If a and b are positive real numbers such that $ab = x$, then*

$$\sum_{\substack{q, d \\ qd \le x}} f(d) g(q) = \sum_{n \le a} f(n) G\left(\frac{x}{n}\right) + \sum_{n \le b} g(n) F\left(\frac{x}{n}\right) - F(a) G(b). \tag{24}$$

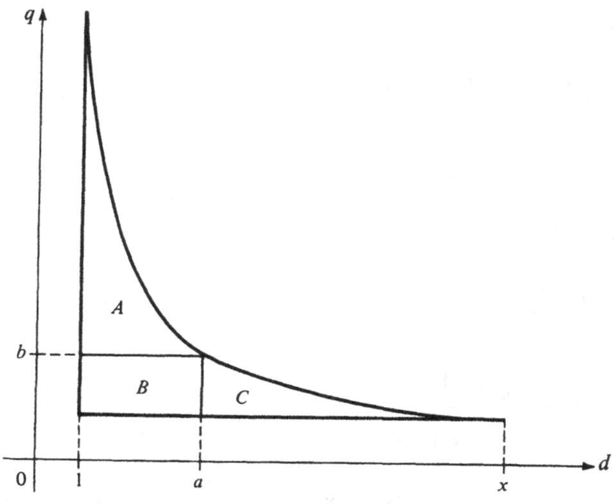

Figure 3.4

PROOF. The sum $H(x)$ on the left of (24) is extended over the lattice points in the hyperbolic region shown in Figure 3.4. We split the sum into two parts, one over the lattice points in $A \cup B$ and the other over those in $B \cup C$. The lattice points in B are covered twice, so we have

$$H(x) = \sum_{d \le a} \sum_{q \le x/d} f(d)g(q) + \sum_{q \le b} \sum_{d \le x/q} f(d)g(q) - \sum_{d \le a} \sum_{q \le b} f(d)g(q),$$

which is the same as (24). □

Note. Taking $a = 1$ and $b = 1$, respectively, we obtain the two equations in Theorem 3.10, since $f(1) = F(1)$ and $g(1) = G(1)$.

Exercises for Chapter 3

1. Use Euler's summation formula to deduce the following for $x \ge 2$:

(a) $\displaystyle\sum_{n \le x} \frac{\log n}{n} = \frac{1}{2}\log^2 x + A + O\left(\frac{\log x}{x}\right)$, where A is a constant.

(b) $\displaystyle\sum_{2 \le n \le x} \frac{1}{n \log n} = \log(\log x) + B + O\left(\frac{1}{x \log x}\right)$, where B is a constant.

2. If $x \ge 2$ prove that

$$\sum_{n \le x} \frac{d(n)}{n} = \frac{1}{2}\log^2 x + 2C \log x + O(1), \text{ where } C \text{ is Euler's constant.}$$

3. If $x \ge 2$ and $\alpha > 0$, $\alpha \ne 1$, prove that

$$\sum_{n \le x} \frac{d(n)}{n^\alpha} = \frac{x^{1-\alpha}\log x}{1-\alpha} + \zeta(\alpha)^2 + O(x^{1-\alpha}).$$

4. If $x \ge 2$ prove that:

(a) $\displaystyle\sum_{n \le x} \mu(n)\left[\frac{x}{n}\right]^2 = \frac{x^2}{\zeta(2)} + O(x \log x)$.

(b) $\displaystyle\sum_{n \le x} \frac{\mu(n)}{n}\left[\frac{x}{n}\right] = \frac{x}{\zeta(2)} + O(\log x)$.

5. If $x \ge 1$ prove that:

(a) $\displaystyle\sum_{n \le x} \varphi(n) = \frac{1}{2}\sum_{n \le x} \mu(n)\left[\frac{x}{n}\right]^2 + \frac{1}{2}$.

(b) $\displaystyle\sum_{n \le x} \frac{\varphi(n)}{n} = \sum_{n \le x} \frac{\mu(n)}{n}\left[\frac{x}{n}\right]$.

These formulas, together with those in Exercise 4, show that, for $x \ge 2$,

$$\sum_{n \le x} \varphi(n) = \frac{1}{2}\frac{x^2}{\zeta(2)} + O(x \log x) \text{ and } \sum_{n \le x} \frac{\varphi(n)}{n} = \frac{x}{\zeta(2)} + O(\log x).$$

6. If $x \geq 2$ prove that

$$\sum_{n \leq x} \frac{\varphi(n)}{n^2} = \frac{1}{\zeta(2)} \log x + \frac{C}{\zeta(2)} - A + O\left(\frac{\log x}{x}\right),$$

where C is Euler's constant and

$$A = \sum_{n=1}^{\infty} \frac{\mu(n) \log n}{n^2}.$$

7. In a later chapter we will prove that $\sum_{n=1}^{\infty} \mu(n) n^{-\alpha} = 1/\zeta(\alpha)$ if $\alpha > 1$. Assuming this, prove that for $x \geq 2$ and $\alpha > 1$, $\alpha \neq 2$, we have

$$\sum_{n \leq x} \frac{\varphi(n)}{n^\alpha} = \frac{x^{2-\alpha}}{2 - \alpha} \frac{1}{\zeta(2)} + \frac{\zeta(\alpha - 1)}{\zeta(\alpha)} + O(x^{1-\alpha} \log x).$$

8. If $\alpha \leq 1$ and $x \geq 2$ prove that

$$\sum_{n \leq x} \frac{\varphi(n)}{n^\alpha} = \frac{x^{2-\alpha}}{2 - \alpha} \frac{1}{\zeta(2)} + O(x^{1-\alpha} \log x).$$

9. In a later chapter we will prove that the infinite product $\prod_p (1 - p^{-2})$, extended over all primes, converges to the value $1/\zeta(2) = 6/\pi^2$. Assuming this result, prove that

(a) $\dfrac{\sigma(n)}{n} < \dfrac{n}{\varphi(n)} < \dfrac{\pi^2}{6} \dfrac{\sigma(n)}{n}$ if $n \geq 2$.

[*Hint*: Use the formula $\varphi(n) = n \prod_{p|n} (1 - p^{-1})$ and the relation

$$1 + x + x^2 + \cdots = \frac{1}{1 - x} = \frac{1 + x}{1 - x^2} \quad \text{with } x = \frac{1}{p}.]$$

(b) If $x \geq 2$ prove that

$$\sum_{n \leq x} \frac{n}{\varphi(n)} = O(x).$$

10. If $x \geq 2$ prove that

$$\sum_{n \leq x} \frac{1}{\varphi(n)} = O(\log x).$$

11. Let $\varphi_1(n) = n \sum_{d|n} |\mu(d)|/d$.

(a) Prove that φ_1 is multiplicative and that $\varphi_1(n) = n \prod_{p|n} (1 + p^{-1})$.
(b) Prove that

$$\varphi_1(n) = \sum_{d^2|n} \mu(d) \sigma\left(\frac{n}{d^2}\right)$$

where the sum is over those divisors of n for which $d^2 | n$.
(c) Prove that

$$\sum_{n \leq x} \varphi_1(n) = \sum_{d \leq \sqrt{x}} \mu(d) S\left(\frac{x}{d^2}\right), \text{ where } S(x) = \sum_{k \leq x} \sigma(k),$$

then use Theorem 3.4 to deduce that, for $x \geq 2$,

$$\sum_{n \leq x} \varphi_1(n) = \frac{\zeta(2)}{2\zeta(4)} x^2 + O(x \log x).$$

As in Exercise 7, you may assume the result $\sum_{n=1}^{\infty} \mu(n)n^{-\alpha} = 1/\zeta(\alpha)$ for $\alpha > 1$.

12. For real $s > 0$ and integer $k \geq 1$ find an asymptotic formula for the partial sums

$$\sum_{\substack{n \leq x \\ (n,k)=1}} \frac{1}{n^s}.$$

with an error term that tends to 0 as $x \to \infty$. Be sure to include the case $s = 1$.

Properties of the greatest-integer function

For each real x the symbol $[x]$ denotes the greatest integer $\leq x$. Exercises 13 through 26 describe some properties of the greatest-integer function. In these exercises x and y denote real numbers, n denotes an integer.

13. Prove each of the following statements:

(a) If $x = k + y$ where k is an integer and $0 \leq y < 1$, then $k = [x]$.
(b) $[x + n] = [x] + n$.
(c) $[-x] = \begin{cases} -[x] & \text{if } x = [x], \\ -[x] - 1 & \text{if } x \neq [x]. \end{cases}$
(d) $[x/n] = [[x]/n]$ if $n \geq 1$.

14. If $0 < y < 1$, what are the possible values of $[x] - [x - y]$?

15. The number $\{x\} = x - [x]$ is called the *fractional part* of x. It satisfies the inequalities $0 \leq \{x\} < 1$, with $\{x\} = 0$ if, and only if, x is an integer. What are the possible values of $\{x\} + \{-x\}$?

16. (a) Prove that $[2x] - 2[x]$ is either 0 or 1.
(b) Prove that $[2x] + [2y] \geq [x] + [y] + [x + y]$.

17. Prove that $[x] + [x + \frac{1}{2}] = [2x]$ and, more generally,

$$\sum_{k=0}^{n-1} \left[x + \frac{k}{n} \right] = [nx].$$

18. Let $f(x) = x - [x] - \frac{1}{2}$. Prove that

$$\sum_{k=0}^{n-1} f\left(x + \frac{k}{n} \right) = f(nx)$$

and deduce that

$$\left| \sum_{n=1}^{m} f\left(2^n x + \frac{1}{2} \right) \right| \leq 1 \quad \text{for all } m \geq 1 \text{ and all real } x.$$

19. Given positive odd integers h and k, $(h, k) = 1$, let $a = (k - 1)/2$, $b = (h - 1)/2$.

(a) Prove that $\sum_{r=1}^{a} [hr/k] + \sum_{r=1}^{b} [kr/h] = ab$. *Hint*. Lattice points.
(b) Obtain a corresponding result if $(h, k) = d$.

20. If n is a positive integer prove that $[\sqrt{n} + \sqrt{n+1}] = [\sqrt{4n+2}]$.

21. Determine all positive integers n such that $[\sqrt{n}]$ divides n.

22. If n is a positive integer, prove that

$$\left[\frac{8n+13}{25}\right] - \left[\frac{n - 12 - \left[\frac{n-17}{25}\right]}{3}\right]$$

is independent of n.

23. Prove that

$$\sum_{n \le x} \lambda(n)\left[\frac{x}{n}\right] = [\sqrt{x}].$$

24. Prove that

$$\sum_{n \le x}\left[\sqrt{\frac{x}{n}}\right] = \sum_{n \le \sqrt{x}}\left[\frac{x}{n^2}\right].$$

25. Prove that

$$\sum_{k=1}^{n}\left[\frac{k}{2}\right] = \left[\frac{n^2}{4}\right]$$

and that

$$\sum_{k=1}^{n}\left[\frac{k}{3}\right] = \left[\frac{n(n-1)}{6}\right].$$

26. If $a = 1, 2, \ldots, 7$ prove that there exists an integer b (depending on a) such that

$$\sum_{k=1}^{n}\left[\frac{k}{a}\right] = \left[\frac{(2n+b)^2}{8a}\right].$$

4 Some Elementary Theorems on the Distribution of Prime Numbers

4.1 Introduction

If $x > 0$ let $\pi(x)$ denote the number of primes not exceeding x. Then $\pi(x) \to \infty$ as $x \to \infty$ since there are infinitely many primes. The behavior of $\pi(x)$ as a function of x has been the object of intense study by many celebrated mathematicians ever since the eighteenth century. Inspection of tables of primes led Gauss (1792) and Legendre (1798) to conjecture that $\pi(x)$ is asymptotic to $x/\log x$, that is,

$$\lim_{x \to \infty} \frac{\pi(x)\log x}{x} = 1.$$

This conjecture was first proved in 1896 by Hadamard [28] and de la Vallée Poussin [71] and is known now as the *prime number theorem.*

Proofs of the prime number theorem are often classified as *analytic* or *elementary*, depending on the methods used to carry them out. The proof of Hadamard and de la Vallée Poussin is analytic, using complex function theory and properties of the Riemann zeta function. An elementary proof was discovered in 1949 by A. Selberg and P. Erdös. Their proof makes no use of the zeta function nor of complex function theory but is quite intricate. At the end of this chapter we give a brief outline of the main features of the elementary proof. In Chapter 13 we present a short analytic proof which is more transparent than the elementary proof.

This chapter is concerned primarily with elementary theorems on primes. In particular, we show that the prime number theorem can be expressed in several equivalent forms.

For example, we will show that the prime number theorem is equivalent to the asymptotic formula

$$\sum_{n \le x} \Lambda(n) \sim x \quad \text{as } x \to \infty. \tag{1}$$

The partial sums of the Mangoldt function $\Lambda(n)$ define a function introduced by Chebyshev in 1848.

4.2 Chebyshev's functions $\psi(x)$ and $\vartheta(x)$

Definition For $x > 0$ we define Chebyshev's ψ-function by the formula

$$\psi(x) = \sum_{n \le x} \Lambda(n).$$

Thus, the asymptotic formula in (1) states that

$$\lim_{x \to \infty} \frac{\psi(x)}{x} = 1. \tag{2}$$

Since $\Lambda(n) = 0$ unless n is a prime power we can write the definition of $\psi(x)$ as follows:

$$\psi(x) = \sum_{n \le x} \Lambda(n) = \sum_{m=1}^{\infty} \sum_{\substack{p \\ p^m \le x}} \Lambda(p^m) = \sum_{m=1}^{\infty} \sum_{p \le x^{1/m}} \log p.$$

The sum on m is actually a finite sum. In fact, the sum on p is empty if $x^{1/m} < 2$, that is, if $(1/m)\log x < \log 2$, or if

$$m > \frac{\log x}{\log 2} = \log_2 x.$$

Therefore we have

$$\psi(x) = \sum_{m \le \log_2 x} \sum_{p \le x^{1/m}} \log p.$$

This can be written in a slightly different form by introducing another function of Chebyshev.

Definition If $x > 0$ we define Chebyshev's ϑ-function by the equation

$$\vartheta(x) = \sum_{p \le x} \log p,$$

where p runs over all primes $\le x$.

The last formula for $\psi(x)$ can now be restated as follows:

(3) $$\psi(x) = \sum_{m \le \log_2 x} \vartheta(x^{1/m}).$$

The next theorem relates the two quotients $\psi(x)/x$ and $\vartheta(x)/x$.

Theorem 4.1 *For $x > 0$ we have*

$$0 \le \frac{\psi(x)}{x} - \frac{\vartheta(x)}{x} \le \frac{(\log x)^2}{2\sqrt{x}\log 2}.$$

Note. This inequality implies that

$$\lim_{x \to \infty}\left(\frac{\psi(x)}{x} - \frac{\vartheta(x)}{x}\right) = 0.$$

In other words, if one of $\psi(x)/x$ or $\vartheta(x)/x$ tends to a limit then so does the other, and the two limits are equal.

PROOF. From (3) we find

$$0 \le \psi(x) - \vartheta(x) = \sum_{2 \le m \le \log_2 x} \vartheta(x^{1/m}).$$

But from the definition of $\vartheta(x)$ we have the trivial inequality

$$\vartheta(x) \le \sum_{p \le x} \log x \le x \log x$$

so

$$0 \le \psi(x) - \vartheta(x) \le \sum_{2 \le m \le \log_2 x} x^{1/m} \log(x^{1/m}) \le (\log_2 x)\sqrt{x} \log \sqrt{x}$$

$$= \frac{\log x}{\log 2} \cdot \frac{\sqrt{x}}{2} \log x = \frac{\sqrt{x}(\log x)^2}{2 \log 2}.$$

Now divide by x to obtain the theorem. □

4.3 Relations connecting $\vartheta(x)$ and $\pi(x)$

In this section we obtain two formulas relating $\vartheta(x)$ and $\pi(x)$. These will be used to show that the prime number theorem is equivalent to the limit relation

$$\lim_{x \to \infty} \frac{\vartheta(x)}{x} = 1.$$

Both functions $\pi(x)$ and $\vartheta(x)$ are step functions with jumps at the primes; $\pi(x)$ has a jump 1 at each prime p, whereas $\vartheta(x)$ has a jump of $\log p$ at p. Sums involving step functions of this type can be expressed as integrals by means of the following theorem.

Theorem 4.2 Abel's identity. *For any arithmetical function $a(n)$ let*

$$A(x) = \sum_{n \le x} a(n),$$

where $A(x) = 0$ if $x < 1$. Assume f has a continuous derivative on the interval $[y, x]$, where $0 < y < x$. Then we have

$$\sum_{y < n \le x} a(n)f(n) = A(x)f(x) - A(y)f(y) - \int_y^x A(t)f'(t)\,dt. \tag{4}$$

PROOF. Let $k = [x]$ and $m = [y]$, so that $A(x) = A(k)$ and $A(y) = A(m)$. Then

$$\begin{aligned}
\sum_{y < n \le x} a(n)f(n) &= \sum_{n=m+1}^{k} a(n)f(n) = \sum_{n=m+1}^{k} \{A(n) - A(n-1)\}f(n) \\
&= \sum_{n=m+1}^{k} A(n)f(n) - \sum_{n=m}^{k-1} A(n)f(n+1) \\
&= \sum_{n=m+1}^{k-1} A(n)\{f(n) - f(n+1)\} + A(k)f(k) - A(m)f(m+1) \\
&= -\sum_{n=m+1}^{k-1} A(n)\int_n^{n+1} f'(t)\,dt + A(k)f(k) - A(m)f(m+1) \\
&= -\sum_{n=m+1}^{k-1} \int_n^{n+1} A(t)f'(t)\,dt + A(k)f(k) - A(m)f(m+1) \\
&= -\int_{m+1}^{k} A(t)f'(t)\,dt + A(x)f(x) - \int_k^x A(t)f'(t)\,dt \\
&\quad - A(y)f(y) - \int_y^{m+1} A(t)f'(t)\,dt \\
&= A(x)f(x) - A(y)f(y) - \int_y^x A(t)f'(t)\,dt.
\end{aligned}$$

□

ALTERNATE PROOF. A shorter proof of (4) is available to those readers familiar with Riemann–Stieltjes integration. (See [2], Chapter 7.) Since $A(x)$ is a step function with jump $a(n)$ at each integer n the sum in (4) can be expressed as a Riemann–Stieltjes integral,

$$\sum_{y < n \le x} a(n)f(n) = \int_y^x f(t)\,dA(t).$$

Integration by parts gives us

$$\sum_{y<n\le x} a(n)f(n) = f(x)A(x) - f(y)A(y) - \int_y^x A(t)\,df(t)$$

$$= f(x)A(x) - f(y)A(y) - \int_y^x A(t)f'(t)\,dt. \qquad \square$$

Note. Since $A(t) = 0$ if $t < 1$, when $y < 1$ Equation (4) takes the form

$$(5) \qquad \sum_{n\le x} a(n)f(n) = A(x)f(x) - \int_1^x A(t)f'(t)\,dt.$$

It should also be noted that Euler's summation formula can easily be deduced from (4). In fact, if $a(n) = 1$ for all $n \ge 1$ we find $A(x) = [x]$ and (4) implies

$$\sum_{y<n\le x} f(n) = f(x)[x] - f(y)[y] - \int_y^x [t]f'(t)\,dt.$$

Combining this with the integration by parts formula

$$\int_y^x tf'(t)\,dt = xf(x) - yf(y) - \int_y^x f(t)\,dt$$

we immediately obtain Euler's summation formula (Theorem 3.1).

Now we use (4) to express $\vartheta(x)$ and $\pi(x)$ in terms of integrals.

Theorem 4.3 *For $x \ge 2$ we have*

$$(6) \qquad \vartheta(x) = \pi(x)\log x - \int_2^x \frac{\pi(t)}{t}\,dt$$

and

$$(7) \qquad \pi(x) = \frac{\vartheta(x)}{\log x} + \int_2^x \frac{\vartheta(t)}{t\log^2 t}\,dt.$$

Proof. Let $a(n)$ denote the characteristic function of the primes; that is,

$$a(n) = \begin{cases} 1 & \text{if } n \text{ is prime,} \\ 0 & \text{otherwise.} \end{cases}$$

Then we have

$$\pi(x) = \sum_{p\le x} 1 = \sum_{1<n\le x} a(n) \qquad \text{and } \vartheta(x) = \sum_{p\le x} \log p = \sum_{1<n\le x} a(n)\log n.$$

Taking $f(x) = \log x$ in (4) with $y = 1$ we obtain

$$\vartheta(x) = \sum_{1 < n \le x} a(n)\log n = \pi(x)\log x - \pi(1)\log 1 - \int_1^x \frac{\pi(t)}{t}\,dt,$$

which proves (6) since $\pi(t) = 0$ for $t < 2$.

Next, let $b(n) = a(n) \log n$ and write

$$\pi(x) = \sum_{3/2 < n \le x} b(n)\frac{1}{\log n}, \qquad \vartheta(x) = \sum_{n \le x} b(n).$$

Taking $f(x) = 1/\log x$ in (4) with $y = 3/2$ we obtain

$$\pi(x) = \frac{\vartheta(x)}{\log x} - \frac{\vartheta(3/2)}{\log 3/2} + \int_{3/2}^x \frac{\vartheta(t)}{t \log^2 t}\,dt,$$

which proves (7) since $\vartheta(t) = 0$ if $t < 2$. □

4.4 Some equivalent forms of the prime number theorem

Theorem 4.4 *The following relations are logically equivalent:*

(8) $$\lim_{x\to\infty} \frac{\pi(x)\log x}{x} = 1.$$

(9) $$\lim_{x\to\infty} \frac{\vartheta(x)}{x} = 1.$$

(10) $$\lim_{x\to\infty} \frac{\psi(x)}{x} = 1.$$

PROOF. From (6) and (7) we obtain, respectively,

$$\frac{\vartheta(x)}{x} = \frac{\pi(x)\log x}{x} - \frac{1}{x}\int_2^x \frac{\pi(t)}{t}\,dt$$

and

$$\frac{\pi(x)\log x}{x} = \frac{\vartheta(x)}{x} + \frac{\log x}{x}\int_2^x \frac{\vartheta(t)\,dt}{t \log^2 t}.$$

To show that (8) implies (9) we need only show that (8) implies

$$\lim_{x\to\infty} \frac{1}{x}\int_2^x \frac{\pi(t)}{t}\,dt = 0.$$

But (8) implies $\dfrac{\pi(t)}{t} = O\left(\dfrac{1}{\log t}\right)$ for $t \ge 2$ so

$$\frac{1}{x}\int_2^x \frac{\pi(t)}{t}\,dt = O\left(\frac{1}{x}\int_2^x \frac{dt}{\log t}\right).$$

Now

$$\int_2^x \frac{dt}{\log t} = \int_2^{\sqrt{x}} \frac{dt}{\log t} + \int_{\sqrt{x}}^x \frac{dt}{\log t} \leq \frac{\sqrt{x}}{\log 2} + \frac{x - \sqrt{x}}{\log \sqrt{x}}$$

so

$$\frac{1}{x} \int_2^x \frac{dt}{\log t} \to 0 \quad \text{as } x \to \infty.$$

This shows that (8) implies (9).

To show that (9) implies (8) we need only show that (9) implies

$$\lim_{x \to \infty} \frac{\log x}{x} \int_2^x \frac{\vartheta(t)\, dt}{t \log^2 t} = 0.$$

But (9) implies $\vartheta(t) = O(t)$ so

$$\frac{\log x}{x} \int_2^x \frac{\vartheta(t)\, dt}{t \log^2 t} = O\left(\frac{\log x}{x} \int_2^x \frac{dt}{\log^2 t}\right).$$

Now

$$\int_2^x \frac{dt}{\log^2 t} = \int_2^{\sqrt{x}} \frac{dt}{\log^2 t} + \int_{\sqrt{x}}^x \frac{dt}{\log^2 t} \leq \frac{\sqrt{x}}{\log^2 2} + \frac{x - \sqrt{x}}{\log^2 \sqrt{x}}$$

hence

$$\frac{\log x}{x} \int_2^x \frac{dt}{\log^2 t} \to 0 \quad \text{as } x \to \infty.$$

This proves that (9) implies (8), so (8) and (9) are equivalent. We know already, from Theorem 4.1, that (9) and (10) are equivalent. □

The next theorem relates the prime number theorem to the asymptotic value of the nth prime.

Theorem 4.5 *Let p_n denote the* nth *prime. Then the following asymptotic relations are logically equivalent:*

(11) $$\lim_{x \to \infty} \frac{\pi(x) \log x}{x} = 1.$$

(12) $$\lim_{x \to \infty} \frac{\pi(x) \log \pi(x)}{x} = 1.$$

(13) $$\lim_{n \to \infty} \frac{p_n}{n \log n} = 1.$$

PROOF. We show that (11) implies (12), (12) implies (13), (13) implies (12), and (12) implies (11).

Assume (11) holds. Taking logarithms we obtain

$$\lim_{x\to\infty} [\log \pi(x) + \log\log x - \log x] = 0$$

or

$$\lim_{x\to\infty}\left[\log x\left(\frac{\log \pi(x)}{\log x} + \frac{\log\log x}{\log x} - 1\right)\right] = 0.$$

Since $\log x \to \infty$ as $x \to \infty$ it follows that

$$\lim_{x\to\infty}\left(\frac{\log \pi(x)}{\log x} + \frac{\log\log x}{\log x} - 1\right) = 0$$

from which we obtain

$$\lim_{x\to\infty}\frac{\log \pi(x)}{\log x} = 1.$$

This, together with (11), gives (12).

Now assume (12) holds. If $x = p_n$ then $\pi(x) = n$ and

$$\pi(x)\log \pi(x) = n \log n$$

so (12) implies

$$\lim_{n\to\infty}\frac{n\log n}{p_n} = 1.$$

Thus, (12) implies (13).

Next, assume (13) holds. Given x, define n by the inequalities

$$p_n \le x < p_{n+1},$$

so that $n = \pi(x)$. Dividing by $n \log n$, we get

$$\frac{p_n}{n\log n} \le \frac{x}{n\log n} < \frac{p_{n+1}}{n\log n} = \frac{p_{n+1}}{(n+1)\log(n+1)}\,\frac{(n+1)\log(n+1)}{n\log n}.$$

Now let $n \to \infty$ and use (13) to get

$$\lim_{n\to\infty}\frac{x}{n\log n} = 1, \qquad \text{or } \lim_{x\to\infty}\frac{x}{\pi(x)\log \pi(x)} = 1.$$

Therefore, (13) implies (12).

Finally, we show that (12) implies (11). Taking logarithms in (12) we obtain

$$\lim_{x\to\infty}(\log \pi(x) + \log\log \pi(x) - \log x) = 0$$

or

$$\lim_{x\to\infty}\left[\log \pi(x)\left(1 + \frac{\log\log \pi(x)}{\log \pi(x)} - \frac{\log x}{\log \pi(x)}\right)\right] = 0.$$

Since $\log \pi(x) \to \infty$ it follows that

$$\lim_{x\to\infty}\left(1 + \frac{\log\log \pi(x)}{\log \pi(x)} - \frac{\log x}{\log \pi(x)}\right) = 0$$

or

$$\lim_{x\to\infty} \frac{\log x}{\log \pi(x)} = 1.$$

This, together with (12), gives (11). □

4.5 Inequalities for $\pi(n)$ and p_n

The prime number theorem states that $\pi(n) \sim n/\log n$ as $n \to \infty$. The inequalities in the next theorem show that $n/\log n$ is the correct order of magnitude of $\pi(n)$. Although better inequalities can be obtained with greater effort (see [60]) the following theorem is of interest because of the elementary nature of its proof.

Theorem 4.6 *For every integer $n \geq 2$ we have*

$$\frac{1}{6}\frac{n}{\log n} < \pi(n) < 6\frac{n}{\log n}. \tag{14}$$

Proof. We begin with the inequalities

$$2^n \leq \binom{2n}{n} < 4^n, \tag{15}$$

where $\binom{2n}{n} = \dfrac{(2n)!}{n!\,n!}$ is a binomial coefficient. The rightmost inequality follows from the relation

$$4^n = (1 + 1)^{2n} = \sum_{k=0}^{2n}\binom{2n}{k} > \binom{2n}{n},$$

and the other one is easily verified by induction. Taking logarithms in (15) we find

$$n \log 2 \leq \log(2n)! - 2\log n! < n \log 4. \tag{16}$$

But Theorem 3.14 implies that

$$\log n! = \sum_{p\leq n} \alpha(p)\log p$$

where the sum is extended over primes and $\alpha(p)$ is given by

$$\alpha(p) = \sum_{m=1}^{\left[\frac{\log n}{\log p}\right]}\left[\frac{n}{p^m}\right].$$

Hence

$$\log(2n)! - 2\log n! = \sum_{p \le 2n} \sum_{m=1}^{\left[\frac{\log 2n}{\log p}\right]} \left\{\left[\frac{2n}{p^m}\right] - 2\left[\frac{n}{p^m}\right]\right\}\log p. \tag{17}$$

Since $[2x] - 2[x]$ is either 0 or 1 the leftmost inequality in (16) implies

$$n\log 2 \le \sum_{p \le 2n} \left(\sum_{m=1}^{\left[\frac{\log 2n}{\log p}\right]} 1\right)\log p \le \sum_{p \le 2n} \log 2n = \pi(2n)\log 2n.$$

This gives us

$$\pi(2n) \ge \frac{n\log 2}{\log 2n} = \frac{2n}{\log 2n}\frac{\log 2}{2} > \frac{1}{4}\frac{2n}{\log 2n} \tag{18}$$

since $\log 2 > 1/2$. For odd integers we have

$$\pi(2n+1) \ge \pi(2n) > \frac{1}{4}\frac{2n}{\log 2n} > \frac{1}{4}\frac{2n}{2n+1}\frac{2n+1}{\log(2n+1)} \ge \frac{1}{6}\frac{2n+1}{\log(2n+1)}$$

since $2n/(2n+1) \ge 2/3$. This, together with (18), gives us

$$\pi(n) > \frac{1}{6}\frac{n}{\log n}$$

for all $n \ge 2$, which proves the leftmost inequality in (14).

To prove the other inequality we return to (17) and extract the term corresponding to $m = 1$. The remaining terms are nonnegative so we have

$$\log(2n)! - 2\log n! \ge \sum_{p \le 2n} \left\{\left[\frac{2n}{p}\right] - 2\left[\frac{n}{p}\right]\right\}\log p.$$

For those primes p in the interval $n < p \le 2n$ we have $[2n/p] - 2[n/p] = 1$ so

$$\log(2n)! - 2\log n! \ge \sum_{n < p \le 2n} \log p = \vartheta(2n) - \vartheta(n).$$

Hence (16) implies

$$\vartheta(2n) - \vartheta(n) < n\log 4.$$

In particular, if n is a power of 2, this gives

$$\vartheta(2^{r+1}) - \vartheta(2^r) < 2^r\log 4 = 2^{r+1}\log 2.$$

Summing on $r = 0, 1, 2, \ldots, k$, the sum on the left telescopes and we find

$$\vartheta(2^{k+1}) < 2^{k+2}\log 2.$$

Now we choose k so that $2^k \le n < 2^{k+1}$ and we obtain

$$\vartheta(n) \le \vartheta(2^{k+1}) < 2^{k+2}\log 2 \le 4n\log 2.$$

But if $0 < \alpha < 1$ we have

$$(\pi(n) - \pi(n^\alpha))\log n^\alpha < \sum_{n^\alpha < p \le n} \log p \le \vartheta(n) < 4n \log 2,$$

hence

$$\pi(n) < \frac{4n \log 2}{\alpha \log n} + \pi(n^\alpha) < \frac{4n \log 2}{\alpha \log n} + n^\alpha$$
$$= \frac{n}{\log n}\left(\frac{4 \log 2}{\alpha} + \frac{\log n}{n^{1-\alpha}}\right).$$

Now if $c > 0$ and $x \ge 1$ the function $f(x) = x^{-c} \log x$ attains its maximum at $x = e^{1/c}$, so $n^{-c} \log n \le 1/(ce)$ for $n \ge 1$. Taking $\alpha = 2/3$ in the last inequality for $\pi(n)$ we find

$$\pi(n) < \frac{n}{\log n}\left(6 \log 2 + \frac{3}{e}\right) < 6 \frac{n}{\log n}.$$

This completes the proof. □

Theorem 4.6 can be used to obtain upper and lower bounds on the size of the nth prime.

Theorem 4.7 *For $n \ge 1$ the nth prime p_n satisfies the inequalities*

$$\frac{1}{6} n \log n < p_n < 12\left(n \log n + n \log \frac{12}{e}\right). \tag{19}$$

Proof. If $k = p_n$ then $k \ge 2$ and $n = \pi(k)$. From (14) we have

$$n = \pi(k) < 6 \frac{k}{\log k} = 6 \frac{p_n}{\log p_n}$$

hence

$$p_n > \frac{1}{6} n \log p_n > \frac{1}{6} n \log n.$$

This gives the lower bound in (19).

To obtain the upper bound we again use (14) to write

$$n = \pi(k) > \frac{1}{6} \frac{k}{\log k} = \frac{1}{6} \frac{p_n}{\log p_n},$$

from which we find

$$p_n < 6n \log p_n. \tag{20}$$

Since $\log x \le (2/e)\sqrt{x}$ if $x \ge 1$ we have $\log p_n \le (2/e)\sqrt{p_n}$, so (20) implies

$$\sqrt{p_n} < \frac{12}{e} n.$$

Therefore

$$\frac{1}{2} \log p_n < \log n + \log \frac{12}{e}$$

which, when used in (20), gives us

$$p_n < 6n\left(2 \log n + 2 \log \frac{12}{e}\right).$$

This proves the upper bound in (19). □

Note. The upper bound in (19) shows once more that the series

$$\sum_{n=1}^{\infty} \frac{1}{p_n}$$

diverges, by comparison with $\sum_{n=2}^{\infty} 1/(n \log n)$.

4.6 Shapiro's Tauberian theorem

We have shown that the prime number theorem is equivalent to the asymptotic formula

$$\frac{1}{x} \sum_{n \le x} \Lambda(n) \sim 1 \quad \text{as } x \to \infty. \tag{21}$$

In Theorem 3.15 we derived a related asymptotic formula,

$$\sum_{n \le x} \Lambda(n)\left[\frac{x}{n}\right] = x \log x - x + O(\log x). \tag{22}$$

Both sums in (21) and (22) are weighted averages of the function $\Lambda(n)$. Each term $\Lambda(n)$ is multiplied by a weight factor $1/x$ in (21) and by $[x/n]$ in (22).

Theorems relating different weighted averages of the same function are called *Tauberian theorems.* We discuss next a Tauberian theorem proved in 1950 by H. N. Shapiro [64]. It relates sums of the form $\sum_{n \le x} a(n)$ with those of the form $\sum_{n \le x} a(n)[x/n]$ for nonnegative $a(n)$.

Theorem 4.8 *Let $\{a(n)\}$ be a nonnegative sequence such that*

$$\sum_{n \le x} a(n)\left[\frac{x}{n}\right] = x \log x + O(x) \quad \textit{for all } x \ge 1. \tag{23}$$

Then:

(a) *For $x \geq 1$ we have*

$$\sum_{n \leq x} \frac{a(n)}{n} = \log x + O(1).$$

(In other words, dropping the square brackets in (23) *leads to a correct result.)*

(b) *There is a constant $B > 0$ such that*

$$\sum_{n \leq x} a(n) \leq Bx \quad \textit{for all } x \geq 1.$$

(c) *There is a constant $A > 0$ and an $x_0 > 0$ such that*

$$\sum_{n \leq x} a(n) \geq Ax \quad \textit{for all } x \geq x_0.$$

PROOF. Let

$$S(x) = \sum_{n \leq x} a(n), \qquad T(x) = \sum_{n \leq x} a(n)\left[\frac{x}{n}\right].$$

First we prove (b). To do this we establish the inequality

$$S(x) - S\left(\frac{x}{2}\right) \leq T(x) - 2T\left(\frac{x}{2}\right). \tag{24}$$

We write

$$\begin{aligned} T(x) - 2T\left(\frac{x}{2}\right) &= \sum_{n \leq x} \left[\frac{x}{n}\right] a(n) - 2 \sum_{n \leq x/2} \left[\frac{x}{2n}\right] a(n) \\ &= \sum_{n \leq x/2} \left(\left[\frac{x}{n}\right] - 2\left[\frac{x}{2n}\right]\right) a(n) + \sum_{x/2 < n \leq x} \left[\frac{x}{n}\right] a(n). \end{aligned}$$

Since $[2y] - 2[y]$ is either 0 or 1, the first sum is nonnegative, so

$$T(x) - 2T\left(\frac{x}{2}\right) \geq \sum_{x/2 < n \leq x} \left[\frac{x}{n}\right] a(n) = \sum_{x/2 < n \leq x} a(n) = S(x) - S\left(\frac{x}{2}\right).$$

This proves (24). But (23) implies

$$T(x) - 2T\left(\frac{x}{2}\right) = x \log x + O(x) - 2\left(\frac{x}{2} \log \frac{x}{2} + O(x)\right) = O(x).$$

Hence (24) implies $S(x) - S(x/2) = O(x)$. This means that there is some constant $K > 0$ such that

$$S(x) - S\left(\frac{x}{2}\right) \leq Kx \quad \text{for all } x \geq 1.$$

Replace x successively by $x/2, x/4, \ldots$ to get

$$S\left(\frac{x}{2}\right) - S\left(\frac{x}{4}\right) \le K\frac{x}{2},$$

$$S\left(\frac{x}{4}\right) - S\left(\frac{x}{8}\right) \le K\frac{x}{4},$$

etc. Note that $S(x/2^n) = 0$ when $2^n > x$. Adding these inequalities we get

$$S(x) \le Kx\left(1 + \frac{1}{2} + \frac{1}{4} + \cdots\right) = 2Kx.$$

This proves (b) with $B = 2K$.

Next we prove (a). We write $[x/n] = (x/n) + O(1)$ and obtain

$$T(x) = \sum_{n \le x}\left[\frac{x}{n}\right]a(n) = \sum_{n \le x}\left(\frac{x}{n} + O(1)\right)a(n) = x\sum_{n \le x}\frac{a(n)}{n} + O\left(\sum_{n \le x} a(n)\right)$$
$$= x\sum_{n \le x}\frac{a(n)}{n} + O(x),$$

by part (b). Hence

$$\sum_{n \le x}\frac{a(n)}{n} = \frac{1}{x}T(x) + O(1) = \log x + O(1).$$

This proves (a).

Finally, we prove (c). Let

$$A(x) = \sum_{n \le x}\frac{a(n)}{n}.$$

Then (a) can be written as follows:

$$A(x) = \log x + R(x),$$

where $R(x)$ is the error term. Since $R(x) = O(1)$ we have $|R(x)| \le M$ for some $M > 0$.

Choose α to satisfy $0 < \alpha < 1$ (we shall specify α more exactly in a moment) and consider the difference

$$A(x) - A(\alpha x) = \sum_{\alpha x < n \le x}\frac{a(n)}{n} = \sum_{n \le x}\frac{a(n)}{n} - \sum_{n \le \alpha x}\frac{a(n)}{n}.$$

If $x \ge 1$ and $\alpha x \ge 1$ we can apply the asymptotic formula for $A(x)$ to write

$$\begin{aligned} A(x) - A(\alpha x) &= \log x + R(x) - (\log \alpha x + R(\alpha x)) \\ &= -\log \alpha + R(x) - R(\alpha x) \\ &\ge -\log \alpha - |R(x)| - |R(\alpha x)| \ge -\log \alpha - 2M. \end{aligned}$$

Now choose α so that $-\log \alpha - 2M = 1$. This requires $\log \alpha = -2M - 1$, or $\alpha = e^{-2M-1}$. Note that $0 < \alpha < 1$. For this α, we have the inequality

$$A(x) - A(\alpha x) \geq 1 \quad \text{if } x \geq 1/\alpha.$$

But

$$A(x) - A(\alpha x) = \sum_{\alpha x < n \leq x} \frac{a(n)}{n} \leq \frac{1}{\alpha x} \sum_{n \leq x} a(n) = \frac{S(x)}{\alpha x}.$$

Hence

$$\frac{S(x)}{\alpha x} \geq 1 \quad \text{if } x \geq 1/\alpha.$$

Therefore $S(x) \geq \alpha x$ if $x \geq 1/\alpha$, which proves (c) with $A = \alpha$ and $x_0 = 1/\alpha$. □

4.7 Applications of Shapiro's theorem

Equation (22) implies

$$\sum_{n \leq x} \Lambda(n)\left[\frac{x}{n}\right] = x \log x + O(x).$$

Since $\Lambda(n) \geq 0$ we can apply Shapiro's theorem with $a(n) = \Lambda(n)$ to obtain:

Theorem 4.9 *For all $x \geq 1$ we have*

$$\sum_{n \leq x} \frac{\Lambda(n)}{n} = \log x + O(1). \tag{25}$$

Also, there exist positive constants c_1 and c_2 such that

$$\psi(x) \leq c_1 x \quad \textit{for all } x \geq 1$$

and

$$\psi(x) \geq c_2 x \quad \textit{for all sufficiently large } x.$$

Another application can be deduced from the asymptotic formula

$$\sum_{p \leq x} \left[\frac{x}{p}\right] \log p = x \log x + O(x)$$

proved in Theorem 3.16. This can be written in the form

$$\sum_{n \leq x} \Lambda_1(n)\left[\frac{x}{n}\right] = x \log x + O(x), \tag{26}$$

where Λ_1 is the function defined as follows:

$$\Lambda_1(n) = \begin{cases} \log p & \text{if } n \text{ is a prime } p, \\ 0 & \text{otherwise.} \end{cases}$$

Since $\Lambda_1(n) \ge 0$, Equation (26) shows that the hypothesis of Shapiro's theorem is satisfied with $a(n) = \Lambda_1(n)$. Since $\vartheta(x) = \sum_{n \le x} \Lambda_1(n)$, part (a) of Shapiro's theorem gives us the following asymptotic formula.

Theorem 4.10 *For all $x \ge 1$ we have*

$$\sum_{p \le x} \frac{\log p}{p} = \log x + O(1). \tag{27}$$

Also, there exist positive constants c_1 and c_2 such that

$$\vartheta(x) \le c_1 x \quad \textit{for all } x \ge 1$$

and

$$\vartheta(x) \ge c_2 x \quad \textit{for all sufficiently large } x.$$

In Theorem 3.11 we proved that

$$\sum_{n \le x} f(n)\left[\frac{x}{n}\right] = \sum_{n \le x} F\left(\frac{x}{n}\right)$$

for any arithmetical function $f(n)$ with partial sums $F(x) = \sum_{n \le x} f(n)$. Since $\psi(x) = \sum_{n \le x} \Lambda(n)$ and $\vartheta(x) = \sum_{n \le x} \Lambda_1(n)$ the asymptotic formulas in (22) and (26) can be expressed directly in terms of $\psi(x)$ and $\vartheta(x)$. We state these as a formal theorem.

Theorem 4.11 *For all $x \ge 1$ we have*

$$\sum_{n \le x} \psi\left(\frac{x}{n}\right) = x \log x - x + O(\log x) \tag{28}$$

and

$$\sum_{n \le x} \vartheta\left(\frac{x}{n}\right) = x \log x + O(x).$$

4.8 An asymptotic formula for the partial sums $\sum_{p \le x} (1/p)$

In Chapter 1 we proved that the series $\sum (1/p)$ diverges. Now we obtain an asymptotic formula for its partial sums. The result is an application of Theorem 4.10, Equation (27).

Theorem 4.12 *There is a constant A such that*

$$\sum_{p \le x} \frac{1}{p} = \log\log x + A + O\left(\frac{1}{\log x}\right) \quad \textit{for all } x \ge 2. \tag{29}$$

Proof. Let

$$A(x) = \sum_{p \le x} \frac{\log p}{p}$$

and let

$$a(n) = \begin{cases} 1 & \text{if } n \text{ is prime,} \\ 0 & \text{otherwise.} \end{cases}$$

Then

$$\sum_{p \le x} \frac{1}{p} = \sum_{n \le x} \frac{a(n)}{n} \qquad \text{and } A(x) = \sum_{n \le x} \frac{a(n)}{n} \log n.$$

Therefore if we take $f(t) = 1/\log t$ in Theorem 4.2 we find, since $A(t) = 0$ for $t < 2$,

$$\sum_{p \le x} \frac{1}{p} = \frac{A(x)}{\log x} + \int_2^x \frac{A(t)}{t \cdot \log^2 t}\, dt. \tag{30}$$

From (27) we have $A(x) = \log x + R(x)$, where $R(x) = O(1)$. Using this on the right of (30) we find

$$\begin{aligned} \sum_{p \le x} \frac{1}{p} &= \frac{\log x + O(1)}{\log x} + \int_2^x \frac{\log t + R(t)}{t \log^2 t}\, dt \\ &= 1 + O\left(\frac{1}{\log x}\right) + \int_2^x \frac{dt}{t \log t} + \int_2^x \frac{R(t)}{t \log^2 t}\, dt. \end{aligned} \tag{31}$$

Now

$$\int_2^x \frac{dt}{t \log t} = \log\log x - \log\log 2$$

and

$$\int_2^x \frac{R(t)}{t \log^2 t}\, dt = \int_2^\infty \frac{R(t)}{t \log^2 t}\, dt - \int_x^\infty \frac{R(t)}{t \log^2 t}\, dt,$$

the existence of the improper integral being assured by the condition $R(t) = O(1)$. But

$$\int_x^\infty \frac{R(t)}{t \log^2 t}\, dt = O\left(\int_x^\infty \frac{dt}{t \log^2 t}\right) = O\left(\frac{1}{\log x}\right).$$

Hence Equation (31) can be written as follows:

$$\sum_{p \le x} \frac{1}{p} = \log\log x + 1 - \log\log 2 + \int_2^{\infty} \frac{R(t)}{t \log^2 t}\, dt + O\left(\frac{1}{\log x}\right).$$

This proves the theorem with

$$A = 1 - \log\log 2 + \int_2^{\infty} \frac{R(t)}{t \log^2 t}\, dt. \qquad \square$$

4.9 The partial sums of the Möbius function

Definition If $x \ge 1$ we define

$$M(x) = \sum_{n \le x} \mu(n).$$

The exact order of magnitude of $M(x)$ is not known. For many years it was thought that

$$|M(x)| < \sqrt{x} \quad \text{if } x > 1, \qquad \text{(Mertens' conjecture)}$$

but A. M. Odlyzko and H. J. J. te Riele (J. Reine Angew. Math. 357 (1985), 138–160) showed that this inequality is violated infinitely often. The best O-result obtained to date is

$$M(x) = O(x\delta(x))$$

where $\delta(x) = \exp\{-A \log^{3/5} x(\log\log x)^{-1/5}\}$ for some positive constant A. (A proof is given in Walfisz [75].)

In this section we prove that the weaker statement

$$\lim_{x \to \infty} \frac{M(x)}{x} = 0$$

is equivalent to the prime number theorem. First we relate $M(x)$ to another weighted average of $\mu(n)$.

Definition If $x \ge 1$ we define

$$H(x) = \sum_{n \le x} \mu(n) \log n.$$

The next theorem shows that the behavior of $M(x)/x$ is determined by that of $H(x)/(x \log x)$.

Theorem 4.13 *We have*

$$\lim_{x \to \infty} \left(\frac{M(x)}{x} - \frac{H(x)}{x \log x}\right) = 0. \tag{32}$$

PROOF. Taking $f(t) = \log t$ in Theorem 4.2 we obtain

$$H(x) = \sum_{n \leq x} \mu(n)\log n = M(x)\log x - \int_1^x \frac{M(t)}{t}\,dt.$$

Hence if $x > 1$ we have

$$\frac{M(x)}{x} - \frac{H(x)}{x \log x} = \frac{1}{x \log x}\int_1^x \frac{M(t)}{t}\,dt.$$

Therefore to prove the theorem we must show that

$$\lim_{x\to\infty} \frac{1}{x \log x}\int_1^x \frac{M(t)}{t}\,dt = 0. \tag{33}$$

But we have the trivial estimate $M(x) = O(x)$ so

$$\int_1^x \frac{M(t)}{t}\,dt = O\left(\int_1^x dt\right) = O(x),$$

from which we obtain (33), and hence (32). □

Theorem 4.14 *The prime number theorem implies*

$$\lim_{x\to\infty} \frac{M(x)}{x} = 0.$$

PROOF. We use the prime number theorem in the form $\psi(x) \sim x$ and prove that $H(x)/(x \log x) \to 0$ as $x \to \infty$. For this purpose we shall require the identity

$$-H(x) = -\sum_{n \leq x} \mu(n)\log n = \sum_{n \leq x} \mu(n)\psi\left(\frac{x}{n}\right). \tag{34}$$

To prove (34) we begin with Theorem 2.11, which states that

$$\Lambda(n) = -\sum_{d|n} \mu(d)\log d$$

and apply Möbius inversion to get

$$-\mu(n)\log n = \sum_{d|n} \mu(d)\Lambda\left(\frac{n}{d}\right).$$

Summing over all $n \leq x$ and using Theorem 3.10 with $f = \mu$, $g = \Lambda$, we obtain (34).

Since $\psi(x) \sim x$, if $\varepsilon > 0$ is given there is a constant $A > 0$ such that

$$\left|\frac{\psi(x)}{x} - 1\right| < \varepsilon \quad \text{whenever } x \geq A.$$

In other words, we have

$$|\psi(x) - x| < \varepsilon x \quad \text{whenever } x \geq A. \tag{35}$$

Choose $x > A$ and split the sum on the right of (34) into two parts,

$$\sum_{n \leq y} + \sum_{y < n \leq x},$$

where $y = [x/A]$. In the first sum we have $n \leq y$ so $n \leq x/A$, and hence $x/n \geq A$. Therefore we can use (35) to write

$$\left|\psi\left(\frac{x}{n}\right) - \frac{x}{n}\right| < \varepsilon \frac{x}{n} \quad \text{if } n \leq y.$$

Thus,

$$\begin{aligned}\sum_{n \leq y} \mu(n)\psi\left(\frac{x}{n}\right) &= \sum_{n \leq y} \mu(n)\left(\frac{x}{n} + \psi\left(\frac{x}{n}\right) - \frac{x}{n}\right) \\ &= x \sum_{n \leq y} \frac{\mu(n)}{n} + \sum_{n \leq y} \mu(n)\left(\psi\left(\frac{x}{n}\right) - \frac{x}{n}\right),\end{aligned}$$

so

$$\begin{aligned}\left|\sum_{n \leq y} \mu(n)\psi\left(\frac{x}{n}\right)\right| &\leq x\left|\sum_{n \leq y} \frac{\mu(n)}{n}\right| + \sum_{n \leq y}\left|\psi\left(\frac{x}{n}\right) - \frac{x}{n}\right| \\ &< x + \varepsilon \sum_{n \leq y} \frac{x}{n} < x + \varepsilon x(1 + \log y) \\ &< x + \varepsilon x + \varepsilon x \log x.\end{aligned}$$

In the second sum we have $y < n \leq x$ so $n \geq y + 1$. Hence

$$\frac{x}{n} \leq \frac{x}{y + 1} < A$$

because

$$y \leq \frac{x}{A} < y + 1.$$

The inequality $(x/n) < A$ implies $\psi(x/n) \leq \psi(A)$. Therefore the second sum is dominated by $x\psi(A)$. Hence the full sum in (34) is dominated by

$$(1 + \varepsilon)x + \varepsilon x \log x + x\psi(A) < (2 + \psi(A))x + \varepsilon x \log x$$

if $\varepsilon < 1$. In other words, given any ε such that $0 < \varepsilon < 1$ we have

$$|H(x)| < (2 + \psi(A))x + \varepsilon x \log x \quad \text{if } x > A,$$

or

$$\frac{|H(x)|}{x \log x} < \frac{2 + \psi(A)}{\log x} + \varepsilon.$$

Now choose $B > A$ so that $x > B$ implies $(2 + \psi(A))/\log x < \varepsilon$. Then for $x > B$ we have

$$\frac{|H(x)|}{x \log x} < 2\varepsilon,$$

which shows that $H(x)/(x \log x) \to 0$ as $x \to \infty$. □

We turn next to the converse of Theorem 4.14 and prove that the relation

$$\lim_{x \to \infty} \frac{M(x)}{x} = 0 \tag{36}$$

implies the prime number theorem. First we introduce the "little oh" notation.

Definition The notation

$$f(x) = o(g(x)) \quad \text{as } x \to \infty \qquad \text{(read: } f(x) \text{ is little oh of } g(x))$$

means that

$$\lim_{x \to \infty} \frac{f(x)}{g(x)} = 0.$$

An equation of the form

$$f(x) = h(x) + o(g(x)) \quad \text{as } x \to \infty$$

means that $f(x) - h(x) = o(g(x))$ as $x \to \infty$.

Thus, (36) states that

$$M(x) = o(x) \quad \text{as } x \to \infty,$$

and the prime number theorem, expressed in the form $\psi(x) \sim x$, can also be written as

$$\psi(x) = x + o(x) \quad \text{as } x \to \infty.$$

More generally, an asymptotic relation

$$f(x) \sim g(x) \quad \text{as } x \to \infty$$

is equivalent to

$$f(x) = g(x) + o(g(x)) \quad \text{as } x \to \infty.$$

We also note that $f(x) = O(1)$ implies $f(x) = o(x)$ as $x \to \infty$.

Theorem 4.15 *The relation*

$$M(x) = o(x) \quad \textit{as } x \to \infty \tag{37}$$

implies $\psi(x) \sim x$ *as* $x \to \infty$.

PROOF. First we express $\psi(x)$ by a formula of the type

$$\psi(x) = x - \sum_{\substack{q,d \\ qd \le x}} \mu(d) f(q) + O(1) \tag{38}$$

and then use (37) to show that the sum is $o(x)$ as $x \to \infty$. The function f in (38) is given by

$$f(n) = \sigma_0(n) - \log n - 2C,$$

where C is Euler's constant and $\sigma_0(n) = d(n)$ is the number of divisors of n. To obtain (38) we start with the identities

$$[x] = \sum_{n \le x} 1, \qquad \psi(x) = \sum_{n \le x} \Lambda(n), \qquad 1 = \sum_{n \le x} \left[\frac{1}{n}\right]$$

and express each summand as a Dirichlet product involving the Möbius function,

$$1 = \sum_{d|n} \mu(d)\sigma_0\left(\frac{n}{d}\right), \qquad \Lambda(n) = \sum_{d|n} \mu(d) \log \frac{n}{d}, \qquad \left[\frac{1}{n}\right] = \sum_{d|n} \mu(d).$$

Then

$$\begin{aligned}
[x] - \psi(x) - 2C &= \sum_{n \le x} \left\{1 - \Lambda(n) - 2C\left[\frac{1}{n}\right]\right\} \\
&= \sum_{n \le x} \sum_{d|n} \mu(d) \left\{\sigma_0\left(\frac{n}{d}\right) - \log \frac{n}{d} - 2C\right\} \\
&= \sum_{\substack{q,d \\ qd \le x}} \mu(d)\{\sigma_0(q) - \log q - 2C\} \\
&= \sum_{\substack{q,d \\ qd \le x}} \mu(d) f(q).
\end{aligned}$$

This implies (38). Therefore the proof of the theorem will be complete if we show that

$$\sum_{\substack{q,d \\ qd \le x}} \mu(d) f(q) = o(x) \quad \text{as } x \to \infty. \tag{39}$$

For this purpose we use Theorem 3.17 to write

$$\sum_{\substack{q,d \\ qd \le x}} \mu(d) f(q) = \sum_{n \le b} \mu(n) F\left(\frac{x}{n}\right) + \sum_{n \le a} f(n) M\left(\frac{x}{n}\right) - F(a)M(b) \tag{40}$$

where a and b are any positive numbers such that $ab = x$ and

$$F(x) = \sum_{n \le x} f(n).$$

We show next that $F(x) = O(\sqrt{x})$ by using Dirichlet's formula (Theorem 3.3)

$$\sum_{n \le x} \sigma_0(n) = x \log x + (2C - 1)x + O(\sqrt{x})$$

together with the relation

$$\sum_{n \le x} \log n = \log[x]! = x \log x - x + O(\log x).$$

These give us

$$\begin{aligned} F(x) &= \sum_{n \le x} \sigma_0(n) - \sum_{n \le x} \log n - 2C \sum_{n \le x} 1 \\ &= x \log x + (2C - 1)x + O(\sqrt{x}) - (x \log x - x + O(\log x)) \\ &\quad -2Cx + O(1) \\ &= O(\sqrt{x}) + O(\log x) + O(1) = O(\sqrt{x}). \end{aligned}$$

Therefore there is a constant $B > 0$ such that

$$|F(x)| \le B\sqrt{x} \quad \text{for all } x \ge 1.$$

Using this in the first sum on the right of (40) we obtain

$$\left| \sum_{n \le b} \mu(n) F\left(\frac{x}{n}\right) \right| \le B \sum_{n \le b} \sqrt{\frac{x}{n}} \le A\sqrt{xb} = \frac{Ax}{\sqrt{a}} \tag{41}$$

for some constant $A > B > 0$.

Now let $\varepsilon > 0$ be arbitrary and choose $a > 1$ such that

$$\frac{A}{\sqrt{a}} < \varepsilon.$$

Then (41) becomes

$$\left| \sum_{n \le b} \mu(n) F\left(\frac{x}{n}\right) \right| < \varepsilon x \tag{42}$$

for all $x \ge 1$. Note that a depends on ε and not on x.

Since $M(x) = o(x)$ as $x \to \infty$, for the same ε there exists $c > 0$ (depending only on ε) such that

$$x > c \text{ implies } \frac{|M(x)|}{x} < \frac{\varepsilon}{K},$$

where K is any positive number. (We will specify K presently.) The second sum on the right of (40) satisfies

$$\left| \sum_{n \le a} f(n) M\left(\frac{x}{n}\right) \right| \le \sum_{n \le a} |f(n)| \frac{\varepsilon}{K} \frac{x}{n} = \frac{\varepsilon x}{K} \sum_{n \le a} \frac{|f(n)|}{n} \tag{43}$$

provided $x/n > c$ for all $n \leq a$. Therefore (43) holds if $x > ac$. Now take

$$K = \sum_{n \leq a} \frac{|f(n)|}{n}.$$

Then (43) implies

$$\left| \sum_{n \leq a} f(n) M\left(\frac{x}{n}\right) \right| < \varepsilon x \quad \text{provided } x > ac. \tag{44}$$

The last term on the right of (40) is dominated by

$$|F(a)M(b)| \leq A\sqrt{a}\,|M(b)| < A\sqrt{a}\,b < \varepsilon\sqrt{a}\sqrt{a}\,b = \varepsilon x$$

since $ab = x$. Combining this with (44) and (42) we find that (40) implies

$$\left| \sum_{\substack{q,d \\ qd \leq x}} \mu(d) f(q) \right| < 3\varepsilon x$$

provided $x > ac$, where a and c depend only on ε. This proves (39). □

Theorem 4.16 *If*

$$A(x) = \sum_{n \leq x} \frac{\mu(n)}{n}$$

the relation

$$A(x) = o(1) \quad \text{as } x \to \infty \tag{45}$$

implies the prime number theorem. In other words, the prime number theorem is a consequence of the statement that the series

$$\sum_{n=1}^{\infty} \frac{\mu(n)}{n}$$

converges and has sum 0.

Note. It can also be shown (see [3]) that the prime number theorem implies convergence of this series to 0, so (45) is actually equivalent to the prime number theorem.

PROOF. We will show that (45) implies $M(x) = o(x)$. By Abel's identity we have

$$M(x) = \sum_{n \leq x} \mu(n) = \sum_{n \leq x} \frac{\mu(n)}{n} n = xA(x) - \int_1^x A(t)\,dt,$$

so

$$\frac{M(x)}{x} = A(x) - \frac{1}{x}\int_1^x A(t)\,dt.$$

Therefore, to complete the proof it suffices to show that

$$\lim_{x\to\infty} \frac{1}{x}\int_1^x A(t)\,dt = 0. \tag{46}$$

Now if $\varepsilon > 0$ is given there exists a c (depending only on ε) such that $|A(x)| < \varepsilon$ if $x \geq c$. Since $|A(x)| \leq 1$ for all $x \geq 1$ we have

$$\left|\frac{1}{x}\int_1^x A(t)\,dt\right| \leq \left|\frac{1}{x}\int_1^c A(t)\,dt\right| + \left|\frac{1}{x}\int_c^x A(t)\,dt\right| \leq \frac{c-1}{x} + \frac{\varepsilon(x-c)}{x}.$$

Letting $x \to \infty$ we find

$$\limsup_{x\to\infty}\left|\frac{1}{x}\int_1^x A(t)\,dt\right| \leq \varepsilon,$$

and since ε is arbitrary this proves (46). □

4.10 Brief sketch of an elementary proof of the prime number theorem

This section gives a very brief sketch of an elementary proof of the prime number theorem. Complete details can be found in [31] or in [46]. The key to this proof is an asymptotic formula of Selberg which states that

$$\psi(x)\log x + \sum_{n\leq x}\Lambda(n)\psi\left(\frac{x}{n}\right) = 2x\log x + O(x).$$

The proof of Selberg's formula is relatively simple and is given in the next section. This section outlines the principal steps used to deduce the prime number theorem from Selberg's formula.

First, Selberg's formula is cast in a more convenient form which involves the function

$$\sigma(x) = e^{-x}\psi(e^x) - 1.$$

Selberg's formula implies an integral inequality of the form

$$|\sigma(x)|x^2 \leq 2\int_0^x\int_0^y |\sigma(u)|\,du\,dy + O(x), \tag{47}$$

and the prime number theorem is equivalent to showing that $\sigma(x) \to 0$ as $x \to \infty$. Therefore, if we let

$$C = \limsup_{x\to\infty}|\sigma(x)|,$$

the prime number theorem is equivalent to showing that $C = 0$. This is proved by assuming that $C > 0$ and obtaining a contradiction as follows. From the definition of C we have

$$|\sigma(x)| \leq C + g(x), \tag{48}$$

where $g(x) \to 0$ as $x \to \infty$. Let $D = \limsup_{x\to\infty} (1/x)\int_0^x |\sigma(t)|\,dt$. Inequality (48), together with (47), implies $C \le D$. If $C > 0$ a clever application of Selberg's asymptotic formula produces a companion inequality resembling (48),

$$\frac{1}{x}\int_0^x |\sigma(t)|\,dt \le C' + h(x), \tag{49}$$

where $0 < C' < C$ and $h(x) \to 0$ as $x \to \infty$. Letting $x \to \infty$ in (49) we find $D \le C' < C$, a contradiction that completes the proof.

4.11 Selberg's asymptotic formula

We deduce Selberg's formula by a method given by Tatuzawa and Iseki [68] in 1951. It is based on the following theorem which has the nature of an inversion formula.

Theorem 4.17 *Let F be a real- or complex-valued function defined on* $(0, \infty)$, *and let*

$$G(x) = \log x \sum_{n\le x} F\left(\frac{x}{n}\right).$$

Then

$$F(x)\log x + \sum_{n\le x} F\left(\frac{x}{n}\right)\Lambda(n) = \sum_{d\le x} \mu(d)G\left(\frac{x}{d}\right).$$

Proof. First we write $F(x)\log x$ as a sum,

$$F(x)\log x = \sum_{n\le x}\left[\frac{1}{n}\right]F\left(\frac{x}{n}\right)\log\frac{x}{n} = \sum_{n\le x}F\left(\frac{x}{n}\right)\log\frac{x}{n}\sum_{d|n}\mu(d).$$

Then we use the identity of Theorem 2.11,

$$\Lambda(n) = \sum_{d|n}\mu(d)\log\frac{n}{d}$$

to write

$$\sum_{n\le x}F\left(\frac{x}{n}\right)\Lambda(n) = \sum_{n\le x}F\left(\frac{x}{n}\right)\sum_{d|n}\mu(d)\log\frac{n}{d}.$$

Adding these equations we find

$$\begin{aligned} F(x)\log x + \sum_{n\le x}F\left(\frac{x}{n}\right)\Lambda(n) &= \sum_{n\le x}F\left(\frac{x}{n}\right)\sum_{d|n}\mu(d)\left\{\log\frac{x}{n} + \log\frac{n}{d}\right\} \\ &= \sum_{n\le x}\sum_{d|n}F\left(\frac{x}{n}\right)\mu(d)\log\frac{x}{d}. \end{aligned}$$

In the last sum we write $n = qd$ to obtain

$$\sum_{n \le x} \sum_{d|n} F\left(\frac{x}{n}\right)\mu(d)\log\frac{x}{d} = \sum_{d \le x} \mu(d)\log\frac{x}{d} \sum_{q \le x/d} F\left(\frac{x}{qd}\right) = \sum_{d \le x} \mu(d)G\left(\frac{x}{d}\right),$$

which proves the theorem. □

Theorem 4.18 Selberg's asymptotic formula. *For $x > 0$ we have*

$$\psi(x)\log x + \sum_{n \le x} \Lambda(n)\psi\left(\frac{x}{n}\right) = 2x\log x + O(x).$$

PROOF. We apply Theorem 4.17 to the function $F_1(x) = \psi(x)$ and also to $F_2(x) = x - C - 1$, where C is Euler's constant. Corresponding to F_1 we have

$$G_1(x) = \log x \sum_{n \le x} \psi\left(\frac{x}{n}\right) = x\log^2 x - x\log x + O(\log^2 x),$$

where we have used Theorem 4.11. Corresponding to F_2 we have

$$\begin{aligned} G_2(x) &= \log x \sum_{n \le x} F_2\left(\frac{x}{n}\right) = \log x \sum_{n \le x}\left(\frac{x}{n} - C - 1\right) \\ &= x\log x \sum_{n \le x} \frac{1}{n} - (C + 1)\log x \sum_{n \le x} 1 \\ &= x\log x\left(\log x + C + O\left(\frac{1}{x}\right)\right) - (C + 1)\log x(x + O(1)) \\ &= x\log^2 x - x\log x + O(\log x). \end{aligned}$$

Comparing the formulas for $G_1(x)$ and $G_2(x)$ we see that $G_1(x) - G_2(x) = O(\log^2 x)$. Actually, we shall only use the weaker estimate

$$G_1(x) - G_2(x) = O(\sqrt{x}).$$

Now we apply Theorem 4.17 to each of F_1 and F_2 and subtract the two relations so obtained. The difference of the two right members is

$$\sum_{d \le x} \mu(d)\left\{G_1\left(\frac{x}{d}\right) - G_2\left(\frac{x}{d}\right)\right\} = O\left(\sum_{d \le x}\sqrt{\frac{x}{d}}\right) = O\left(\sqrt{x}\sum_{d \le x}\frac{1}{\sqrt{d}}\right) = O(x)$$

by Theorem 3.2(b). Therefore the difference of the two left members is also $O(x)$. In other words, we have

$$\{\psi(x) - (x - C - 1)\}\log x + \sum_{n \le x}\left\{\psi\left(\frac{x}{n}\right) - \left(\frac{x}{n} - C - 1\right)\right\}\Lambda(n) = O(x).$$

Rearranging terms and using Theorem 4.9 we find that

$$\psi(x)\log x + \sum_{n \le x} \psi\left(\frac{x}{n}\right)\Lambda(n) = (x - C - 1)\log x$$

$$+ \sum_{n \le x} \left(\frac{x}{n} - C - 1\right)\Lambda(n) + O(x)$$

$$= 2x \log x + O(x). \qquad \square$$

Exercises for Chapter 4

1. Let $S = \{1, 5, 9, 13, 17, \ldots\}$ denote the set of all positive integers of the form $4n + 1$. An element p of S is called an *S-prime* if $p > 1$ and if the only divisors of p, among the elements of S, are 1 and p. (For example, 49 is an S-prime.) An element $n > 1$ in S which is not an S-prime is called an *S-composite*.
   (a) Prove that every S-composite is a product of S-primes.
   (b) Find the smallest S-composite that can be expressed in more than one way as a product of S-primes
   This example shows that unique factorization does not hold in S.

2. Consider the following finite set of integers:

$$T = \{1, 7, 11, 13, 17, 19, 23, 29\}.$$

   (a) For each prime p in the interval $30 < p < 100$ determine a pair of integers m, n, where $m \ge 0$ and $n \in T$, such that $p = 30m + n$.
   (b) Prove the following statement or exhibit a counter example:
   Every prime $p > 5$ can be expressed in the form $30m + n$, where $m \ge 0$ and $n \in T$.

3. Let $f(x) = x^2 + x + 41$. Find the smallest integer $x \ge 0$ for which $f(x)$ is composite.

4. Let $f(x) = a_0 + a_1 x + \cdots + a_n x^n$ be a polynomial with integer coefficients, where $a_n > 0$ and $n \ge 1$. Prove that $f(x)$ is composite for infinitely many integers x.

5. Prove that for every $n > 1$ there exist n consecutive composite numbers.

6. Prove that there do not exist polynomials P and Q such that

$$\pi(x) = \frac{P(x)}{Q(x)} \text{ for } x = 1, 2, 3, \ldots$$

7. Let $a_1 < a_2 < \cdots < a_n \le x$ be a set of positive integers such that no a_i divides the product of the others. Prove that $n \le \pi(x)$.

8. Calculate the highest power of 10 that divides 1000!.

9. Given an arithmetic progression of integers

$$h, h + k, h + 2k, \ldots, h + nk, \ldots,$$

   where $0 < k < 2000$. If $h + nk$ is prime for $n = t, t + 1, \ldots, t + r$ prove that $r \le 9$. In other words, at most 10 consecutive terms of this progression can be primes.

10. Let s_n denote the nth partial sum of the series

$$\sum_{r=1}^{\infty} \frac{1}{r(r+1)}.$$

Prove that for every integer $k > 1$ there exist integers m and n such that $s_m - s_n = 1/k$.

11. Let s_n denote the sum of the first n primes. Prove that for each n there exists an integer whose square lies between s_n and s_{n+1}.

Prove each of the statements in Exercises 12 through 16. In this group of exercises you may use the prime number theorem.

12. If $a > 0$ and $b > 0$, then $\pi(ax)/\pi(bx) \sim a/b$ as $x \to \infty$.

13. If $0 < a < b$, there exists an x_0 such that $\pi(ax) < \pi(bx)$ if $x \geq x_0$.

14. If $0 < a < b$, there exists an x_0 such that for $x \geq x_0$ there is at least one prime between ax and bx.

15. Every interval $[a, b]$ with $0 < a < b$, contains a rational number of the form p/q, where p and q are primes.

16. (a) Given a positive integer n there exists a positive integer k and a prime p such that $10^k n < p < 10^k(n + 1)$.
(b) Given m integers $a_1, \ldots, a_m$ such that $0 \leq a_i \leq 9$ for $i = 1, 2, \ldots, m$, there exists a prime p whose decimal expansion has $a_1, \ldots, a_m$ for its first m digits.

17. Given an integer $n > 1$ with two factorizations $n = \prod_{i=1}^{r} p_i$ and $n = \prod_{i=1}^{t} q_i$, where the p_i are primes (not necessarily distinct) and the q_i are arbitrary integers > 1. Let α be a nonnegative real number.
(a) If $\alpha \geq 1$ prove that

$$\sum_{i=1}^{r} p_i^{\alpha} \leq \sum_{i=1}^{t} q_i^{\alpha}.$$

(b) Obtain a corresponding inequality relating these sums if $0 \leq \alpha < 1$.

18. Prove that the following two relations are equivalent:

(a)
$$\pi(x) = \frac{x}{\log x} + O\left(\frac{x}{\log^2 x}\right).$$

(b)
$$\vartheta(x) = x + O\left(\frac{x}{\log x}\right).$$

19. If $x \geq 2$, let

$$\text{Li}(x) = \int_2^x \frac{dt}{\log t} \text{ (the } \textit{logarithmic integral} \text{ of } x).$$

(a) Prove that

$$\text{Li}(x) = \frac{x}{\log x} + \int_2^x \frac{dt}{\log^2 t} - \frac{2}{\log 2},$$

and that, more generally,

$$\operatorname{Li}(x) = \frac{x}{\log x}\left(1 + \sum_{k=1}^{n-1} \frac{k!}{\log^k x}\right) + n! \int_2^x \frac{dt}{\log^{n+1} t} + C_n,$$

where C_n is independent of x.

(b) If $x \geq 2$ prove that

$$\int_2^x \frac{dt}{\log^n t} = O\left(\frac{x}{\log^n x}\right).$$

20. Let f be an arithmetical function such that

$$\sum_{p \leq x} f(p)\log p = (ax + b)\log x + cx + O(1) \quad \text{for } x \geq 2.$$

Prove that there is a constant A (depending on f) such that, if $x \geq 2$,

$$\sum_{p \leq x} f(p) = ax + (a + c)\left(\frac{x}{\log x} + \int_2^x \frac{dt}{\log^2 t}\right) + b \log(\log x) + A + O\left(\frac{1}{\log x}\right).$$

21. Given two real-valued functions $S(x)$ and $T(x)$ such that

$$T(x) = \sum_{n \leq x} S\left(\frac{x}{n}\right) \quad \text{for all } x \geq 1.$$

If $S(x) = O(x)$ and if c is a positive constant, prove that the relation

$$S(x) \sim cx \quad \text{as } x \to \infty$$

implies

$$T(x) \sim cx \log x \quad \text{as } x \to \infty.$$

22. Prove that Selberg's formula, as expressed in Theorem 4.18, is equivalent to each of the following relations:

(a)
$$\psi(x)\log x + \sum_{p \leq x} \psi\left(\frac{x}{p}\right)\log p = 2x \log x + O(x).$$

(b)
$$\vartheta(x)\log x + \sum_{p \leq x} \vartheta\left(\frac{x}{p}\right)\log p = 2x \log x + O(x).$$

23. Let $M(x) = \sum_{n \leq x} \mu(n)$. Prove that

$$M(x)\log x + \sum_{n \leq x} M\left(\frac{x}{n}\right)\Lambda(n) = O(x)$$

and that

$$M(x)\log x + \sum_{p \leq x} M\left(\frac{x}{p}\right)\log p = O(x).$$

[*Hint:* Theorem 4.17.]

24. Let $A(x)$ be defined for all $x > 0$ and assume that

$$T(x) = \sum_{n \le x} A\left(\frac{x}{n}\right) = ax \log x + bx + o\left(\frac{x}{\log x}\right) \quad \text{as } x \to \infty,$$

where a and b are constants. Prove that

$$A(x)\log x + \sum_{n \le x} A\left(\frac{x}{n}\right)\Lambda(n) = 2ax \log x + o(x \log x) \quad \text{as } x \to \infty.$$

Verify that Selberg's formula of Theorem 4.18 is a special case.

25. Prove that the prime number theorem in the form $\psi(x) \sim x$ implies Selberg's asymptotic formula in Theorem 4.18 with an error term $o(x \log x)$ as $x \to \infty$.

26. In 1851 Chebyshev proved that if $\psi(x)/x$ tends to a limit as $x \to \infty$ then this limit equals 1. This exercise outlines a simple proof of this result based on the formula

$$\sum_{n \le x} \psi\left(\frac{x}{n}\right) = x \log x + O(x) \tag{50}$$

which follows from Theorem 4.11.

(a) Let $\delta = \limsup_{x \to \infty}(\psi(x)/x)$. Given $\varepsilon > 0$ choose $N = N(\varepsilon)$ so that $x \ge N$ implies $\psi(x) \le (\delta + \varepsilon)x$. Split the sum in (50) into two parts, one with $n \le x/N$, the other with $n > x/N$, and estimate each part to obtain the inequality

$$\sum_{n \le x} \psi\left(\frac{x}{n}\right) \le (\delta + \varepsilon)x \log x + x\psi(N).$$

Comparing this with (50), deduce that $\delta \ge 1$.

(b) Let $\gamma = \liminf_{x \to \infty}(\psi(x)/x)$ and use an argument similar to that in (a) to deduce that $\gamma \le 1$. Therefore, if $\psi(x)/x$ has a limit as $x \to \infty$ then $\gamma = \delta = 1$.

In Exercises 27 through 30, let $A(x) = \sum_{n \le x} a(n)$, where $a(n)$ satisfies

$$a(n) \ge 0 \quad \text{for all } n \ge 1, \tag{51}$$

and

$$\sum_{n \le x} A\left(\frac{x}{n}\right) = \sum_{n \le x} a(n)\left[\frac{x}{n}\right] = ax \log x + bx + o\left(\frac{x}{\log x}\right) \quad \text{as } x \to \infty. \tag{52}$$

When $a(n) = \Lambda(n)$ these relations hold with $a = 1$ and $b = -1$. The following exercises show that (51) and (52), together with the prime number theorem, $\psi(x) \sim x$, imply $A(x) \sim ax$, a result due to Basil Gordon. This should be compared with Theorem 4.8 (Shapiro's Tauberian theorem) which assumes only (51) and the weaker condition $\sum_{n \le x} A(x/n) = ax \log x + O(x)$ and concludes that $Cx \le A(x) \le Bx$ for some positive constants C and B.

27. Prove that

(a) $$\sum_{n \le x} A\left(\frac{x}{n}\right)\Lambda(n) = \sum_{n \le \sqrt{x}} A\left(\frac{x}{n}\right)\Lambda(n) + \sum_{n \le \sqrt{x}} \psi\left(\frac{x}{n}\right)a(n) + O(x)$$

and use this to deduce the relation

(b) $$\frac{A(x)}{x} + \frac{1}{x \log x} \sum_{n \le \sqrt{x}} A\left(\frac{x}{n}\right)\Lambda(n) + \frac{1}{x \log x} \sum_{n \le \sqrt{x}} \psi\left(\frac{x}{n}\right)a(n) = 2a + o(1).$$

28. Let $\alpha = \liminf_{x \to \infty}(A(x)/x)$ and let $\beta = \limsup_{x \to \infty}(A(x)/x)$.

(*a*) Choose any $\varepsilon > 0$ and use the fact that

$$A\left(\frac{x}{t}\right) < (\beta + \varepsilon)\frac{x}{t} \quad \text{and } \psi\left(\frac{x}{t}\right) < (1 + \varepsilon)\frac{x}{t}$$

for all sufficiently large x/t to deduce, from Exercise 27(b), that

$$\alpha + \frac{\beta}{2} + \frac{a}{2} + \frac{\varepsilon}{2} + \frac{a\varepsilon}{2} > 2a.$$

Since ε is arbitrary this implies

$$\alpha + \frac{\beta}{2} + \frac{a}{2} \ge 2a.$$

[*Hint:* Let $x \to \infty$ in such a way that $A(x)/x \to \alpha$.]

(b) By a similar argument, prove that

$$\beta + \frac{\alpha}{2} + \frac{a}{2} \le 2a$$

and deduce that $\alpha = \beta = a$. In other words, $A(x) \sim ax$ as $x \to \infty$.

29. Take $a(n) = 1 + \mu(n)$ and verify that (52) is satisfied with $a = 1$ and $b = 2C - 1$, where C is Euler's constant. Show that the result of Exercise 28 implies

$$\lim_{x \to \infty} \frac{1}{x} \sum_{n \le x} \mu(n) = 0.$$

This gives an alternate proof of Theorem 4.14.

30. Suppose that, in Exercise 28, we do not assume the prime number theorem. Instead, let

$$\gamma = \liminf_{x \to \infty} \frac{\psi(x)}{x}, \qquad \delta = \limsup_{x \to \infty} \frac{\psi(x)}{x}.$$

(a) Show that the argument suggested in Exercise 28 leads to the inequalities

$$\alpha + \frac{\beta}{2} + \frac{a\delta}{2} \ge 2a, \qquad \beta + \frac{\alpha}{2} + \frac{a\gamma}{2} \le 2a.$$

(b) Use part (a) to show that $\beta - \alpha \le a\delta - a\gamma$, and deduce that

$$a\gamma \le \alpha \le \beta \le a\delta.$$

This shows that among all numbers $a(n)$ satisfying (51) and (52) with a fixed a, the most widely separated limits of indetermination,

$$\liminf_{x \to \infty} \frac{A(x)}{x} \text{ and } \limsup_{x \to \infty} \frac{A(x)}{x},$$

occur when $a(n) = a\Lambda(n)$. Hence to deduce $A(x) \sim ax$ from (51) and (52) it suffices to treat only the special case $a(n) = a\Lambda(n)$.

5 Congruences

5.1 Definition and basic properties of congruences

Gauss introduced a remarkable notation which simplifies many problems concerning divisibility of integers. In so doing he created a new branch of number theory called the *theory of congruences*, the foundations of which are discussed in this chapter.

Unless otherwise indicated, small latin and Greek letters will denote integers (positive, negative, or zero).

Definition Given integers a, b, m with $m > 0$. We say that a is congruent to b modulo m, and we write

$$a \equiv b \pmod{m}, \tag{1}$$

if m divides the difference $a - b$. The number m is called the modulus of the congruence.

In other words, the congruence (1) is equivalent to the divisibility relation

$$m \mid (a - b).$$

In particular, $a \equiv 0 \pmod{m}$ if, and only if, $m \mid a$. Hence $a \equiv b \pmod{m}$ if, and only if, $a - b \equiv 0 \pmod{m}$. If $m \nmid (a - b)$ we write $a \not\equiv b \pmod{m}$ and say that a and b are *incongruent* mod m.

EXAMPLES

1. $19 \equiv 7 \pmod{12}$, $1 \equiv -1 \pmod 2$, $3^2 \equiv -1 \pmod 5$.
2. n is even if, and only if, $n \equiv 0 \pmod 2$.
3. n is odd if, and only if, $n \equiv 1 \pmod 2$.
4. $a \equiv b \pmod 1$ for every a and b.
5. If $a \equiv b \pmod m$ then $a \equiv b \pmod d$ when $d \mid m$, $d > 0$.

The congruence symbol $\equiv$ was chosen by Gauss to suggest analogy with the equals sign $=$. The next two theorems show that congruences do indeed possess many of the formal properties of equations.

Theorem 5.1 *Congruence is an equivalence relation. That is, we have:*

(a) $a \equiv a \pmod m$ *(reflexivity)*
(b) $a \equiv b \pmod m$ *implies* $b \equiv a \pmod m$ *(symmetry)*
(c) $a \equiv b \pmod m$ *and* $b \equiv c \pmod m$ *imply* $a \equiv c \pmod m$ *(transitivity).*

PROOF. The proof follows at once from the following properties of divisibility:

(a) $m \mid 0$.
(b) If $m \mid (a - b)$ then $m \mid (b - a)$.
(c) If $m \mid (a - b)$ and $m \mid (b - c)$ then $m \mid (a - b) + (b - c) = a - c$. □

Theorem 5.2 *If* $a \equiv b \pmod m$ *and* $\alpha \equiv \beta \pmod m$, *then we have:*

(a) $ax + \alpha y \equiv bx + \beta y \pmod m$ *for all integers* x *and* y.
(b) $a\alpha \equiv b\beta \pmod m$.
(c) $a^n \equiv b^n \pmod m$ *for every positive integer* n.
(d) $f(a) \equiv f(b) \pmod m$ *for every polynomial* f *with integer coefficients.*

PROOF. (a) Since $m \mid (a - b)$ and $m \mid (\alpha - \beta)$ we have

$$m \mid x(a - b) + y(\alpha - \beta) = (ax + \alpha y) - (bx + \beta y).$$

(b) Note that $a\alpha - b\beta = \alpha(a - b) + b(\alpha - \beta) \equiv 0 \pmod m$ by part (a).
(c) Take $\alpha = a$ and $\beta = b$ in part (b) and use induction on n.
(d) Use part (c) and induction on the degree of f. □

Theorem 5.2 tells us that two congruences with the same modulus can be added, subtracted, or multiplied, member by member, as though they were equations. The same holds true for any finite number of congruences with the same modulus.

Before developing further properties of congruences we give two examples to illustrate their usefulness.

Example 1 Test for divisibility by 9. An integer $n > 0$ is divisible by 9 if, and only if, the sum of its digits in its decimal expansion is divisible by 9. This property is easily proved using congruences. If the digits of n in decimal notation are $a_0, a_1, \ldots, a_k$, then

$$n = a_0 + 10a_1 + 10^2a_2 + \cdots + 10^ka_k.$$

Using Theorem 5.2 we have, modulo 9,

$$10 \equiv 1, \qquad 10^2 \equiv 1, \ldots, \qquad 10^k \equiv 1 \pmod 9$$

so

$$n \equiv a_0 + a_1 + \cdots + a_k \pmod 9.$$

Note that all these congruences hold modulo 3 as well, so a number is divisible by 3 if, and only if, the sum of its digits is divisible by 3.

Example 2 The Fermat numbers $F_n = 2^{2^n} + 1$ were mentioned in the Historical Introduction. The first five are primes:

$$F_0 = 3, \qquad F_1 = 5, \qquad F_2 = 17, \qquad F_3 = 257, \qquad \text{and } F_4 = 65{,}537.$$

We now show that F_5 is divisible by 641 without explicitly calculating F_5. To do this we consider the successive powers 2^{2^n} modulo 641. We have

$$2^2 = 4, \qquad 2^4 = 16, \qquad 2^8 = 256, \qquad 2^{16} = 65{,}536 \equiv 154 \pmod{641},$$

so

$$2^{32} \equiv (154)^2 = 23{,}716 \equiv 640 \equiv -1 \pmod{641}.$$

Therefore $F_5 = 2^{32} + 1 \equiv 0 \pmod{641}$, so F_5 is composite.

We return now to general properties of congruences. Common nonzero factors cannot always be cancelled from both members of a congruence as they can in equations. For example, both members of the congruence

$$48 \equiv 18 \pmod{10}$$

are divisible by 6, but if we cancel the common factor 6 we get an incorrect result, $8 \equiv 3 \pmod{10}$. The next theorem shows that a common factor can be cancelled if the modulus is also divisible by this factor.

Theorem 5.3 *If $c > 0$ then*

$$a \equiv b \pmod m \quad \textit{if, and only if,} \ ac \equiv bc \pmod{mc}.$$

Proof. We have $m \mid (b - a)$ if, and only if, $cm \mid c(b - a)$. □

The next theorem describes a cancellation law which can be used when the modulus is not divisible by the common factor.

Theorem 5.4 Cancellation law. *If* $ac \equiv bc \pmod{m}$ *and if* $d = (m, c)$, *then*

$$a \equiv b \left(\text{mod}\, \frac{m}{d}\right).$$

In other words, a common factor c can be cancelled provided the modulus is divided by $d = (m, c)$. *In particular, a common factor which is relatively prime to the modulus can always be cancelled.*

PROOF. Since $ac \equiv bc \pmod{m}$ we have

$$m \mid c(a - b) \qquad \text{so}\ \frac{m}{d} \,\Big|\, \frac{c}{d}(a - b).$$

But $(m/d, c/d) = 1$, hence $m/d \mid (a - b)$. □

Theorem 5.5 *Assume* $a \equiv b \pmod{m}$. *If* $d \mid m$ *and* $d \mid a$ *then* $d \mid b$.

PROOF. It suffices to assume that $d > 0$. If $d \mid m$ then $a \equiv b \pmod{m}$ implies $a \equiv b \pmod{d}$. But if $d \mid a$ then $a \equiv 0 \pmod{d}$ so $b \equiv 0 \pmod{d}$. □

Theorem 5.6 *If* $a \equiv b \pmod{m}$ *then* $(a, m) = (b, m)$. *In other words, numbers which are congruent* mod m *have the same* gcd *with* m.

PROOF. Let $d = (a, m)$ and $e = (b, m)$. Then $d \mid m$ and $d \mid a$ so $d \mid b$; hence $d \mid e$. Similarly, $e \mid m$, $e \mid b$, so $e \mid a$; hence $e \mid d$. Therefore $d = e$. □

Theorem 5.7 *If* $a \equiv b \pmod{m}$ *and if* $0 \leq |b - a| < m$, *then* $a = b$.

PROOF. Since $m \mid (a - b)$ we have $m \leq |a - b|$ unless $a - b = 0$. □

Theorem 5.8 *We have* $a \equiv b \pmod{m}$ *if, and only if, a and b give the same remainder when divided by m.*

PROOF. Write $a = mq + r$, $b = mQ + R$, where $0 \leq r < m$ and $0 \leq R < m$. Then $a - b \equiv r - R \pmod{m}$ and $0 \leq |r - R| < m$. Now use Theorem 5.7. □

Theorem 5.9 *If* $a \equiv b \pmod{m}$ *and* $a \equiv b \pmod{n}$ *where* $(m, n) = 1$, *then* $a \equiv b \pmod{mn}$.

PROOF. Since both m and n divide $a - b$ so does their product since $(m, n) = 1$. □

5.2 Residue classes and complete residue systems

Definition Consider a fixed modulus $m > 0$. We denote by $\hat{a}$ the set of all integers x such that $x \equiv a \pmod{m}$ and we call $\hat{a}$ the residue class a modulo m.

Thus, $\hat{a}$ consists of all integers of the form $a + mq$, where $q = 0, \pm 1, \pm 2, \ldots$

The following properties of residue classes are easy consequences of this definition.

Theorem 5.10 *For a given modulus m we have:*

(a) $\hat{a} = \hat{b}$ *if, and only if,* $a \equiv b \pmod{m}$.
(b) *Two integers x and y are in the same residue class if, and only if,* $x \equiv y \pmod{m}$.
(c) *The m residue classes* $\hat{1}, \hat{2}, \ldots, \hat{m}$ *are disjoint and their union is the set of all integers.*

PROOF. Parts (a) and (b) follow at once from the definition. To prove (c) we note that the numbers $0, 1, 2, \ldots, m - 1$ are incongruent modulo m (by Theorem 5.7). Hence by part (b) the residue classes

$$\hat{0}, \hat{1}, \hat{2}, \ldots, \widehat{m-1}$$

are disjoint. But every integer x must be in exactly one of these classes because $x = qm + r$ where $0 \le r < m$, so $x \equiv r \pmod{m}$ and hence $x \in \hat{r}$. Since $\hat{0} = \hat{m}$ this proves (c). □

Definition A set of m representatives, one from each of the residue classes $\hat{1}, \hat{2}, \ldots, \hat{m}$, is called a complete residue system modulo m.

EXAMPLES Any set consisting of m integers, incongruent mod m, is a complete residue system mod m. For example,

$$\{1, 2, \ldots, m\}; \qquad \{0, 1, 2, \ldots, m - 1\};$$

$$\{1, m + 2, 2m + 3, 3m + 4, \ldots, m^2\}.$$

Theorem 5.11 *Assume* $(k, m) = 1$. *If* $\{a_1, \ldots, a_m\}$ *is a complete residue system modulo m, so is* $\{ka_1, \ldots, ka_m\}$.

PROOF. If $ka_i \equiv ka_j \pmod{m}$ then $a_i \equiv a_j \pmod{m}$ since $(k, m) = 1$. Therefore no two elements in the set $\{ka_1, \ldots, ka_m\}$ are congruent modulo m. Since there are m elements in this set it forms a complete residue system. □

5.3 Linear congruences

Polynomial congruences can be studied in much the same way that polynomial equations are studied in algebra. Here, however, we deal with polynomials $f(x)$ with integer coefficients so that the values of these polynomials will be integers when x is an integer. An integer x satisfying a polynomial congruence

(2) $$f(x) \equiv 0 \pmod{m}$$

is called a *solution* of the congruence. Of course, if $x \equiv y \pmod{m}$ then $f(x) \equiv f(y) \pmod{m}$ so every congruence having one solution has infinitely many. Therefore we make the convention that solutions belonging to the same residue class will not be counted as distinct. And when we speak of the *number* of solutions of a congruence such as (2) we shall mean the number of *incongruent* solutions, that is, the number of solutions contained in the set $\{1, 2, \ldots, m\}$ or in any other complete residue system modulo m. Therefore every polynomial congruence modulo m has at most m solutions.

EXAMPLE 1 The linear congruence $2x \equiv 3 \pmod{4}$ has no solutions, since $2x - 3$ is odd for every x and therefore cannot be divisible by 4.

EXAMPLE 2 The quadratic congruence $x^2 \equiv 1 \pmod{8}$ has exactly four solutions given by $x \equiv 1, 3, 5, 7 \pmod{8}$.

The theory of linear congruences is completely described by the next three theorems.

Theorem 5.12 *Assume* $(a, m) = 1$. *Then the linear congruence*

$$ax \equiv b \pmod{m} \tag{3}$$

has exactly one solution.

PROOF. We need only test the numbers $1, 2, \ldots, m$, since they constitute a complete residue system. Therefore we form the products $a, 2a, \ldots, ma$. Since $(a, m) = 1$ these numbers also constitute a complete residue system. Hence exactly one of these products is congruent to b modulo m. That is, there is exactly one x satisfying (3). □

Although Theorem 5.12 tells us that the linear congruence (3) has a unique solution if $(a, m) = 1$, it does not tell us how to determine this solution except by testing all the numbers in a complete residue system. There are more expeditious methods known for determining the solution; some of them are discussed later in this chapter.

Note. If $(a, m) = 1$ the unique solution of the congruence $ax \equiv 1 \pmod{m}$ is called the *reciprocal* of a modulo m. If a' is the reciprocal of a then ba' is the solution of (3).

Theorem 5.13 *Assume* $(a, m) = d$. *Then the linear congruence*

$$ax \equiv b \pmod{m} \tag{4}$$

has solutions if, and only if, $d \mid b$.

PROOF. If a solution exists then $d|b$ since $d|m$ and $d|a$. Conversely, if $d|b$ the congruence

$$\frac{a}{d}x \equiv \frac{b}{d}\left(\text{mod } \frac{m}{d}\right)$$

has a solution since $(a/d, m/d) = 1$, and this solution is also a solution of (4). □

Theorem 5.14 *Assume $(a, m) = d$ and suppose that $d|b$. Then the linear congruence*

$$(5) \qquad ax \equiv b \pmod{m}$$

has exactly d solutions modulo m. These are given by

$$(6) \qquad t, t + \frac{m}{d}, t + 2\frac{m}{d}, \ldots, t + (d - 1)\frac{m}{d},$$

where t is the solution, unique modulo m/d, of the linear congruence

$$(7) \qquad \frac{a}{d}x \equiv \frac{b}{d}\left(\text{mod } \frac{m}{d}\right).$$

PROOF. Every solution of (7) is also a solution of (5). Conversely, every solution of (5) satisfies (7). Now the d numbers listed in (6) are solutions of (7) hence of (5). No two of these are congruent modulo m since the relations

$$t + r\frac{m}{d} \equiv t + s\frac{m}{d} \pmod{m}, \quad \text{with } 0 \le r < d, 0 \le s < d$$

imply

$$r\frac{m}{d} \equiv s\frac{m}{d} \pmod{m}, \qquad \text{and hence } r \equiv s \pmod{d}.$$

But $0 \le |r - s| < d$ so $r = s$.

It remains to show that (5) has no solutions except those listed in (6). If y is a solution of (5) then $ay \equiv at \pmod{m}$ so $y \equiv t \pmod{m/d}$. Hence $y = t + km/d$ for some k. But $k \equiv r \pmod{d}$ for some r satisfying $0 \le r < d$. Therefore

$$k\frac{m}{d} \equiv r\frac{m}{d} \pmod{m} \qquad \text{so } y \equiv t + r\frac{m}{d} \pmod{m}.$$

Therefore y is congruent modulo m to one of the numbers in (6). This completes the proof. □

In Chapter 1 we proved that the gcd of two numbers a and b is a linear combination of a and b. The same result can be deduced as a consequence of Theorem 5.14.

Theorem 5.15 *If* $(a, b) = d$ *there exist integers* x *and* y *such that*

$$ax + by = d. \tag{8}$$

PROOF. The linear congruence $ax \equiv d \pmod{b}$ has a solution. Hence there is an integer y such that $d - ax = by$. This gives us $ax + by = d$, as required. □

Note. Geometrically, the pairs (x, y) satisfying (8) are lattice points lying on a straight line. The x-coordinate of each of these points is a solution of the congruence $ax \equiv d \pmod{b}$.

5.4 Reduced residue systems and the Euler–Fermat theorem

Definition By a reduced residue system modulo m we mean any set of $\varphi(m)$ integers, incongruent modulo m, each of which is relatively prime to m.

Note. $\varphi(m)$ is Euler's totient, introduced in Chapter 2.

Theorem 5.16 *If* $\{a_1, a_2, \ldots, a_{\varphi(m)}\}$ *is a reduced residue system modulo* m *and if* $(k, m) = 1$, *then* $\{ka_1, ka_2, \ldots, ka_{\varphi(m)}\}$ *is also a reduced residue system modulo* m.

PROOF. No two of the numbers ka_i are congruent modulo m. Also, since $(a_i, m) = (k, m) = 1$ we have $(ka_i, m) = 1$ so each ka_i is relatively prime to m. □

Theorem 5.17 Euler–Fermat theorem. *Assume* $(a, m) = 1$. *Then we have*

$$a^{\varphi(m)} \equiv 1 \pmod{m}.$$

PROOF. Let $\{b_1, b_2, \ldots, b_{\varphi(m)}\}$ be a reduced residue system modulo m. Then $\{ab_1, ab_2, \ldots, ab_{\varphi(m)}\}$ is also a reduced residue system. Hence the product of all the integers in the first set is congruent to the product of those in the second set. Therefore

$$b_1 \cdots b_{\varphi(m)} \equiv a^{\varphi(m)} b_1 \cdots b_{\varphi(m)} \pmod{m}.$$

Each b_i is relatively prime to m so we can cancel each b_i to obtain the theorem. □

Theorem 5.18 *If a prime* p *does not divide* a *then*

$$a^{p-1} \equiv 1 \pmod{p}.$$

PROOF. This is a corollary of the foregoing theorem since $\varphi(p) = p - 1$. □

Theorem 5.19 Little Fermat theorem. *For any integer a and any prime p we have*

$$a^p \equiv a \pmod{p}.$$

PROOF. If $p \nmid a$ this is Theorem 5.18. If $p \mid a$ then both a^p and a are congruent to 0 mod p. □

The Euler–Fermat theorem can be used to calculate the solutions of a linear congruence.

Theorem 5.20 *If* $(a, m) = 1$ *the solution* (*unique* mod m) *of the linear congruence*

$$ax \equiv b \pmod{m} \tag{9}$$

is given by

$$x \equiv ba^{\varphi(m)-1} \pmod{m}. \tag{10}$$

PROOF. The number x given by (10) satisfies (9) because of the Euler–Fermat theorem. The solution is unique mod m since $(a, m) = 1$. □

EXAMPLE 1 Solve the congruence $5x \equiv 3 \pmod{24}$.

Solution

Since $(5, 24) = 1$ there is a unique solution. Using (10) we find

$$x \equiv 3 \cdot 5^{\varphi(24)-1} \equiv 3 \cdot 5^7 \pmod{24}$$

since $\varphi(24) = \varphi(3)\varphi(8) = 2 \cdot 4$. Modulo 24 we have $5^2 \equiv 1$, and

$$5^4 \equiv 5^6 \equiv 1, \qquad 5^7 \equiv 5, \qquad \text{so } x \equiv 15 \pmod{24}.$$

EXAMPLE 2 Solve the congruence $25x \equiv 15 \pmod{120}$.

Solution

Since $d = (25, 120) = 5$ and $d \mid 15$ the congruence has exactly five solutions modulo 120. To find them we divide by 5 and solve the congruence $5x \equiv 3 \pmod{24}$. Using Example 1 and Theorem 5.14 we find that the five solutions are given by $x = 15 + 24k$, $k = 0, 1, 2, 3, 4$, or

$$x \equiv 15, 39, 63, 87, 111 \pmod{120}.$$

5.5 Polynomial congruences modulo p. Lagrange's theorem

The fundamental theorem of algebra states that for every polynomial f of degree $n \geq 1$ the equation $f(x) = 0$ has n solutions among the complex numbers. There is no direct analog of this theorem for polynomial congruences. For example, we have seen that some linear congruences have no solutions, some have exactly one solution, and some have more than one. Thus, even in this special case, there appears to be no simple relation between the number of solutions and the degree of the polynomial. However, for congruences modulo a *prime* we have the following theorem of Lagrange.

Theorem 5.21 (Lagrange). *Given a prime p, let*

$$f(x) = c_0 + c_1x + \cdots + c_nx^n$$

be a polynomial of degree n with integer coefficients such that $c_n \not\equiv 0 \pmod{p}$. *Then the polynomial congruence*

$$f(x) \equiv 0 \pmod{p} \tag{11}$$

has at most n solutions.

Note. This result is not true for composite moduli. For example, the quadratic congruence $x^2 \equiv 1 \pmod{8}$ has 4 solutions.

PROOF. We use induction on n, the degree of f. When $n = 1$ the congruence is linear:

$$c_1x + c_0 \equiv \pmod{p}.$$

Since $c_1 \not\equiv 0 \pmod{p}$ we have $(c_1, p) = 1$ and there is exactly one solution. Assume, then, that the theorem is true for polynomials of degree $n - 1$. Assume also that the congruence (11) has $n + 1$ incongruent solutions modulo p, say

$$x_0, x_1, \ldots, x_n,$$

where $f(x_k) \equiv 0 \pmod{p}$ for each $k = 0, 1, \ldots, n$. We shall obtain a contradiction. We have the algebraic identity

$$f(x) - f(x_0) = \sum_{r=1}^{n} c_r(x^r - x_0^r) = (x - x_0)g(x)$$

where $g(x)$ is a polynomial of degree $n - 1$ with integer coefficients and with leading coefficient c_n. Thus we have

$$f(x_k) - f(x_0) = (x_k - x_0)g(x_k) \equiv 0 \pmod{p},$$

since $f(x_k) \equiv f(x_0) \equiv 0 \pmod{p}$. But $x_k - x_0 \not\equiv 0 \pmod{p}$ if $k \neq 0$ so we must have $g(x_k) \equiv 0 \pmod{p}$ for each $k \neq 0$. But this means that the congruence $g(x) \equiv 0 \pmod{p}$ has n incongruent solutions modulo p, contradicting our induction hypothesis. This completes the proof. □

5.6 Applications of Lagrange's theorem

Theorem 5.22 *If* $f(x) = c_0 + c_1x + \cdots + c_nx^n$ *is a polynomial of degree n with integer coefficients, and if the congruence*

$$f(x) \equiv 0 \pmod{p}$$

has more than n solutions, where p is prime, then every coefficient of f is divisible by p.

PROOF. If there is some coefficient not divisible by p, let c_k be the one with largest index. Then $k \leq n$ and the congruence

$$c_0 + c_1 x + \cdots + c_k x^k \equiv 0 \pmod{p}$$

has more than k solutions so, by Lagrange's theorem, $p \mid c_k$, a contradiction. □

Now we apply Theorem 5.22 to a particular polynomial.

Theorem 5.23 *For any prime p all the coefficients of the polynomial*

$$f(x) = (x - 1)(x - 2)\cdots(x - p + 1) - x^{p-1} + 1$$

are divisible by p.

PROOF. Let $g(x) = (x - 1)(x - 2)\cdots(x - p + 1)$. The roots of g are the numbers $1, 2, \ldots, p - 1$, hence they satisfy the congruence

$$g(x) \equiv 0 \pmod{p}.$$

By the Euler–Fermat theorem, these numbers also satisfy the congruence $h(x) \equiv 0 \pmod{p}$, where

$$h(x) = x^{p-1} - 1.$$

The difference $f(x) = g(x) - h(x)$ has degree $p - 2$ but the congruence $f(x) \equiv 0 \pmod{p}$ has $p - 1$ solutions, $1, 2, \ldots, p - 1$. Therefore, by Theorem 5.22, each coefficient of $f(x)$ is divisible by p. □

We obtain the next two theorems by considering two particular coefficients of the polynomial $f(x)$ in Theorem 5.23.

Theorem 5.24 Wilson's theorem. *For any prime p we have*

$$(p - 1)! \equiv -1 \pmod{p}.$$

PROOF. The constant term of the polynomial $f(x)$ in Theorem 5.23 is $(p - 1)! + 1$. □

Note. The converse of Wilson's theorem also holds. That is, if $n > 1$ and $(n - 1)! \equiv -1 \pmod{n}$, then n is prime. (See Exercise 5.7.)

Theorem 5.25 Wolstenholme's theorem. *For any prime $p \geq 5$ we have*

$$\sum_{k=1}^{p-1} \frac{(p - 1)!}{k} \equiv 0 \pmod{p^2}.$$

PROOF. The sum in question is the sum of the products of the numbers 1, $2, \ldots, p-1$ taken $p-2$ at a time. This sum is also equal to the coefficient of $-x$ in the polynomial

$$g(x) = (x-1)(x-2)\cdots(x-p+1).$$

In fact, $g(x)$ can be written in the form

$$g(x) = x^{p-1} - S_1x^{p-2} + S_2x^{p-3} - \cdots + S_{p-3}x^2 - S_{p-2}x + (p-1)!,$$

where the coefficient S_k is the kth elementary symmetric function of the roots, that is, the sum of the products of the numbers $1, 2, \ldots, p-1$, taken k at a time. Theorem 5.23 shows that each of the numbers $S_1, S_2, \ldots, S_{p-2}$ is divisible by p. We wish to show that S_{p-2} is divisible by p^2.

The product for $g(x)$ shows that $g(p) = (p-1)!$ so

$$(p-1)! = p^{p-1} - S_1 p^{p-2} + \cdots + S_{p-3}p^2 - S_{p-2}p + (p-1)!.$$

Canceling $(p-1)!$ and reducing the equation mod p^3 we find, since $p \geq 5$,

$$pS_{p-2} \equiv 0 \pmod{p^3},$$

and hence $S_{p-2} \equiv 0 \pmod{p^2}$, as required. □

5.7 Simultaneous linear congruences. The Chinese remainder theorem

A system of two or more linear congruences need not have a solution, even though each individual congruence has a solution. For example, there is no x which simultaneously satisfies $x \equiv 1 \pmod 2$ and $x \equiv 0 \pmod 4$, even though each of these separately has solutions. In this example the moduli 2 and 4 are not relatively prime. We shall prove next that any system of two or more linear congruences which can be solved separately with unique solutions can also be solved simultaneously if the moduli are relatively prime in pairs. We begin with a special case.

Theorem 5.26 Chinese remainder theorem. *Assume $m_1, \ldots, m_r$ are positive integers, relatively prime in pairs:*

$$(m_i, m_k) = 1 \quad \textit{if } i \neq k.$$

Let $b_1, \ldots, b_r$ be arbitrary integers. Then the system of congruences

$$x \equiv b_1 \pmod{m_1}$$
$$\vdots$$
$$x \equiv b_r \pmod{m_r}$$

has exactly one solution modulo the product $m_1 \cdots m_r$.

Proof. Let $M = m_1 \cdots m_r$ and let $M_k = M/m_k$. Then $(M_k, m_k) = 1$ so each M_k has a unique reciprocal M'_k modulo m_k. Now let

$$x = b_1 M_1 M'_1 + b_2 M_2 M'_2 + \cdots + b_r M_r M'_r.$$

Consider each term in this sum modulo m_k. Since $M_i \equiv 0 \pmod{m_k}$ if $i \neq k$ we have

$$x \equiv b_k M_k M'_k \equiv b_k \pmod{m_k}.$$

Hence x satisfies every congruence in the system. But it is easy to show that the system has only one solution mod M. In fact, if x and y are two solutions of the system we have $x \equiv y \pmod{m_k}$ for each k and, since the m_k are relatively prime in pairs, we also have $x \equiv y \pmod{M}$. This completes the proof. □

The following extension is now easily deduced.

Theorem 5.27 *Assume $m_1, \ldots, m_r$ are relatively prime in pairs. Let $b_1, \ldots, b_r$ be arbitrary integers and let $a_1, \ldots, a_r$ satisfy*

$$(a_k, m_k) = 1 \quad \textit{for } k = 1, 2, \ldots, r.$$

Then the linear system of congruences

$$\begin{aligned} a_1 x &\equiv b_1 \pmod{m_1} \\ &\vdots \\ a_r x &\equiv b_r \pmod{m_r} \end{aligned}$$

has exactly one solution modulo $m_1 m_2 \cdots m_r$.

Proof. Let a'_k denote the reciprocal of a_k modulo m_k. This exists since $(a_k, m_k) = 1$. Then the congruence $a_k x \equiv b_k \pmod{m_k}$ is equivalent to the congruence $x \equiv b_k a'_k \pmod{m_k}$. Now apply Theorem 5.26. □

5.8 Applications of the Chinese remainder theorem

The first application deals with polynomial congruences with composite moduli.

Theorem 5.28 *Let f be a polynomial with integer coefficients, let $m_1, m_2, \ldots, m_r$ be positive integers relatively prime in pairs, and let $m = m_1 m_2 \cdots m_r$. Then the congruence*

$$f(x) \equiv 0 \pmod{m} \tag{12}$$

has a solution if, and only if, each of the congruences

$$f(x) \equiv 0 \pmod{m_i} \qquad (i = 1, 2, \ldots, r) \tag{13}$$

has a solution. Moreover, if $\nu(m)$ *and* $\nu(m_i)$ *denote the number of solutions of* (12) *and* (13), *respectively, then*

$$\nu(m) = \nu(m_1)\nu(m_2)\cdots\nu(m_r). \tag{14}$$

PROOF. If $f(a) \equiv 0 \pmod{m}$ then $f(a) \equiv 0 \pmod{m_i}$ for each i. Hence every solution of (12) is also a solution of (13).

Conversely, let a_i be a solution of (13). Then by the Chinese remainder theorem there exists an integer a such that

$$a \equiv a_i \pmod{m_i} \quad \text{for } i = 1, 2, \ldots, r, \tag{15}$$

so

$$f(a) \equiv f(a_i) \equiv 0 \pmod{m_i}.$$

Since the moduli are relatively prime in pairs we also have $f(a) \equiv 0 \pmod{m}$. Therefore if each of the congruences in (13) has a solution, so does (12).

We also know, by Theorem 5.26, that each r-tuple of solutions $(a_1, \ldots, a_r)$ of the congruences in (13) gives rise to a unique integer a mod m satisfying (15). As each a_i runs through the $\nu(m_i)$ solutions of (13) the number of integers a which satisfy (15) and hence (13) is $\nu(m_1)\cdots\nu(m_r)$. This proves (14). □

Note. If m has the prime power decomposition

$$m = p_1^{\alpha_1}\cdots p_r^{\alpha_r}$$

we can take $m_i = p_i^{\alpha_i}$ in Theorem 5.28 and we see that the problem of solving a polynomial congruence for a composite modulus is reduced to that for prime power moduli. Later we will show that the problem can be reduced further to polynomial congruences with prime moduli plus a set of linear congruences. (See Section 5.9.)

The next application of the Chinese remainder theorem concerns the set of lattice points visible from the origin. (See Section 3.8.)

Theorem 5.29 *The set of lattice points in the plane visible from the origin contains arbitrarily large square gaps. That is, given any integer* $k > 0$ *there exists a lattice point* (a, b) *such that none of the lattice points*

$$(a + r, b + s), \quad 0 < r \le k, 0 < s \le k,$$

is visible from the origin.

PROOF. Let $p_1, p_2, \ldots,$ be the sequence of primes. Given $k > 0$ consider the $k \times k$ matrix whose entries in the first row consist of the first k primes, those in the second row consist of the next k primes, and so on. Let m_i be the product of the primes in the ith row and let M_i be the product of the primes in the ith column. Then the numbers m_i are relatively prime in pairs, as are the M_i.

Next consider the set of congruences

$$x \equiv -1 \pmod{m_1}$$
$$x \equiv -2 \pmod{m_2}$$
$$\vdots$$
$$x \equiv -k \pmod{m_k}.$$

This system has a solution a which is unique mod $m_1 \cdots m_k$. Similarly, the system

$$y \equiv -1 \pmod{M_1}$$
$$\vdots$$
$$y \equiv -k \pmod{M_k}$$

has a solution b which is unique mod $M_1 \cdots M_k = m_1 \cdots m_k$.

Now consider the square with opposite vertices at (a, b) and $(a + k, b + k)$. Any lattice point inside this square has the form

$$(a + r, b + s), \quad \text{where } 0 < r < k, 0 < s < k,$$

and those with $r = k$ or $s = k$ lie on the boundary of the square. We now show that no such point is visible from the origin. In fact,

$$a \equiv -r \pmod{m_r} \qquad \text{and } b \equiv -s \pmod{M_s}$$

so the prime in the intersection of row r and column s divides both $a + r$ and $b + s$. Hence $a + r$ and $b + s$ are not relatively prime, and therefore the lattice point $(a + r, b + s)$ is not visible from the origin. □

5.9 Polynomial congruences with prime power moduli

Theorem 5.28 shows that the problem of solving a polynomial congruence

$$f(x) \equiv 0 \pmod{m}$$

can be reduced to that of solving a system of congruences

$$f(x) \equiv 0 \pmod{p_i^{\alpha_i}} \qquad (i = 1, 2, \ldots, r),$$

where $m = p_1^{\alpha_1} \cdots p_r^{\alpha_r}$. In this section we show that the problem can be further reduced to congruences with prime moduli plus a set of linear congruences.

Let f be a polynomial with integer coefficients, and suppose that for some prime p and some $\alpha \geq 2$ the congruence

$$(16) \qquad f(x) \equiv 0 \pmod{p^\alpha}$$

has a solution, say $x = a$, where a is chosen so that it lies in the interval

$$0 \leq a < p^\alpha.$$

This solution also satisfies each of the congruences $f(x) \equiv 0 \pmod{p^\beta}$ for each $\beta < \alpha$. In particular, a satisfies the congruence

(17) $$f(x) \equiv 0 \pmod{p^{\alpha-1}}.$$

Now divide a by $p^{\alpha-1}$ and write

(18) $$a = qp^{\alpha-1} + r, \quad \text{where } 0 \le r < p^{\alpha-1}.$$

The remainder r determined by (18) is said to be *generated* by a. Since $r \equiv a \pmod{p^{\alpha-1}}$ the number r is also a solution of (17). In other words, every solution a of congruence (16) in the interval $0 \le a < p^\alpha$ generates a solution r of congruence (17) in the interval $0 \le r < p^{\alpha-1}$.

Now suppose we start with a solution r of (17) in the interval $0 \le r < p^{\alpha-1}$ and ask whether there is a solution a of (16) in the interval $0 \le a < p^\alpha$ which generates r. If so, we say that r can be *lifted* from $p^{\alpha-1}$ to p^α. The next theorem shows that the possibility of r being lifted depends on $f(r)$ mod p^α and on the derivative $f'(r)$ mod p.

Theorem 5.30 *Assume $\alpha \ge 2$ and let r be a solution of the congruence*

(19) $$f(x) \equiv 0 \pmod{p^{\alpha-1}}$$

lying in the interval $0 \le r < p^{\alpha-1}$.

(a) *Assume $f'(r) \not\equiv 0 \pmod{p}$. Then r can be lifted in a unique way from $p^{\alpha-1}$ to p^α. That is, there is a unique a in the interval $0 \le a < p^\alpha$ which generates r and which satisfies the congruence*

(20) $$f(x) \equiv 0 \pmod{p^\alpha}.$$

(b) *Assume $f'(r) \equiv 0 \pmod{p}$. Then we have two possibilities:*
(b$_1$) *If $f(r) \equiv 0 \pmod{p^\alpha}$, r can be lifted from $p^{\alpha-1}$ to p^α in p distinct ways.*
(b$_2$) *If $f(r) \not\equiv 0 \pmod{p^\alpha}$, r cannot be lifted from $p^{\alpha-1}$ to p^α.*

PROOF. If n is the degree of f we have the identity (Taylor's formula)

(21) $$f(x + h) = f(x) + f'(x)h + \frac{f''(x)}{2!}h^2 + \cdots + \frac{f^{(n)}(x)}{n!}h^n$$

for every x and h. We note that each polynomial $f^{(k)}(x)/k!$ has integer coefficients. (The reader should verify this.) Now take $x = r$ in (21), where r is a solution of (19) in the interval $0 \le r < p^{\alpha-1}$, and let $h = qp^{\alpha-1}$ where q is an integer to be specified presently. Since $\alpha \ge 2$ the terms in (21) involving h^2 and higher powers of h are integer multiples of p^α. Therefore (21) gives us the congruence

$$f(r + qp^{\alpha-1}) \equiv f(r) + f'(r)qp^{\alpha-1} \pmod{p^\alpha}.$$

Since r satisfies (19) we can write $f(r) = kp^{\alpha-1}$ for some integer k, and the last congruence becomes

$$f(r + qp^{\alpha-1}) \equiv \{qf'(r) + k\}p^{\alpha-1} \pmod{p^\alpha}.$$

Now let

$$a = r + qp^{\alpha-1}. \tag{22}$$

Then a satisfies congruence (20) if, and only if, q satisfies the linear congruence

$$qf'(r) + k \equiv 0 \pmod{p}. \tag{23}$$

If $f'(r) \not\equiv 0 \pmod{p}$ this congruence has a unique solution q mod p, and if we choose q in the interval $0 \le q < p$ then the number a given by (22) will satisfy (20) and will lie in the interval $0 \le a < p^\alpha$.

On the other hand, if $f'(r) \equiv 0 \pmod{p}$ then (23) has a solution q if, and only if, $p \mid k$, that is, if and only if $f(r) \equiv 0 \pmod{p^\alpha}$. If $p \nmid k$ there is no choice of q to make a satisfy (20). But if $p \mid k$ then the p values $q = 0, 1, \ldots, p - 1$ give p solutions a of (20) which generate r and lie in the interval $0 \le a < p^\alpha$. This completes the proof. □

The proof of the foregoing theorem also describes a method for obtaining solutions of congruence (20) if solutions of (19) are known. By applying the method repeatedly the problem is ultimately reduced to that of solving the congruence

$$f(x) \equiv 0 \pmod{p}. \tag{24}$$

If (24) has no solutions, then (20) has no solutions. If (24) has solutions, we choose one, call it r, which lies in the interval $0 \le r < p$. Corresponding to r there will be 0, 1, or p solutions of the congruence

$$f(x) \equiv 0 \pmod{p^2} \tag{25}$$

depending on the numbers $f'(r)$ and $k = f(r)/p$. If $p \nmid k$ and $p \mid f'(r)$ then r cannot be lifted to a solution of (25). In this case we begin anew with a different solution r. If no r can be lifted then (25) has no solution.

If $p \mid k$ for some r, we examine the linear congruence

$$qf'(r) + k \equiv 0 \pmod{p}.$$

This has 1 or p solutions q according as $p \nmid f'(r)$ or $p \mid f'(r)$. For each solution q the number $a = r + qp$ gives a solution of (25). For each solution of (25) a similar procedure can be used to find all solutions of

$$f(x) \equiv 0 \pmod{p^3},$$

and so on, until all solutions of (20) are obtained.

5.10 The principle of cross-classification

Some problems in number theory can be dealt with by applying a general combinatorial theorem about sets called the *principle of cross-classification.* This is a formula which counts the number of elements of a finite set S which do not belong to certain prescribed subsets $S_1, \ldots, S_n$.

Notation *If T is a subset of S we write $N(T)$ for the number of elements of T. We denote by $S - T$ the set of those elements of S which are not in T. Thus,*

$$S - \bigcup_{i=1}^{n} S_i$$

consists of those elements of S which are not in any of the subsets $S_1, \ldots, S_n$. For brevity we write S_iS_j, $S_iS_jS_k, \ldots,$ for the intersections $S_i \cap S_j$, $S_i \cap S_j \cap S_k, \ldots$, respectively.

Theorem 5.31 Principle of cross-classification. *If $S_1, \ldots, S_n$ are given subsets of a finite set S, then*

$$N\left(S - \bigcup_{i=1}^{n} S_i\right) = N(S) - \sum_{1 \le i \le n} N(S_i) + \sum_{1 \le i < j \le n} N(S_iS_j)$$

$$- \sum_{1 \le i < j < k \le n} N(S_iS_jS_k) + \cdots + (-1)^n N(S_1S_2 \cdots S_n).$$

PROOF. If $T \subseteq S$ let $N_r(T)$ denote the number of elements of T which are not in any of the first r subsets $S_1, \ldots, S_r$, with $N_0(T)$ being simply $N(T)$. The elements enumerated by $N_{r-1}(T)$ fall into two disjoint sets, those which are not in S_r and those which are in S_r. Therefore we have

$$N_{r-1}(T) = N_r(T) + N_{r-1}(TS_r).$$

Hence

$$N_r(T) = N_{r-1}(T) - N_{r-1}(TS_r). \tag{26}$$

Now take $T = S$ and use (26) to express each term on the right in terms of N_{r-2}. We obtain

$$\begin{aligned} N_r(S) &= \{N_{r-2}(S) - N_{r-2}(SS_{r-1})\} - \{N_{r-2}(S_r) - N_{r-2}(S_rS_{r-1})\} \\ &= N_{r-2}(S) - N_{r-2}(S_{r-1}) - N_{r-2}(S_r) + N_{r-2}(S_rS_{r-1}). \end{aligned}$$

Applying (26) repeatedly we finally obtain

$$N_r(S) = N_0(S) - \sum_{i=1}^{r} N_0(S_i) + \sum_{1 \le i < j \le r} N_0(S_iS_j) - \cdots + (-1)^r N_0(S_1 \cdots S_r).$$

When $r = n$ this gives the required formula. □

EXAMPLE The product formula for Euler's totient can be derived from the cross-classification principle. Let $p_1, \ldots, p_r$ denote the distinct prime divisors of n. Let $S = \{1, 2, \ldots, n\}$ and let S_k be the subset of S consisting of those integers divisible by p_k. The numbers in S relatively prime to n are those in none of the sets $S_1, \ldots, S_r$, so

$$\varphi(n) = N\left(S - \bigcup_{k=1}^{r} S_k\right).$$

If $d \mid n$ there are n/d multiples of d in the set S. Hence

$$N(S_i) = \frac{n}{p_i},\ N(S_i S_j) = \frac{n}{p_i p_j}, \ldots, N(S_1 \cdots S_r) = \frac{n}{p_1 \cdots p_r},$$

so the cross-classification principle gives us

$$\begin{aligned} \varphi(n) &= n - \sum_{i=1}^{n} \frac{n}{p_i} + \sum_{1 \le i < j \le r} \frac{n}{p_i p_j} - \cdots + (-1)^r \frac{n}{p_1 \cdots p_r} \\ &= n \sum_{d \mid n} \frac{\mu(d)}{d} = n \prod_{p \mid n} \left(1 - \frac{1}{p}\right). \end{aligned}$$

The next application of the cross-classification principle counts the number of elements in a reduced residue system mod k which belong to a given residue class r mod d, where $d \mid k$ and $(r, d) = 1$.

Theorem 5.32 *Given integers r, d and k such that $d \mid k$, $d > 0$, $k \ge 1$ and $(r, d) = 1$. Then the number of elements in the set*

$$S = \{r + td : t = 1, 2, \ldots, k/d\}$$

which are relatively prime to k is $\varphi(k)/\varphi(d)$.

PROOF. If a prime p divides k and $r + td$ then $p \nmid d$, otherwise $p \mid r$, contradicting the hypothesis $(r, d) = 1$. Therefore, the primes which divide k and elements of S are those which divide k but do not divide d. Call them $p_1, \ldots, p_m$ and let

$$k' = p_1 p_2 \cdots p_m.$$

Then the elements of S relatively prime to k are those not divisible by any of these primes. Let

$$S_i = \{x : x \in S \text{ and } p_i \mid x\} \qquad (i = 1, 2, \ldots, m).$$

If $x \in S_i$ and $x = r + td$, then $r + td \equiv 0 \pmod{p_i}$. Since $p_i \nmid d$ there is a unique t mod p_i with this property, therefore exactly one t in each of the intervals $[1, p_i]$, $[p_i + 1, 2p_i], \ldots, [(q-1)p_i + 1, qp_i]$ where $qp_i = k/d$.

Therefore

$$N(S_i) = \frac{k/d}{p_i}.$$

Similarly,

$$N(S_iS_j) = \frac{k/d}{p_ip_j}, \ldots, N(S_1 \cdots S_m) = \frac{k/d}{p_1 \cdots p_m}.$$

Hence by the cross-classification principle the number of integers in S which are relatively prime to k is

$$N\left(S - \bigcup_{i=1}^{m} S_i\right) = \frac{k}{d} \sum_{\delta|k'} \frac{\mu(\delta)}{\delta} = \frac{k}{d} \prod_{p|k'} \left(1 - \frac{1}{p}\right) = \frac{k \prod_{p|k} \left(1 - \frac{1}{p}\right)}{d \prod_{p|d} \left(1 - \frac{1}{p}\right)} = \frac{\varphi(k)}{\varphi(d)}. \quad \square$$

5.11 A decomposition property of reduced residue systems

As an application of the foregoing theorem we discuss a property of reduced residue systems which will be used in a later chapter. We begin with a numerical example.

Let S be a reduced residue system mod 15, say

$$S = \{1, 2, 4, 7, 8, 11, 13, 14\}.$$

We display the 8 elements of S in a 4×2 matrix as follows:

$$\begin{bmatrix} 1 & 2 \\ 4 & 8 \\ 7 & 11 \\ 13 & 14 \end{bmatrix}.$$

Note that each row contains a reduced residue system mod 3, and the numbers in each column are congruent to each other mod 3. This example illustrates a general property of reduced residue systems described in the following theorem.

Theorem 5.33 *Let S be a reduced residue system* mod k, *and let $d > 0$ be a divisor of k. Then we have the following decompositions of S:*

(a) *S is the union of $\varphi(k)/\varphi(d)$ disjoint sets, each of which is a reduced residue system* mod d.
(b) *S is the union of $\varphi(d)$ disjoint sets, each of which consists of $\varphi(k)/\varphi(d)$ numbers congruent to each other* mod d.

Note. In the foregoing example, $k = 15$ and $d = 3$. The rows of the matrix represent the disjoint sets of part (a), and the columns represent the disjoint sets of part (b). If we apply the theorem to the divisor $d = 5$ we obtain the decomposition given by the matrix

$$\begin{bmatrix} 1 & 2 & 4 & 8 \\ 11 & 7 & 14 & 13 \end{bmatrix}.$$

Each row is a reduced residue system mod 5 and each column consists of numbers congruent to each other mod 5.

PROOF. First we prove that properties (a) and (b) are equivalent. If (b) holds we can display the $\varphi(k)$ elements of S as a matrix, using the $\varphi(d)$ disjoint sets of (b) as columns. This matrix has $\varphi(k)/\varphi(d)$ rows. Each row contains a reduced system mod d, and these are the disjoint sets required for part (a). Similarly, it is easy to verify that (a) implies (b).

Now we prove (b). Let S_d be a given reduced residue system mod d, and suppose $r \in S_d$. We will prove that there are at least $\varphi(k)/\varphi(d)$ integers n in S, distinct mod k, such that $n \equiv r \pmod{d}$. Since there are $\varphi(d)$ values of r in S_d and $\varphi(k)$ integers in S, there can't be more than $\varphi(k)/\varphi(d)$ such numbers n, so this will prove part (b).

The required numbers n will be selected from the residue classes mod k represented by the following k/d integers:

$$r, r + d, r + 2d, \ldots, r + \frac{k}{d} d.$$

These numbers are congruent to each other mod d and they are incongruent mod k. Since $(r, d) = 1$, Theorem 5.32 shows that $\varphi(k)/\varphi(d)$ of them are relatively prime to k, so this completes the proof. (For a different proof based on group theory see [1].) □

Exercises for Chapter 5

1. Let S be a set of n integers (not necessarily distinct). Prove that some nonempty subset of S has a sum which is divisible by n.

2. Prove that $5n^3 + 7n^5 \equiv 0 \pmod{12}$ for all integers n.

3. (a) Find all positive integers n for which $n^{13} \equiv n \pmod{1365}$.
   (b) Find all positive integers n for which $n^{17} \equiv n \pmod{4080}$.

4. (a) Prove that $\varphi(n) \equiv 2 \pmod{4}$ when $n = 4$ and when $n = p^a$, where p is a prime, $p \equiv 3 \pmod{4}$.
   (b) Find all n for which $\varphi(n) \equiv 2 \pmod{4}$.

5. A yardstick divided into inches is again divided into 70 equal parts. Prove that among the four shortest divisions two have left endpoints corresponding to 1 and 19 inches. What are the right endpoints of the other two?

6. Find all x which simultaneously satisfy the system of congruences

$$x \equiv 1 \pmod{3}, \qquad x \equiv 2 \pmod{4}, \qquad x \equiv 3 \pmod{5}.$$

7. Prove the converse of Wilson's theorem: *If* $(n - 1)! + 1 \equiv 0 \pmod{n}$, *then* n *is prime if* $n > 1$.

8. Find all positive integers n for which $(n - 1)! + 1$ is a power of n.

9. If p is an odd prime, let $q = (p - 1)/2$. Prove that

$$(q!)^2 + (-1)^q \equiv 0 \pmod{p}.$$

This gives $q!$ as an explicit solution to the congruence $x^2 + 1 \equiv 0 \pmod{p}$ when $p \equiv 1 \pmod{4}$, and it shows that $q! \equiv \pm 1 \pmod{p}$ if $p \equiv 3 \pmod{4}$. No simple general rule is known for determining the sign.

10. If p is an odd prime, prove that

$$1^2 3^2 5^2 \cdots (p - 2)^2 \equiv (-1)^{(p+1)/2} \pmod{p}$$

and

$$2^2 4^2 6^2 \cdots (p - 1)^2 \equiv (-1)^{(p+1)/2} \pmod{p}.$$

11. Let p be a prime, $p \geq 5$, and write

$$1 + \frac{1}{2} + \frac{1}{3} + \cdots + \frac{1}{p} = \frac{r}{ps}.$$

Prove that $p^3 \mid (r - s)$.

12. If p is a prime, prove that

$$\binom{n}{p} \equiv \left[\frac{n}{p}\right] \pmod{p}.$$

Also, if $p^\alpha \mid [n/p]$ prove that

$$p^\alpha \,\Big|\, \binom{n}{p}.$$

13. Let a, b, n be positive integers such that n divides $a^n - b^n$. Prove that n also divides $(a^n - b^n)/(a - b)$.

14. Let a, b, and x_0 be positive integers and define

$$x_n = ax_{n-1} + b \quad \text{for } n = 1, 2, \ldots$$

Prove that not all the x_n can be primes.

15. Let n, r, a denote positive integers. The congruence $n^2 \equiv n \pmod{10^a}$ implies $n^r \equiv n \pmod{10^a}$ for all r. Find all values of r such that $n^r \equiv n \pmod{10^a}$ implies $n^2 \equiv n \pmod{10^a}$.

16. Let n, a, d be given integers with $(a, d) = 1$. Prove that there exists an integer m such that $m \equiv a \pmod{d}$ and $(m, n) = 1$.

17. Let f be an integer-valued arithmetical function such that

$$f(m + n) \equiv f(n) \pmod{m}$$

for all $m \geq 1$, $n \geq 1$. Let $g(n)$ be the number of values (including repetitions) of $f(1), f(2), \ldots, f(n)$ divisible by n, and let $h(n)$ be the number of these values relatively prime to n. Prove that

$$h(n) = n \sum_{d|n} \mu(d) \frac{g(d)}{d}.$$

18. Given an odd integer $n > 3$, let k and t be the smallest positive integers such that both $kn + 1$ and tn are squares. Prove that n is prime if, and only if, both k and t are greater than $n/4$.

19. Prove that each member of the set of $n - 1$ consecutive integers

$$n! + 2, n! + 3, \ldots, n! + n$$

is divisible by a prime which does not divide any other member of the set.

20. Prove that for any positive integers n and k, there exists a set of n consecutive integers such that each member of this set is divisible by k distinct prime factors no one of which divides any other member of the set.

21. Let n be a positive integer which is not a square. Prove that for every integer a relatively prime to n there exist integers x and y satisfying

$$ax \equiv y \pmod{n} \quad \text{with } 0 < x < \sqrt{n} \text{ and } 0 < |y| < \sqrt{n}.$$

22. Let p be a prime, $p \equiv 1 \pmod 4$, let $q = (p - 1)/2$, and let $a = q!$.

(a) Prove that there exist positive integers x and y satisfying $0 < x < \sqrt{p}$ and $0 < y < \sqrt{p}$ such that

$$a^2x^2 - y^2 \equiv 0 \pmod{p}.$$

(b) For the x and y in part (a), prove that $p = x^2 + y^2$. This shows that every prime $p \equiv 1 \pmod 4$ is the sum of two squares.

(c) Prove that no prime $p \equiv 3 \pmod 4$ is the sum of two squares.

Finite Abelian Groups and Their Characters 6

6.1 Definitions

In Chapter 2 we had occasion to mention groups but made no essential use of their properties. Now we wish to discuss some elementary aspects of group theory in more detail. In Chapter 7 our discussion of Dirichlet's theorem on primes in arithmetical progressions will require a knowledge of certain arithmetical functions called *Dirichlet characters*. Although the study of Dirichlet characters can be undertaken without any knowledge of groups, the introduction of a minimal amount of group theory places the theory of Dirichlet characters in a more natural setting and simplifies some of the discussion.

Definition Postulates for a group. A group G is a nonempty set of elements together with a binary operation, which we denote by $\cdot$, such that the following postulates are satisfied:

(a) *Closure*. For every a and b in G, $a \cdot b$ is also in G.
(b) *Associativity*. For every a, b, c in G, we have $(a \cdot b) \cdot c = a \cdot (b \cdot c)$.
(c) *Existence of identity*. There is a unique element e in G, called the identity, such that $a \cdot e = e \cdot a = a$ for every a in G.
(d) *Existence of inverses*. For every a in G there is a unique element b in G such that $a \cdot b = b \cdot a = e$. This b is denoted by a^{-1} and is called the inverse of a.

Note. We usually omit the dot and write ab for $a \cdot b$.

Definition Abelian group. A group G is called abelian if every pair of elements commute; that is, if $ab = ba$ for all a and b in G.

Definition Finite group. A group G is called finite if G is a finite set. In this case the number of elements in G is called the order of G and is denoted by $|G|$.

Definition Subgroup. A nonempty subset G' of a group G which is itself a group, under the same operation, is called a subgroup of G.

6.2 Examples of groups and subgroups

EXAMPLE 1 Trivial subgroups. Every group G has at least two subgroups, G itself and the set $\{e\}$ consisting of the identity element alone.

EXAMPLE 2 Integers under addition. The set of all integers is an abelian group with $+$ as the operation and 0 as the identity. The inverse of n is $-n$.

EXAMPLE 3 Complex numbers under multiplication. The set of all non-zero complex numbers is an abelian group with ordinary multiplication of complex numbers as the operation and 1 as the identity. The inverse of z is the reciprocal $1/z$. The set of all complex numbers of absolute value 1 is a subgroup.

EXAMPLE 4 The nth roots of unity. The groups in Examples 2 and 3 are infinite groups. An example of a finite group is the set $\{1, \varepsilon, \varepsilon^2, \ldots, \varepsilon^{n-1}\}$, where $\varepsilon = e^{2\pi i/n}$ and the operation $\cdot$ is ordinary multiplication of complex numbers. This group, of order n, is called the group of nth roots of unity. It is a subgroup of both groups in Example 3.

6.3 Elementary properties of groups

The following elementary theorems concern an arbitrary group G. Unless otherwise stated, G is not required to be abelian nor finite.

Theorem 6.1 Cancellation laws. *If elements a, b, c in G satisfy*

$$ac = bc \quad or \quad ca = cb,$$

then $a = b$.

PROOF. In the first case multiply each member on the right by c^{-1} and use associativity. In the second case multiply on the left by c^{-1}. □

Theorem 6.2 Properties of inverses. *In any group G we have*:

(a) $e^{-1} = e$.
(b) *For every a in G, $(a^{-1})^{-1} = a$.*
(c) *For all a and b in G, $(ab)^{-1} = b^{-1}a^{-1}$. (Note reversal of order.)*
(d) *For all a and b in G the equation $ax = b$ has the unique solution $x = a^{-1}b$; the equation $ya = b$ has the unique solution $y = ba^{-1}$.*

PROOF.

(a) Since $ee = ee^{-1}$ we cancel e to obtain $e = e^{-1}$.
(b) Since $aa^{-1} = e$ and inverses are unique, a is the inverse of a^{-1}.
(c) By associativity we have

$$(ab)(b^{-1}a^{-1}) = a(bb^{-1})a^{-1} = aea^{-1} = aa^{-1} = e$$

so $b^{-1}a^{-1}$ is the inverse of ab.
(d) Again by associativity we have

$$a(a^{-1}b) = (aa^{-1})b = b \qquad \text{and } (ba^{-1})a = b(a^{-1}a) = b.$$

The solutions are unique because of the cancellation laws. □

Definition Powers of an element. If $a \in G$ we define a^n for any integer n by the following relations:

$$a^0 = e, \qquad a^n = aa^{n-1}, \qquad a^{-n} = (a^{-1})^n \quad \text{for } n > 0.$$

The following laws of exponents can be proved by induction. We omit the proofs.

Theorem 6.3 *If $a \in G$, any two powers of a commute, and for all integers m and n we have*

$$a^m a^n = a^{m+n} = a^n a^m \qquad \text{and } (a^m)^n = a^{mn} = (a^n)^m.$$

Moreover, if a and b commute we have

$$a^n b^n = (ab)^n.$$

Theorem 6.4 Subgroup criterion. *If G' is a nonempty subset of a group G, then G' is a subgroup if, and only if, G' satisfies group postulates* (a) *and* (d):
(a) *Closure: If $a, b \in G'$, then $ab \in G'$.*
(d) *Existence of inverse. If $a \in G'$, then $a^{-1} \in G'$.*

PROOF. Every subgroup G' certainly has these properties. Conversely, if G' satisfies (a) and (d) it is easy to show that G' also satisfies postulates (b) and (c). Postulate (b), associativity, holds in G' because it holds for all elements in G. To prove that (c) holds in G' we note that there is an element a in G' (since G' is nonempty) whose inverse $a^{-1} \in G'$ (by (d)) hence $aa^{-1} \in G'$ by (a). But $aa^{-1} = e$ so $e \in G'$. □

6.4 Construction of subgroups

A subgroup of a given group G can always be constructed by choosing any element a in G and forming the set of all its powers a^n, $n = 0, \pm 1, \pm 2, \ldots$ This set clearly satisfies postulates (a) and (d) so is a subgroup of G. It is called the *cyclic subgroup generated by* a and is denoted by $\langle a \rangle$.

Note that $\langle a \rangle$ is abelian, even if G is not. If $a^n = e$ for some positive integer n there will be a smallest $n > 0$ with this property and the subgroup $\langle a \rangle$ will be a finite group of order n,

$$\langle a \rangle = \{a, a^2, \ldots, a^{n-1}, a^n = e\}.$$

The integer n is also called the *order of the element a*. An example of a cyclic subgroup of order n is the group of nth roots of unity mentioned in Section 6.2.

The next theorem shows that every element of a finite group has finite order.

Theorem 6.5 *If G is finite and $a \in G$, then there is a positive integer $n \le |G|$ such that $a^n = e$.*

PROOF. Let $g = |G|$. Then at least two of the following $g + 1$ elements of G must be equal:

$$e, a, a^2, \ldots, a^g.$$

Suppose that $a^r = a^s$, where $0 \le s < r \le g$. Then we have

$$e = a^r(a^s)^{-1} = a^{r-s}.$$

This proves the theorem with $n = r - s$. □

As noted in Section 6.2, every group G has two trivial subgroups, $\{e\}$ and G itself. When G is a finite abelian group there is a simple process for constructing an increasing collection of subgroups intermediate to $\{e\}$ and G. The process, which will be described in Theorem 6.8, is based on the following observation.

If G' is a subgroup of a finite group G, then for any element a in G there is an integer n such that $a^n \in G'$. If a is already in G' we simply take $n = 1$. If $a \notin G'$ we can take n to be the order of a, since $a^n = e \in G'$. However, there may be a smaller positive power of a which lies in G'. By the well-ordering principle there is a *smallest* positive integer n such that $a^n \in G'$. We call this integer the *indicator* of a in G'.

Theorem 6.6 *Let G' be a subgroup of a finite abelian group G, where $G' \neq G$. Choose an element a in G, $a \notin G'$, and let h be the indicator of a in G'. Then the set of products*

$$G'' = \{xa^k : x \in G' \text{ and } k = 0, 1, 2, \ldots, h - 1\}$$

is a subgroup of G which contains G'. Moreover, the order of G'' is h times that of G',

$$|G''| = h|G'|.$$

Proof. To show G'' is a subgroup we use the subgroup criterion. First we test closure. Choose two elements in G'', say xa^k and ya^j, where $x, y \in G'$ and $0 \le k < h$, $0 \le j < h$. Since G is abelian the product of the elements is

$$(xy)a^{k+j}. \tag{1}$$

Now $k + j = qh + r$ where $0 \le r < h$. Hence

$$a^{k+j} = a^{qh+r} = a^{qh}a^r = za^r,$$

where $z = a^{qh} = (a^h)^q \in G'$ since $a^h \in G'$. Therefore the element in (1) is $(xyz)a^r = wa^r$, where $w \in G'$ and $0 \le r < h$. This proves that G'' satisfies the closure postulate.

Next we show that the inverse of each element in G'' is also in G''. Choose an arbitrary element in G'', say xa^k. If $k = 0$ then the inverse is x^{-1} which is in G''. If $0 < k < h$ the inverse is the element

$$ya^{h-k}, \quad \text{where } y = x^{-1}(a^h)^{-1},$$

which again is in G''. This shows that G'' is indeed a subgroup of G. Clearly G'' contains G'.

Next we determine the order of G''. Let $m = |G'|$. As x runs through the m elements of G' and k runs through the h integers $0, 1, 2, \ldots, h - 1$ we obtain mh products xa^k. If we show that all these are *distinct*, then G'' has order mh. Consider two of these products, say xa^k and ya^j and assume that

$$xa^k = ya^j \quad \text{with } 0 \le j \le k < h.$$

Then $a^{k-j} = x^{-1}y$ and $0 \le k - j < h$. Since $x^{-1}y \in G'$ we must have a^{k-j} in G' so $k = j$ and hence $x = y$. This completes the proof. □

6.5 Characters of finite abelian groups

Definition Let G be an arbitrary group. A complex-valued function f defined on G is called a character of G if f has the multiplicative property

$$f(ab) = f(a)f(b)$$

for all a, b in G, and if $f(c) \ne 0$ for some c in G.

Theorem 6.7 *If f is a character of a finite group G with identity element e, then $f(e) = 1$ and each function value $f(a)$ is a root of unity. In fact, if $a^n = e$ then $f(a)^n = 1$.*

Proof. Choose c in G such that $f(c) \ne 0$. Since $ce = c$ we have

$$f(c)f(e) = f(c)$$

so $f(e) = 1$. If $a^n = e$ then $f(a)^n = f(a^n) = f(e) = 1$. □

EXAMPLE Every group G has at least one character, namely the function which is identically 1 on G. This is called the *principal* character. The next theorem tells us that there are further characters if G is abelian and has finite order > 1.

Theorem 6.8 *A finite abelian group G of order n has exactly n distinct characters.*

PROOF. In Theorem 6.6 we learned how to construct, from a given subgroup $G' \neq G$, a new subgroup G'' containing G' and at least one more element a not in G'. We use the symbol $\langle G'; a\rangle$ to denote the subgroup G'' constructed in Theorem 6.6. Thus

$$\langle G'; a\rangle = \{xa^k : x \in G' \text{ and } 0 \leq k < h\}$$

where h is the indicator of a in G'.

Now we apply this construction repeatedly, starting with the subgroup $\{e\}$ which we denote by G_1. If $G_1 \neq G$ we let a_1 be an element of G other than e and define $G_2 = \langle G_1; a_1\rangle$. If $G_2 \neq G$ let a_2 be an element of G which is not in G_2 and define $G_3 = \langle G_2; a_2\rangle$. Continue the process to obtain a finite set of elements $a_1, a_2, \ldots, a_t$ and a corresponding set of subgroups $G_1, G_2, \ldots, G_{t+1}$ such that

$$G_{r+1} = \langle G_r; a_r\rangle$$

with

$$G_1 \subset G_2 \subset \cdots \subset G_{t+1} = G.$$

The process must terminate in a finite number of steps since the given group G is finite and each G_{r+1} contains more elements than its predecessor G_r. We consider such a chain of subgroups and prove the theorem by induction, showing that if it is true for G_r it must also be true for G_{r+1}.

It is clear that there is only one character for G_1, namely the function which is identically 1. Assume, therefore, that G_r has order m and that there are exactly m distinct characters for G_r. Consider $G_{r+1} = \langle G_r; a_r\rangle$ and let h be the indicator of a_r in G_r, that is, the smallest positive integer such that $a_r^h \in G_r$. We shall show that there are exactly h different ways to extend each character of G_r to obtain a character of G_{r+1}, and that each character of G_{r+1} is the extension of some character of G_r. This will prove that G_{r+1} has exactly mh characters, and since mh is also the order of G_{r+1} this will prove the theorem by induction on r.

A typical element in G_{r+1} has the form

$$xa_r^k, \quad \text{where } x \in G_r \text{ and } 0 \leq k < h.$$

Suppose for the moment that it is possible to extend a character f of G_r to G_{r+1}. Call this extension $\tilde{f}$ and let us see what can be said about $\tilde{f}(xa_r^k)$. The multiplicative property requires

$$\tilde{f}(xa_r^k) = \tilde{f}(x)\tilde{f}(a_r)^k.$$

But $x \in G_r$ so $\tilde{f}(x) = f(x)$ and the foregoing equation implies

$$\tilde{f}(xa_r^k) = f(x)\tilde{f}(a_r)^k.$$

This tells us that $\tilde{f}(xa_r^k)$ is determined as soon as $\tilde{f}(a_r)$ is known.

What are the possible values for $\tilde{f}(a_r)$? Let $c = a_r^h$. Since $c \in G_r$ we have $\tilde{f}(c) = f(c)$, and since $\tilde{f}$ is multiplicative we also have $\tilde{f}(c) = \tilde{f}(a_r)^h$. Hence

$$\tilde{f}(a_r)^h = f(c),$$

so $\tilde{f}(a_r)$ is one of the hth roots of $f(c)$. Therefore there are at most h choices for $\tilde{f}(a_r)$.

These observations tell us how to define $\tilde{f}$. If f is a given character of G_r, we choose one of the hth roots of $f(c)$, where $c = a_r^h$, and define $\tilde{f}(a_r)$ to be this root. Then we define $\tilde{f}$ on the rest of G_{r+1} by the equation

$$\tilde{f}(xa_r^k) = f(x)\tilde{f}(a_r)^k. \tag{2}$$

The h choices for $\tilde{f}(a_r)$ are all different so this gives us h different ways to define $\tilde{f}(xa_r^k)$. Now we verify that the function $\tilde{f}$ so defined has the required multiplicative property. From (2) we find

$$\begin{aligned}\tilde{f}(xa_r^k \cdot ya_r^j) &= \tilde{f}(xy \cdot a_r^{k+j}) = f(xy)\tilde{f}(a_r)^{k+j} \\ &= f(x)f(y)\tilde{f}(a_r)^k\tilde{f}(a_r)^j \\ &= \tilde{f}(xa_r^k)\tilde{f}(ya_r^j),\end{aligned}$$

so $\tilde{f}$ is a character of G_{r+1}. No two of the extensions $\tilde{f}$ and $\tilde{g}$ can be identical on G_{r+1} because the functions f and g which they extend would then be identical on G_r. Therefore each of the m characters of G_r can be extended in h different ways to produce a character of G_{r+1}. Moreover, if φ is any character of G_{r+1} then its restriction to G_r is also a character of G_r, so the extension process produces all the characters of G_{r+1}. This completes the proof. □

6.6 The character group

In this section G is a finite abelian group of order n. The principal character of G is denoted by f_1. The others, denoted by $f_2, f_3, \ldots, f_n$, are called nonprincipal characters. They have the property that $f(a) \neq 1$ for some a in G.

Theorem 6.9 *If multiplication of characters is defined by the relation*

$$(f_i f_j)(a) = f_i(a)f_j(a)$$

for each a in G, then the set of characters of G forms an abelian group of order n. We denote this group by $\hat{G}$. The identity element of $\hat{G}$ is the principal character f_1. The inverse of f_i is the reciprocal $1/f_i$.

PROOF. Verification of the group postulates is a straightforward exercise and we omit the details.

Note. For each character f we have $|f(a)| = 1$. Hence the reciprocal $1/f(a)$ is equal to the complex conjugate $\overline{f(a)}$. Thus, the function $\bar{f}$ defined by $\bar{f}(a) = \overline{f(a)}$ is also a character of G. Moreover, we have

$$\bar{f}(a) = \frac{1}{f(a)} = f(a^{-1})$$

for every a in G.

6.7 The orthogonality relations for characters

Let G be a finite abelian group of order n with elements $a_1, a_2, \ldots, a_n$, and let $f_1, f_2, \ldots, f_n$ be the characters of G, with f_1 the principal character.

Notation *We denote by $A = A(G)$ the $n \times n$ matrix $[a_{ij}]$ whose element a_{ij} in the ith row and jth column is*

$$a_{ij} = f_i(a_j).$$

We will prove that the matrix A has an inverse and then use this fact to deduce the so-called orthogonality relations for characters. First we determine the sum of the entries in each row of A.

Theorem 6.10 *The sum of the entries in the* ith *row of A is given by*

$$\sum_{r=1}^{n} f_i(a_r) = \begin{cases} n & \text{if } f_i \text{ is the principal character } (i = 1), \\ 0 & \text{otherwise.} \end{cases}$$

PROOF. Let S denote the sum in question. If $f_i = f_1$ each term of the sum is 1 and $S = n$. If $f_i \neq f_1$, there is an element b in G for which $f_i(b) \neq 1$. As a_r runs through the elements of G so does the product ba_r. Hence

$$S = \sum_{r=1}^{n} f_i(ba_r) = f_i(b) \sum_{r=1}^{n} f_i(a_r) = f_i(b)S.$$

Therefore $S(1 - f_i(b)) = 0$. Since $f_i(b) \neq 1$ it follows that $S = 0$. □

Now we use this theorem to show that A has an inverse.

Theorem 6.11 *Let A^* denote the conjugate transpose of the matrix A. Then we have*

$$AA^* = nI,$$

where I is the $n \times n$ identity matrix. Hence $n^{-1}A^*$ is the inverse of A.

PROOF. Let $B = AA^*$. The entry b_{ij} in the ith row and jth column of B is given by

$$b_{ij} = \sum_{r=1}^{n} f_i(a_r)\bar{f}_j(a_r) = \sum_{r=1}^{n} (f_i \bar{f}_j)(a_r) = \sum_{r=1}^{n} f_k(a_r),$$

where $f_k = f_i \bar{f}_j = f_i/f_j$. Now $f_i/f_j = f_1$ if, and only if, $i = j$. Hence by Theorem 6.10 we have

$$b_{ij} = \begin{cases} n & \text{if } i = j, \\ 0 & \text{if } i \neq j. \end{cases}$$

In other words, $B = nI$. □

Next we use the fact that a matrix commutes with its inverse to deduce the orthogonality relations for characters.

Theorem 6.12 Orthogonality relations for characters. *We have*

$$\sum_{r=1}^{n} \bar{f}_r(a_i) f_r(a_j) = \begin{cases} n & \text{if } a_i = a_j, \\ 0 & \text{if } a_i \neq a_j. \end{cases} \tag{3}$$

PROOF. The relation $AA^* = nI$ implies $A^*A = nI$. But the element in the ith row and jth column of A^*A is the sum on the left of (3). This completes the proof. □

Note. Since $\bar{f}_r(a_i) = f_r(a_i)^{-1} = f_r(a_i^{-1})$, the general term of the sum in (3) is equal to $f_r(a_i^{-1})f_r(a_j) = f_r(a_i^{-1}a_j)$. Therefore the orthogonality relations can also be expressed as follows:

$$\sum_{r=1}^{n} f_r(a_i^{-1}a_j) = \begin{cases} n & \text{if } a_i = a_j, \\ 0 & \text{if } a_i \neq a_j. \end{cases}$$

When a_i is the identity element e we obtain:

Theorem 6.13 *The sum of the entries in the* jth *column of* A *is given by*

$$\sum_{r=1}^{n} f_r(a_j) = \begin{cases} n & \text{if } a_j = e, \\ 0 & \text{otherwise}. \end{cases} \tag{4}$$

6.8 Dirichlet characters

The foregoing discussion dealt with characters of an arbitrary finite abelian group G. Now we specialize G to be the group of reduced residue classes modulo a fixed positive integer k. First we prove that these residue classes do, indeed, form a group if multiplication is suitably defined.

We recall that a reduced residue system modulo k is a set of $\varphi(k)$ integers $\{a_1, a_2, \ldots, a_{\varphi(k)}\}$ incongruent modulo k, each of which is relatively prime to

k. For each integer a the corresponding residue class $\hat{a}$ is the set of all integers congruent to a modulo k:

$$\hat{a} = \{x: x \equiv a \pmod{k}\}.$$

Multiplication of residue classes is defined by the relation

$$\hat{a} \cdot \hat{b} = \widehat{ab}. \tag{5}$$

That is, the product of two residue classes $\hat{a}$ and $\hat{b}$ is the residue class of the product ab.

Theorem 6.14 *With multiplication defined by* (5), *the set of reduced residue classes modulo k is a finite abelian group of order $\varphi(k)$. The identity is the residue class $\hat{1}$. The inverse of $\hat{a}$ is the residue class $\hat{b}$ where $ab \equiv 1 \pmod{k}$.*

PROOF. The closure property is automatically satisfied because of the way multiplication of residue classes was defined. The class $\hat{1}$ is clearly the identity element. If $(a, k) = 1$ there is a unique b such that $ab \equiv 1 \pmod{k}$. Hence the inverse of $\hat{a}$ is $\hat{b}$. Finally, it is clear that the group is abelian and that its order is $\varphi(k)$. □

Definition Dirichlet characters. Let G be the group of reduced residue classes modulo k. Corresponding to each character f of G we define an arithmetical function $\chi = \chi_f$ as follows:

$$\chi(n) = f(\hat{n}) \quad \text{if } (n, k) = 1,$$

$$\chi(n) = 0 \quad \text{if } (n, k) > 1.$$

The function χ is called a Dirichlet character modulo k. The principal character χ_1 is that which has the properties

$$\chi_1(n) = \begin{cases} 1 & \text{if } (n, k) = 1, \\ 0 & \text{if } (n, k) > 1. \end{cases}$$

Theorem 6.15 *There are $\varphi(k)$ distinct Dirichlet characters modulo k, each of which is completely multiplicative and periodic with period k. That is, we have*

$$\chi(mn) = \chi(m)\chi(n) \quad \textit{for all } m, n \tag{6}$$

and

$$\chi(n + k) = \chi(n) \quad \textit{for all } n.$$

Conversely, if χ is completely multiplicative and periodic with period k, and if $\chi(n) = 0$ if $(n, k) > 1$, then χ is one of the Dirichlet characters mod k.

PROOF. There are $\varphi(k)$ characters f for the group G of reduced residue classes modulo k, hence $\varphi(k)$ characters χ_f modulo k. The multiplicative property (6) of χ_f follows from that of f when both m and n are relatively prime to k. If one of m or n is not relatively prime to k then neither is mn, hence both members of (6) are zero. The periodicity property follows from the fact that $\chi_f(n) = f(\hat{n})$ and that $a \equiv b \pmod{k}$ implies $(a, k) = (b, k)$.

To prove the converse we note that the function f defined on the group G by the equation

$$f(\hat{n}) = \chi(n) \quad \text{if } (n, k) = 1$$

is a character of G, so χ is a Dirichlet character mod k. □

EXAMPLE When $k = 1$ or $k = 2$ then $\varphi(k) = 1$ and the only Dirichlet character is the principal character χ_1. For $k \geq 3$, there are at least two Dirichlet characters since $\varphi(k) \geq 2$. The following tables display all the Dirichlet characters for $k = 3$, 4 and 5.

n	1	2	3
$\chi_1(n)$	1	1	0
$\chi_2(n)$	1	-1	0

$k = 3, \varphi(k) = 2$

n	1	2	3	4
$\chi_1(n)$	1	0	1	0
$\chi_2(n)$	1	0	-1	0

$k = 4, \varphi(k) = 2$

n	1	2	3	4	5
$\chi_1(n)$	1	1	1	1	0
$\chi_2(n)$	1	-1	-1	1	0
$\chi_3(n)$	1	i	$-i$	-1	0
$\chi_4(n)$	1	$-i$	i	-1	0

$k = 5, \varphi(k) = 4$

To fill these tables we use the fact that $\chi(n)^{\varphi(k)} = 1$ whenever $(n, k) = 1$, so $\chi(n)$ is a $\varphi(k)$th root of unity. We also note that if χ is a character mod k so is the complex conjugate $\bar{\chi}$. This information suffices to complete the tables for $k = 3$ and $k = 4$.

When $k = 5$ we have $\varphi(5) = 4$ so the possible values of $\chi(n)$ are ± 1 and $\pm i$ when $(n, 5) = 1$. Also, $\chi(2)\chi(3) = \chi(6) = \chi(1) = 1$ so $\chi(2)$ and $\chi(3)$ are reciprocals. Since $\chi(4) = \chi(2)^2$ this information suffices to fill the table for $k = 5$. As a check we can use Theorems 6.10 and 6.13 which tell us that the sum of the entries is 0 in each row and column except for the first. The following tables display all the Dirichlet characters mod 6 and 7.

n	1	2	3	4	5	6
$\chi_1(n)$	1	0	0	0	1	0
$\chi_2(n)$	1	0	0	0	-1	0

$k = 6, \varphi(k) = 2$

n	1	2	3	4	5	6	7
$\chi_1(n)$	1	1	1	1	1	1	0
$\chi_2(n)$	1	1	-1	1	-1	-1	0
$\chi_3(n)$	1	ω^2	ω	$-\omega$	$-\omega^2$	-1	0
$\chi_4(n)$	1	ω^2	$-\omega$	$-\omega$	ω^2	1	0
$\chi_5(n)$	1	$-\omega$	ω^2	ω^2	$-\omega$	1	0
$\chi_6(n)$	1	$-\omega$	$-\omega^2$	ω^2	ω	-1	0

$\omega = e^{\pi i/3}$

$k = 7, \varphi(k) = 6$

In our discussion of Dirichlet's theorem on primes in an arithmetic progression we shall make use of the following orthogonality relation for characters modulo k.

Theorem 6.16 *Let $\chi_1, \ldots, \chi_{\varphi(k)}$ denote the $\varphi(k)$ Dirichlet characters modulo k. Let m and n be two integers, with $(n, k) = 1$. Then we have*

$$\sum_{r=1}^{\varphi(k)} \chi_r(m)\bar{\chi}_r(n) = \begin{cases} \varphi(k) & \text{if } m \equiv n \pmod{k}, \\ 0 & \text{if } m \not\equiv n \pmod{k}. \end{cases}$$

PROOF. If $(m, k) = 1$ take $a_i = \hat{n}$ and $a_j = \hat{m}$ in the orthogonality relations of Theorem 6.12 and note that $\hat{m} = \hat{n}$ if, and only if, $m \equiv n \pmod{k}$. If $(m, k) > 1$ each term in the sum vanishes and $m \not\equiv n \pmod{k}$. □

6.9 Sums involving Dirichlet characters

This section discusses certain sums which occur in the proof of Dirichlet's theorem on primes in arithmetical progressions.

The first theorem refers to a nonprincipal character χ mod k, but the proof is also valid if χ is any arithmetical function with bounded partial sums.

Theorem 6.17 *Let χ be any nonprincipal character modulo k, and let f be a nonnegative function which has a continuous negative derivative $f'(x)$ for all $x \geq x_0$. Then if $y \geq x \geq x_0$ we have*

$$\sum_{x < n \leq y} \chi(n) f(n) = O(f(x)). \tag{7}$$

If, in addition, $f(x) \to 0$ as $x \to \infty$, then the infinite series

$$\sum_{n=1}^{\infty} \chi(n) f(n)$$

converges and we have, for $x \geq x_0$,

$$\sum_{n \leq x} \chi(n) f(n) = \sum_{n=1}^{\infty} \chi(n) f(n) + O(f(x)). \tag{8}$$

PROOF. Let $A(x) = \sum_{n \leq x} \chi(n)$. Since χ is nonprincipal we have

$$A(k) = \sum_{n=1}^{k} \chi(n) = 0.$$

By periodicity it follows that $A(nk) = 0$ for $n = 2, 3, \ldots$, hence $|A(x)| < \varphi(k)$ for all x. In other words, $A(x) = O(1)$.

Now we use Abel's identity (Theorem 4.2) to express the sum in (7) as an integral. This gives us

$$\sum_{x<n\leq y} \chi(n)f(n) = f(y)A(y) - f(x)A(x) - \int_x^y A(t)f'(t)\,dt$$

$$= O(f(y)) + O(f(x)) + O\left(\int_x^y (-f'(t))\,dt\right) = O(f(x)).$$

This proves (7). If $f(x) \to 0$ as $x \to \infty$ then (7) shows that the series

$$\sum_{n=1}^{\infty} \chi(n)f(n)$$

converges because of the Cauchy convergence criterion. To prove (8) we simply note that

$$\sum_{n=1}^{\infty} \chi(n)f(n) = \sum_{n\leq x} \chi(n)f(n) + \lim_{y\to\infty} \sum_{x<n\leq y} \chi(n)f(n).$$

Because of (7) the limit on the right is $O(f(x))$. This completes the proof. □

Now we apply Theorem 6.17 successively with $f(x) = 1/x$, $f(x) = (\log x)/x$, and $f(x) = 1/\sqrt{x}$ for $x \geq 1$ to obtain:

Theorem 6.18 *If χ is any nonprincipal character* mod k *and if $x \geq 1$ we have*

$$\sum_{n\leq x} \frac{\chi(n)}{n} = \sum_{n=1}^{\infty} \frac{\chi(n)}{n} + O\left(\frac{1}{x}\right), \tag{9}$$

$$\sum_{n\leq x} \frac{\chi(n)\log n}{n} = \sum_{n=1}^{\infty} \frac{\chi(n)\log n}{n} + O\left(\frac{\log x}{x}\right), \tag{10}$$

$$\sum_{n\leq x} \frac{\chi(n)}{\sqrt{n}} = \sum_{n=1}^{\infty} \frac{\chi(n)}{\sqrt{n}} + O\left(\frac{1}{\sqrt{x}}\right). \tag{11}$$

6.10 The nonvanishing of $L(1, \chi)$ for real nonprincipal χ

We denote by $L(1, \chi)$ the sum of the series in (9). Thus,

$$L(1, \chi) = \sum_{n=1}^{\infty} \frac{\chi(n)}{n}.$$

In the proof of Dirichlet's theorem we need to know that $L(1, \chi) \neq 0$ when χ is a nonprincipal character. We prove this here for real nonprincipal characters. First we consider the divisor sum of $\chi(n)$.

Theorem 6.19 *Let χ be any real-valued character* mod k *and let*

$$A(n) = \sum_{d|n} \chi(d).$$

Then $A(n) \geq 0$ for all n, and $A(n) \geq 1$ if n is a square.

PROOF. For prime powers we have

$$A(p^a) = \sum_{t=0}^{a} \chi(p^t) = 1 + \sum_{t=1}^{a} \chi(p)^t.$$

Since χ is real-valued the only possible values for $\chi(p)$ are 0, 1 and -1. If $\chi(p) = 0$ then $A(p^a) = 1$; if $\chi(p) = 1$ then $A(p^a) = a + 1$; and if $\chi(p) = -1$ then

$$A(p^a) = \begin{cases} 0 & \text{if } a \text{ is odd,} \\ 1 & \text{if } a \text{ is even.} \end{cases}$$

In any case, $A(p^a) \geq 1$ if a is even.

Now if $n = p_1^{a_1} \cdots p_r^{a_r}$ then $A(n) = A(p_1^{a_1}) \cdots A(p_r^{a_r})$ since A is multiplicative. Each factor $A(p_i^{a_i}) \geq 0$ hence $A(n) \geq 0$. Also, if n is a square then each exponent a_i is even, so each factor $A(p_i^{a_i}) \geq 1$ hence $A(n) \geq 1$. This proves the theorem. □

Theorem 6.20 *For any real-valued nonprincipal character χ* mod k, *let*

$$A(n) = \sum_{d|n} \chi(d) \qquad \text{and } B(x) = \sum_{n \leq x} \frac{A(n)}{\sqrt{n}}.$$

Then we have:

(a) $B(x) \to \infty$ *as* $x \to \infty$.
(b) $B(x) = 2\sqrt{x}L(1, \chi) + O(1)$ *for all* $x \geq 1$.

Therefore $L(1, \chi) \neq 0$.

PROOF. To prove part (a) we use Theorem 6.19 to write

$$B(x) \geq \sum_{\substack{n \leq x \\ n = m^2}} \frac{1}{\sqrt{n}} = \sum_{m \leq \sqrt{x}} \frac{1}{m}.$$

The last sum tends to ∞ as $x \to \infty$ since the harmonic series $\sum 1/m$ diverges.

To prove part (b) we write

$$B(x) = \sum_{n \leq x} \frac{1}{\sqrt{n}} \sum_{d|n} \chi(d) = \sum_{\substack{q,d \\ qd \leq x}} \frac{\chi(d)}{\sqrt{qd}}.$$

Now we invoke Theorem 3.17 which states that

$$\sum_{\substack{q,d \\ qd \le x}} f(d)g(q) = \sum_{n \le a} f(n)G\left(\frac{x}{n}\right) + \sum_{n \le b} g(n)F\left(\frac{x}{n}\right) - F(a)G(b)$$

where $ab = x$, $F(x) = \sum_{n \le x} f(n)$, and $G(x) = \sum_{n \le x} g(n)$. We take $a = b = \sqrt{x}$ and let $f(n) = \chi(n)/\sqrt{n}$, $g(n) = 1/\sqrt{n}$ to obtain

$$(12) \quad B(x) = \sum_{\substack{q,d \\ qd \le x}} \frac{\chi(d)}{\sqrt{qd}} = \sum_{n \le \sqrt{x}} \frac{\chi(n)}{\sqrt{n}} G\left(\frac{x}{n}\right) + \sum_{n \le \sqrt{x}} \frac{1}{\sqrt{n}} F\left(\frac{x}{n}\right) - F(\sqrt{x})G(\sqrt{x}).$$

By Theorem 3.2 we have

$$G(x) = \sum_{n \le x} \frac{1}{\sqrt{n}} = 2\sqrt{x} + A + O\left(\frac{1}{\sqrt{x}}\right)$$

where A is a constant, and by Theorem 6.18, Equation (11), we have

$$F(x) = \sum_{n \le x} \frac{\chi(n)}{\sqrt{n}} = B + O\left(\frac{1}{\sqrt{x}}\right),$$

where $B = \sum_{n=1}^{\infty} \chi(n)/\sqrt{n}$. Since $F(\sqrt{x})G(\sqrt{x}) = 2Bx^{1/4} + O(1)$, Equation (12) gives us

$$\begin{aligned} B(x) &= \sum_{n \le \sqrt{x}} \frac{\chi(n)}{\sqrt{n}} \left\{2\sqrt{\frac{x}{n}} + A + O\left(\sqrt{\frac{n}{x}}\right)\right\} \\ &\quad + \sum_{n \le \sqrt{x}} \frac{1}{\sqrt{n}} \left\{B + O\left(\sqrt{\frac{n}{x}}\right)\right\} - 2Bx^{1/4} + O(1) \\ &= 2\sqrt{x} \sum_{n \le \sqrt{x}} \frac{\chi(n)}{n} + A \sum_{n \le \sqrt{x}} \frac{\chi(n)}{\sqrt{n}} + O\left(\frac{1}{\sqrt{x}} \sum_{n \le \sqrt{x}} |\chi(n)|\right) \\ &\quad + B \sum_{n \le \sqrt{x}} \frac{1}{\sqrt{n}} + O\left(\frac{1}{\sqrt{x}} \sum_{n \le \sqrt{x}} 1\right) - 2Bx^{1/4} + O(1) \\ &= 2\sqrt{x}\, L(1, \chi) + O(1). \end{aligned}$$

This proves part (b). Now it is clear that parts (a) and (b) together imply that $L(1, \chi) \ne 0$. □

Exercises for Chapter 6

1. Let G be a set of nth roots of a nonzero complex number. If G is a group under multiplication, prove that G is the group of nth roots of unity.
2. Let G be a finite group of order n with identity element e. If $a_1, \ldots, a_n$ are n elements of G, not necessarily distinct, prove that there are integers p and q with $1 \le p \le q \le n$ such that $a_p a_{p+1} \cdots a_q = e$.

3. Let G be the set of all 2×2 matrices $\begin{pmatrix} a & b \\ c & d \end{pmatrix}$, where a, b, c, d are integers with $ad - bc = 1$. Prove that G is a group under matrix multiplication. This group is sometimes called the *modular group*.

4. Let $G = \langle a \rangle$ be a cyclic group generated by a. Prove that every subgroup of G is cyclic. (It is not assumed that G is finite.)

5. Let G be a finite group of order n and let G' be a subgroup of order m. Prove that $m \mid n$ (Lagrange's theorem). Deduce that the order of every element of G divides n.

6. Let G be a group of order 6 with identity element e. Prove that either G is cyclic, or else there are two elements a and b in G such that

$$G = \{a, a^2, a^3, b, ab, a^2b\},$$

with $a^3 = b^2 = e$. Which of these elements is ba?

7. A group table for a finite group $G = \{a_1, \ldots, a_n\}$ of order n is an $n \times n$ matrix whose ij-entry is $a_i a_j$. If $a_i a_j = e$ prove that $a_j a_i = e$. In other words, the identity element is symmetrically located in the group table. Deduce that if n is even the equation $x^2 = e$ has an even number of solutions.

8. Generalizing Exercise 7, let $f(p)$ denote the number of solutions of the equation $x^p = e$, where p is a prime divisor of n, the order of G. Prove that $p \mid f(p)$ (Cauchy's theorem). [*Hint*: Consider the set S of ordered p-tuples $(a_1, \ldots, a_p)$ such that $a_i \in G$ and $a_1 \cdots a_p = e$. There are n^{p-1} p-tuples in S. Call two such p-tuples equivalent if one is a cyclic permutation of the other. Show that $f(p)$ equivalence classes contain exactly one member and that each of the others contains exactly p members. Count the number of members of S in two ways and deduce that $p \mid f(p)$.]

9. Let G be a finite group of order n. Prove that n is odd if, and only if, each element of G is a square. That is, for each a in G there is an element b in G such that $a = b^2$.

10. State and prove a generalization of Exercise 9 in which the condition "n is odd" is replaced by "n is relatively prime to k" for some $k \geq 2$.

11. Let G be a finite group of order n, and let S be a subset containing more than $n/2$ elements of G. Prove that for each g in G there exist elements a and b in S such that $ab = g$.

12. Let G be a group and let S be a subset of n distinct elements of G with the property that $a \in S$ implies $a^{-1} \notin S$. Consider the n^2 products (not necessarily distinct) of the form ab, where $a \in S$ and $b \in S$. Prove that at most $n(n-1)/2$ of these products belong to S.

13. Let $f_1, \ldots, f_m$ be the characters of a finite group G of order m, and let a be an element of G of order n. Theorem 6.7 shows that each number $f_r(a)$ is an nth root of unity. Prove that every nth root of unity occurs equally often among the numbers $f_1(a)$, $f_2(a), \ldots, f_m(a)$. [*Hint*: Evaluate the sum

$$\sum_{r=1}^{m} \sum_{k=1}^{n} f_r(a^k) e^{-2\pi i k/n}$$

in two ways to determine the number of times $e^{2\pi i/n}$ occurs.]

14. Construct tables showing the values of all the Dirichlet characters mod k for $k = 8, 9$, and 10.

15. Let χ be any nonprincipal character mod k. Prove that for all integers $a < b$ we have

$$\left| \sum_{n=a}^{b} \chi(n) \right| \le \frac{1}{2} \varphi(k).$$

16. If χ is a real-valued character mod k then $\chi(n) = \pm 1$ or 0 for each n, so the sum

$$S = \sum_{n=1}^{k} n\chi(n)$$

is an integer. This exercise shows that $12S \equiv 0 \pmod{k}$.

(a) If $(a, k) = 1$ prove that $a\chi(a)S \equiv S \pmod{k}$.
(b) Write $k = 2^{\alpha}q$ where q is odd. Show that there is an integer a with $(a, k) = 1$ such that $a \equiv 3 \pmod{2^{\alpha}}$ and $a \equiv 2 \pmod{q}$. Then use (a) to deduce that $12S \equiv 0 \pmod{k}$.

17. An arithmetical function f is called *periodic* mod k if $k > 0$ and $f(m) = f(n)$ whenever $m \equiv n \pmod{k}$. The integer k is called a *period* of f.

(a) If f is periodic mod k, prove that f has a smallest positive period k_0 and that $k_0 \mid k$.
(b) Let f be periodic and completely multiplicative, and let k be the smallest positive period of f. Prove that $f(n) = 0$ if $(n, k) > 1$. This shows that f is a Dirichlet character mod k.

18. (a) Let f be a Dirichlet character mod k. If k is squarefree, prove that k is the smallest positive period of f.
(b) Give an example of a Dirichlet character mod k for which k is not the smallest positive period of f.

7 Dirichlet's Theorem on Primes in Arithmetical Progressions

7.1 Introduction

The arithmetic progression of odd numbers $1, 3, 5, \ldots, 2n + 1, \ldots$ contains infinitely many primes. It is natural to ask whether other arithmetic progressions have this property. An arithmetic progression with first term h and common difference k consists of all numbers of the form

$$kn + h, \quad n = 0, 1, 2, \ldots \tag{1}$$

If h and k have a common factor d, each term of the progression is divisible by d and there can be no more than one prime in the progression if $d > 1$. In other words, a necessary condition for the existence of infinitely many primes in the arithmetic progression (1) is that $(h, k) = 1$. Dirichlet was the first to prove that this condition is also sufficient. That is, if $(h, k) = 1$ the arithmetic progression (1) contains infinitely many primes. This result, now known as *Dirichlet's theorem*, will be proved in this chapter.

We recall that Euler proved the existence of infinitely many primes by showing that the series $\sum p^{-1}$, extended over all primes, diverges. Dirichlet's idea was to prove a corresponding statement when the primes are restricted to lie in the given progression (1). In a famous memoir [15] published in 1837 Dirichlet carried out this plan by ingenious analytic methods. The proof was later simplified by several authors. The version given in this chapter is based on a proof published in 1950 by Harold N. Shapiro [65] and deals with the series $\sum p^{-1} \log p$ rather than $\sum p^{-1}$.

First we show that for certain special progressions it is easy to prove Dirichlet's theorem by a modification of Euclid's proof of the infinitude of primes.

7.2 Dirichlet's theorem for primes of the form $4n - 1$ and $4n + 1$

Theorem 7.1 *There are infinitely many primes of the form* $4n - 1$.

Proof. We argue by contradiction. Assume there are only a finite number of such primes, let p be the largest, and consider the integer

$$N = 2^2 \cdot 3 \cdot 5 \cdots p - 1.$$

The product $3 \cdot 5 \cdots p$ contains all the odd primes $\leq p$ as factors. Since N is of the form $4n - 1$ it cannot be prime because $N > p$. No prime $\leq p$ divides N, so all the prime factors of N must exceed p. But all of the prime factors of N cannot be of the form $4n + 1$ because the product of two such numbers is again of the same form. Hence some prime factor of N must be of the form $4n - 1$. This is a contradiction. □

A different type of argument can be used for primes of the form $4n + 1$.

Theorem 7.2 *There are infinitely many primes of the form* $4n + 1$.

Proof. Let N be any integer > 1. We will show that there is a prime $p > N$ such that $p \equiv 1 \pmod 4$. Let

$$m = (N!)^2 + 1.$$

Note that m is odd, $m > 1$. Let p be the smallest prime factor of m. None of the numbers $2, 3, \ldots, N$ divides m, so $p > N$. Also, we have

$$(N!)^2 \equiv -1 \pmod p.$$

Raising both members to the $(p - 1)/2$ power we find

$$(N!)^{p-1} \equiv (-1)^{(p-1)/2} \pmod p.$$

But $(N!)^{p-1} \equiv 1 \pmod p$ by the Euler-Fermat theorem, so

$$(-1)^{(p-1)/2} \equiv 1 \pmod p.$$

Now the difference $(-1)^{(p-1)/2} - 1$ is either 0 or -2, and it cannot be -2, because it is divisible by p, so it must be 0. That is,

$$(-1)^{(p-1)/2} = 1.$$

But this means that $(p - 1)/2$ is even, so $p \equiv 1 \pmod 4$. In other words, we have shown that for each integer $N > 1$ there is a prime $p > N$ such that $p \equiv 1 \pmod 4$. Therefore there are infinitely many primes of the form $4n + 1$. □

Simple arguments like those just given for primes of the form $4n - 1$ and $4n + 1$ can also be adapted to treat other special arithmetic progressions, such as $5n - 1$, $8n - 1$, $8n - 3$ and $8n + 3$ (see Sierpinski [67]), but no one has yet found such a simple argument that works for the general progression $kn + h$.

7.3 The plan of the proof of Dirichlet's theorem

In Theorem 4.10 we derived the asymptotic formula

$$\sum_{p \le x} \frac{\log p}{p} = \log x + O(1), \tag{2}$$

where the sum is extended over all primes $p \le x$. We shall prove Dirichlet's theorem as a consequence of the following related asymptotic formula.

Theorem 7.3 *If $k > 0$ and $(h, k) = 1$ we have, for all $x > 1$,*

$$\sum_{\substack{p \le x \\ p \equiv h \pmod{k}}} \frac{\log p}{p} = \frac{1}{\varphi(k)} \log x + O(1), \tag{3}$$

where the sum is extended over those primes $p \le x$ which are congruent to h mod k.

Since $\log x \to \infty$ as $x \to \infty$ this relation implies that there are infinitely many primes $p \equiv h \pmod{k}$, hence infinitely many in the progression $nk + h$, $n = 0, 1, 2, \ldots$

Note that the principal term on the right of (3) is independent of h. Therefore (3) not only implies Dirichlet's theorem but it also shows that the primes in each of the $\varphi(k)$ reduced residue classes mod k make the same contribution to the principal term in (2).

The proof of Theorem 7.3 will be presented through a sequence of lemmas which we have collected together in this section to reveal the plan of the proof. Throughout the chapter we adopt the following notation.

The positive integer k represents a fixed modulus, and h is a fixed integer relatively prime to k. The $\varphi(k)$ Dirichlet characters mod k are denoted by

$$\chi_1, \chi_2, \ldots, \chi_{\varphi(k)}$$

with χ_1 denoting the principal character. For $\chi \ne \chi_1$ we write $L(1, \chi)$ and $L'(1, \chi)$ for the sums of the following series:

$$L(1, \chi) = \sum_{n=1}^{\infty} \frac{\chi(n)}{n},$$

$$L'(1, \chi) = -\sum_{n=1}^{\infty} \frac{\chi(n) \log n}{n}.$$

The convergence of each of these series was shown in Theorem 6.18. Moreover, in Theorem 6.20 we proved that $L(1, \chi) \neq 0$ if χ is real-valued. The symbol p denotes a prime, and $\sum_{p \le x}$ denotes a sum extended over all primes $p \le x$.

Lemma 7.4 *For $x > 1$ we have*

$$\sum_{\substack{p \le x \\ p \equiv h \pmod{k}}} \frac{\log p}{p} = \frac{1}{\varphi(k)} \log x + \frac{1}{\varphi(k)} \sum_{r=2}^{\varphi(k)} \bar{\chi}_r(h) \sum_{p \le x} \frac{\chi_r(p) \log p}{p} + O(1).$$

It is clear that Lemma 7.4 will imply Theorem 7.3 if we show that

$$\sum_{p \le x} \frac{\chi(p) \log p}{p} = O(1) \tag{4}$$

for each $\chi \neq \chi_1$. The next lemma expresses this sum in a form which is not extended over primes.

Lemma 7.5 *For $x > 1$ and $\chi \neq \chi_1$ we have*

$$\sum_{p \le x} \frac{\chi(p) \log p}{p} = -L'(1, \chi) \sum_{n \le x} \frac{\mu(n)\chi(n)}{n} + O(1).$$

Therefore Lemma 7.5 will imply (4) if we show that

$$\sum_{n \le x} \frac{\mu(n)\chi(n)}{n} = O(1). \tag{5}$$

This, in turn, will be deduced from the following lemma.

Lemma 7.6 *For $x > 1$ and $\chi \neq \chi_1$ we have*

$$L(1, \chi) \sum_{n \le x} \frac{\mu(n)\chi(n)}{n} = O(1). \tag{6}$$

If $L(1, \chi) \neq 0$ we can cancel $L(1, \chi)$ in (6) to obtain (5). Therefore, the proof of Dirichlet's theorem depends ultimately on the nonvanishing of $L(1, \chi)$ for all $\chi \neq \chi_1$. As already remarked, this was proved for real $\chi \neq \chi_1$ in Theorem 6.20, so it remains to prove that $L(1, \chi) \neq 0$ for all $\chi \neq \chi_1$ which take complex as well as real values.

For this purpose we let $N(k)$ denote the number of nonprincipal characters χ mod k such that $L(1, \chi) = 0$. If $L(1, \chi) = 0$ then $L(1, \bar{\chi}) = 0$ and $\chi \neq \bar{\chi}$ since χ is not real. Therefore the characters χ for which $L(1, \chi) = 0$ occur in conjugate pairs, so $N(k)$ is *even*. Our goal is to prove that $N(k) = 0$, and this will be deduced from the following asymptotic formula.

Lemma 7.7 *For $x > 1$ we have*

$$\sum_{\substack{p \le x \\ p \equiv 1 \pmod{k}}} \frac{\log p}{p} = \frac{1 - N(k)}{\varphi(k)} \log x + O(1). \tag{7}$$

If $N(k) \ne 0$ then $N(k) \ge 2$ since $N(k)$ is even, hence the coefficient of $\log x$ in (7) is negative and the right member $\to -\infty$ as $x \to \infty$. This is a contradiction since all the terms on the left are positive. Therefore Lemma 7.7 implies that $N(k) = 0$. The proof of Lemma 7.7, in turn, will be based on the following asymptotic formula.

Lemma 7.8 *If $\chi \ne \chi_1$ and $L(1, \chi) = 0$ we have*

$$L'(1, \chi) \sum_{n \le x} \frac{\mu(n)\chi(n)}{n} = \log x + O(1).$$

7.4 Proof of Lemma 7.4

To prove Lemma 7.4 we begin with the asymptotic formula mentioned earlier,

$$\sum_{p \le x} \frac{\log p}{p} = \log x + O(1) \tag{2}$$

and extract those terms in the sum arising from primes $p \equiv h \pmod{k}$. The extraction is done with the aid of the orthogonality relation for Dirichlet characters, as expressed in Theorem 6.16:

$$\sum_{r=1}^{\varphi(k)} \chi_r(m)\bar{\chi}_r(n) = \begin{cases} \varphi(k) & \text{if } m \equiv n \pmod{k}, \\ 0 & \text{if } m \not\equiv n \pmod{k}. \end{cases}$$

This is valid for $(n, k) = 1$. We take $m = p$ and $n = h$, where $(h, k) = 1$, then multiply both members by $p^{-1} \log p$ and sum over all $p \le x$ to obtain

$$\sum_{p \le x} \sum_{r=1}^{\varphi(k)} \chi_r(p)\bar{\chi}_r(h) \frac{\log p}{p} = \varphi(k) \sum_{\substack{p \le x \\ p \equiv h \pmod{k}}} \frac{\log p}{p}. \tag{8}$$

In the sum on the left we isolate those terms involving only the principal character χ_1 and rewrite (8) in the form

$$\varphi(k) \sum_{\substack{p \le x \\ p \equiv h \pmod{k}}} \frac{\log p}{p} = \bar{\chi}_1(h) \sum_{p \le x} \frac{\chi_1(p)\log p}{p} + \sum_{r=2}^{\varphi(k)} \bar{\chi}_r(h) \sum_{p \le x} \frac{\chi_r(p)\log p}{p}. \tag{9}$$

Now $\bar{\chi}_1(h) = 1$ and $\chi_1(p) = 0$ unless $(p, k) = 1$, in which case $\chi_1(p) = 1$. Hence the first term on the right of (9) is given by

$$\sum_{\substack{p \le x \\ (p, k) = 1}} \frac{\log p}{p} = \sum_{p \le x} \frac{\log p}{p} - \sum_{\substack{p \le x \\ p | k}} \frac{\log p}{p} = \sum_{p \le x} \frac{\log p}{p} + O(1), \tag{10}$$

since there are only a finite number of primes which divide k. Combining (10) with (9) we obtain

$$\varphi(k) \sum_{\substack{p \le x \\ p \equiv h \pmod{k}}} \frac{\log p}{p} = \sum_{p \le x} \frac{\log p}{p} + \sum_{r=2}^{\varphi(k)} \bar{\chi}_r(h) \sum_{p \le x} \frac{\chi_r(p)\log p}{p} + O(1).$$

Using (2) and dividing by $\varphi(k)$ we obtain Lemma 7.4. □

7.5 Proof of Lemma 7.5

We begin with the sum

$$\sum_{n \le x} \frac{\chi(n)\Lambda(n)}{n},$$

where $\Lambda(n)$ is Mangoldt's function, and express this sum in two ways. First we note that the definition of $\Lambda(n)$ gives us

$$\sum_{n \le x} \frac{\chi(n)\Lambda(n)}{n} = \sum_{p \le x} \sum_{\substack{a=1 \\ p^a \le x}}^{\infty} \frac{\chi(p^a)\log p}{p^a}.$$

We separate the terms with $a = 1$ and write

$$\sum_{n \le x} \frac{\chi(n)\Lambda(n)}{n} = \sum_{p \le x} \frac{\chi(p)\log p}{p} + \sum_{p \le x} \sum_{\substack{a=2 \\ p^a \le x}}^{\infty} \frac{\chi(p^a)\log p}{p^a}. \tag{11}$$

The second sum on the right is majorized by

$$\sum_{p} \log p \sum_{a=2}^{\infty} \frac{1}{p^a} = \sum_{p} \frac{\log p}{p(p-1)} < \sum_{n=2}^{\infty} \frac{\log n}{n(n-1)} = O(1),$$

so (11) gives us

$$\sum_{p \le x} \frac{\chi(p)\log p}{p} = \sum_{n \le x} \frac{\chi(n)\Lambda(n)}{n} + O(1). \tag{12}$$

Now we recall that $\Lambda(n) = \sum_{d|n} \mu(d)\log(n/d)$, hence

$$\sum_{n \le x} \frac{\chi(n)\Lambda(n)}{n} = \sum_{n \le x} \frac{\chi(n)}{n} \sum_{d|n} \mu(d)\log\frac{n}{d}.$$

In the last sum we write $n = cd$ and use the multiplicative property of χ to obtain

$$\sum_{n \le x} \frac{\chi(n)\Lambda(n)}{n} = \sum_{d \le x} \frac{\mu(d)\chi(d)}{d} \sum_{c \le x/d} \frac{\chi(c)\log c}{c}. \tag{13}$$

Since $x/d \ge 1$, in the sum over c we may use formula (10) of Theorem 6.18 to obtain

$$\sum_{c \le x/d} \frac{\chi(c)\log c}{c} = -L'(1, \chi) + O\left(\frac{\log x/d}{x/d}\right).$$

Equation (13) now becomes

$$\sum_{n\le x}\frac{\chi(n)\Lambda(n)}{n} = -L'(1,\chi)\sum_{d\le x}\frac{\mu(d)\chi(d)}{d} + O\left(\sum_{d\le x}\frac{1}{d}\frac{\log x/d}{x/d}\right). \tag{14}$$

The sum in the O-term is

$$\frac{1}{x}\sum_{d\le x}(\log x - \log d) = \frac{1}{x}\left([x]\log x - \sum_{d\le x}\log d\right) = O(1)$$

since

$$\sum_{d\le x}\log d = \log[x]! = x\log x + O(x).$$

Therefore (14) becomes

$$\sum_{n\le x}\frac{\chi(n)\Lambda(n)}{n} = -L'(1,\chi)\sum_{d\le x}\frac{\mu(d)\chi(d)}{d} + O(1)$$

which, with (12), proves Lemma 7.5. □

7.6 Proof of Lemma 7.6

We use the generalized Möbius inversion formula proved in Theorem 2.23 which states that if α is completely multiplicative we have

$$G(x) = \sum_{n\le x}\alpha(n)F\left(\frac{x}{n}\right) \quad \text{if, and only if,}\quad F(x) = \sum_{n\le x}\mu(n)\alpha(n)G\left(\frac{x}{n}\right). \tag{15}$$

We take $\alpha(n) = \chi(n)$ and $F(x) = x$ to obtain

$$x = \sum_{n\le x}\mu(n)\chi(n)G\left(\frac{x}{n}\right) \tag{16}$$

where

$$G(x) = \sum_{n\le x}\chi(n)\frac{x}{n} = x\sum_{n\le x}\frac{\chi(n)}{n}.$$

By Equation (9) of Theorem 6.18 we can write $G(x) = xL(1,\chi) + O(1)$. Using this in (16) we find

$$x = \sum_{n\le x}\mu(n)\chi(n)\left\{\frac{x}{n}L(1,\chi) + O(1)\right\} = xL(1,\chi)\sum_{n\le x}\frac{\mu(n)\chi(n)}{n} + O(x).$$

Now we divide by x to obtain Lemma 7.6. □

7.7 Proof of Lemma 7.8

We prove Lemma 7.8 and then use it to prove Lemma 7.7. Once again we make use of the generalized Möbius inversion formula (15). This time we take $F(x) = x \log x$ to obtain

$$x \log x = \sum_{n \le x} \mu(n)\chi(n)G\left(\frac{x}{n}\right) \tag{17}$$

where

$$G(x) = \sum_{n \le x} \chi(n)\frac{x}{n} \log \frac{x}{n} = x \log x \sum_{n \le x} \frac{\chi(n)}{n} - x \sum_{n \le x} \frac{\chi(n)\log n}{n}.$$

Now we use formulas (9) and (10) of Theorem 6.18 to get

$$\begin{aligned} G(x) &= x \log x\left\{L(1, \chi) + O\left(\frac{1}{x}\right)\right\} + x\left\{L'(1, \chi) + O\left(\frac{\log x}{x}\right)\right\} \\ &= xL'(1, \chi) + O(\log x) \end{aligned}$$

since we are assuming that $L(1, \chi) = 0$. Hence (17) gives us

$$\begin{aligned} x \log x &= \sum_{n \le x} \mu(n)\chi(n)\left\{\frac{x}{n} L'(1, \chi) + O\left(\log \frac{x}{n}\right)\right\} \\ &= xL'(1, \chi) \sum_{n \le x} \frac{\mu(n)\chi(n)}{n} + O\left(\sum_{n \le x} (\log x - \log n)\right). \end{aligned}$$

We have already noted that the O-term on the right is $O(x)$ (see the proof of Lemma 7.5). Hence we have

$$x \log x = xL'(1, \chi) \sum_{n \le x} \frac{\mu(n)\chi(n)}{n} + O(x),$$

and when we divide by x we obtain Lemma 7.8. □

7.8 Proof of Lemma 7.7

We use Lemma 7.4 with $h = 1$ to get

$$\sum_{\substack{p \le x \\ p \equiv 1 \pmod k}} \frac{\log p}{p} = \frac{1}{\varphi(k)} \log x + \frac{1}{\varphi(k)} \sum_{r=2}^{\varphi(k)} \sum_{p \le x} \frac{\chi_r(p)\log p}{p} + O(1). \tag{18}$$

In the sum over p on the right we use Lemma 7.5 which states that

$$\sum_{p \le x} \frac{\chi_r(p)\log p}{p} = -L'(1, \chi_r) \sum_{n \le x} \frac{\mu(n)\chi_r(n)}{n} + O(1).$$

If $L(1, \chi_r) \neq 0$, Lemma 7.6 shows that the right member of the foregoing equation is $O(1)$. But if $L(1, \chi_r) = 0$ then Lemma 7.8 implies

$$-L'(1, \chi_r) \sum_{n \leq x} \frac{\mu(n)\chi_r(n)}{n} = -\log x + O(1).$$

Therefore the sum on the right of (18) is

$$\frac{1}{\varphi(k)} \{-N(k)\log x + O(1)\},$$

so (18) becomes

$$\sum_{\substack{p \leq x \\ p \equiv 1 \pmod{k}}} \frac{\log p}{p} = \frac{1 - N(k)}{\varphi(k)} \log x + O(1).$$

This proves Lemma 7.7 and therefore also Theorem 7.3. □

As remarked earlier, Theorem 7.3 implies Dirichlet's theorem:

Theorem 7.9 *If $k > 0$ and $(h, k) = 1$ there are infinitely many primes in the arithmetic progression $nk + h$, $n = 0, 1, 2, \ldots$*

7.9 Distribution of primes in arithmetic progressions

If $k > 0$ and $(a, k) = 1$, let

$$\pi_a(x) = \sum_{\substack{p \leq x \\ p \equiv a \pmod{k}}} 1.$$

The function $\pi_a(x)$ counts the number of primes $\leq x$ in the progression $nk + a$, $n = 0, 1, 2, \ldots$ Dirichlet's theorem shows that $\pi_a(x) \to \infty$ as $x \to \infty$. There is also a prime number theorem for arithmetic progressions which states that

$$\pi_a(x) \sim \frac{\pi(x)}{\varphi(k)} \sim \frac{1}{\varphi(k)} \frac{x}{\log x} \quad \text{as } x \to \infty, \tag{19}$$

if $(a, k) = 1$. A proof of (19) is outlined in [44].

The prime number theorem for progressions is suggested by the formula of Theorem 7.3,

$$\sum_{\substack{p \leq x \\ p \equiv h \pmod{k}}} \frac{\log p}{p} = \frac{1}{\varphi(k)} \log x + O(1).$$

Since the principal term is independent of h, the primes seem to be equally distributed among the $\varphi(k)$ reduced residue classes mod k, and (19) is a precise statement of this fact.

We conclude this chapter by giving an alternate formulation of the prime number theorem for arithmetic progressions.

Theorem 7.10 *If the relation*

$$\pi_a(x) \sim \frac{\pi(x)}{\varphi(k)} \quad \text{as } x \to \infty \tag{20}$$

holds for every integer a relatively prime to k, then

$$\pi_a(x) \sim \pi_b(x) \quad \text{as } x \to \infty \tag{21}$$

whenever $(a, k) = (b, k) = 1$. *Conversely,* (21) *implies* (20).

Proof. It is clear that (20) implies (21). To prove the converse we assume (21) and let $A(k)$ denote the number of primes that divide k. If $x > k$ we have

$$\begin{aligned}\pi(x) &= \sum_{p \le x} 1 = A(k) + \sum_{\substack{p \le x \\ p \nmid k}} 1 \\ &= A(k) + \sum_{\substack{a=1 \\ (a,k)=1}}^{k} \sum_{\substack{p \le x \\ p \equiv a \pmod k}} 1 = A(k) + \sum_{\substack{a=1 \\ (a,k)=1}}^{k} \pi_a(x).\end{aligned}$$

Therefore

$$\frac{\pi(x) - A(k)}{\pi_b(x)} = \sum_{\substack{a=1 \\ (a,k)=1}}^{k} \frac{\pi_a(x)}{\pi_b(x)}.$$

By (21) each term in the sum tends to 1 as $x \to \infty$ so the sum tends to $\varphi(k)$. Hence

$$\frac{\pi(x)}{\pi_b(x)} - \frac{A(k)}{\pi_b(x)} \to \varphi(k) \quad \text{as } x \to \infty.$$

But $A(k)/\pi_b(x) \to 0$ so $\pi(x)/\pi_b(x) \to \varphi(k)$, which proves (20). □

Exercises for Chapter 7

In Exercises 1 through 4, h and k are given positive integers, $(h, k) = 1$, and $A(h, k)$ is the arithmetic progression $A(h, k) = \{h + kx : x = 0, 1, 2, \ldots\}$. Exercises 1 through 4 are to be solved without using Dirichlet's theorem.

1. Prove that, for every integer $n \ge 1$, $A(h, k)$ contains infinitely many numbers relatively prime to n.

2. Prove that $A(h, k)$ contains an infinite subset $\{a_1, a_2, \ldots\}$ such that $(a_i, a_j) = 1$ if $i \ne j$.

3. Prove that $A(h, k)$ contains an infinite subset which forms a geometric progression (a set of numbers of the form ar^n, $n = 0, 1, 2, \ldots$). This implies that $A(h, k)$ contains infinitely many numbers having the same prime factors.

4. Let S be any infinite subset of $A(h, k)$. Prove that for every positive integer n there is a number in $A(h, k)$ which can be expressed as a product of more than n different elements of S.

5. Dirichlet's theorem implies the following statement: If h and $k > 0$ are any two integers with $(h, k) = 1$, then there exists at least one prime number of the form $kn + h$. Prove that this statement also implies Dirichlet's theorem.

6. If $(h, k) = 1$, $k > 0$, prove that there is a constant A (depending on h and on k) such that, if $x \geq 2$,

$$\sum_{\substack{p \leq x \\ p \equiv h \pmod{k}}} \frac{1}{p} = \frac{1}{\varphi(k)} \log \log x + A + O\left(\frac{1}{\log x}\right).$$

7. Construct an infinite set S of primes with the following property: If $p \in S$ and $q \in S$ then $(\frac{1}{2}(p - 1), \frac{1}{2}(q - 1)) = (p, q - 1) = (p - 1, q) = 1$.

8. Let f be an integer-coefficient polynomial of degree $n \geq 1$ with the following property: For each prime p there exists a prime q and an integer m such that $f(p) = q^m$. Prove that $q = p$, $m = n$ and $f(x) = x^n$ for all x. [*Hint:* If $q \neq p$ then q^{m+1} divides $f(p + tq^{m+1}) - f(p)$ for each $t = 1, 2, \ldots$]

Periodic Arithmetical Functions and Gauss Sums 8

8.1 Functions periodic modulo k

Let k be a positive integer. An arithmetical function f is said to be *periodic with period* k (or *periodic modulo* k) if

$$f(n + k) = f(n)$$

for all integers n. If k is a period so is mk for any integer $m > 0$. The smallest positive period of f is called the *fundamental period*.

Periodic functions have already been encountered in the earlier chapters. For example, the Dirichlet characters mod k are periodic mod k. A simpler example is the greatest common divisor (n, k) regarded as a function of n. Periodicity enters through the relation

$$(n + k, k) = (n, k).$$

Another example is the exponential function

$$f(n) = e^{2\pi imn/k}$$

where m and k are fixed integers. The number $e^{2\pi im/k}$ is a kth root of unity and $f(n)$ is its nth power. Any finite linear combination of such functions, say

$$\sum_m c(m)e^{2\pi imn/k}$$

is also periodic mod k for every choice of coefficients $c(m)$. Our first goal is to show that every arithmetical function which is periodic mod k can be expressed as a linear combination of this type. These sums are called *finite Fourier series*. We begin the discussion with a simple but important example known as the *geometric sum*.

Theorem 8.1 *For fixed $k \geq 1$ let*

$$g(n) = \sum_{m=0}^{k-1} e^{2\pi imn/k}.$$

Then

$$g(n) = \begin{cases} 0 & \text{if } k \nmid n, \\ k & \text{if } k \mid n. \end{cases}$$

PROOF. Since $g(n)$ is the sum of terms in a geometric progression,

$$g(n) = \sum_{m=0}^{k-1} x^m,$$

where $x = e^{2\pi in/k}$, we have

$$g(n) = \begin{cases} \dfrac{x^k - 1}{x - 1} & \text{if } x \neq 1, \\ k & \text{if } x = 1. \end{cases}$$

But $x^k = 1$, and $x = 1$ if and only if $k \mid n$, so the theorem is proved. □

8.2 Existence of finite Fourier series for periodic arithmetical functions

We shall use Lagrange's polynomial interpolation formula to show that every periodic arithmetical function has a finite Fourier expansion.

Theorem 8.2 Lagrange's interpolation theorem. *Let $z_0, z_1, \ldots, z_{k-1}$ be k distinct complex numbers, and let $w_0, w_1, \ldots, w_{k-1}$ be k complex numbers which need not be distinct. Then there is a unique polynomial $P(z)$ of degree $\leq k - 1$ such that*

$$P(z_m) = w_m \quad \text{for } m = 0, 1, 2, \ldots, k - 1.$$

PROOF. The required polynomial $P(z)$, called the Lagrange interpolation polynomial, can be constructed explicitly as follows. Let

$$A(z) = (z - z_0)(z - z_1) \cdots (z - z_{k-1})$$

and let

$$A_m(z) = \frac{A(z)}{z - z_m}.$$

Then $A_m(z)$ is a polynomial of degree $k - 1$ with the following properties:

$$A_m(z_m) \neq 0, \qquad A_m(z_j) = 0 \quad \text{if } j \neq m.$$

Hence $A_m(z)/A_m(z_m)$ is a polynomial of degree $k - 1$ which vanishes at each z_j for $j \neq m$, and has the value 1 at z_m. Therefore the linear combination

$$P(z) = \sum_{m=0}^{k-1} w_m \frac{A_m(z)}{A_m(z_m)}$$

is a polynomial of degree $\leq k - 1$ with $P(z_j) = w_j$ for each j. If there were another such polynomial, say $Q(z)$, the difference $P(z) - Q(z)$ would vanish at k distinct points, hence $P(z) = Q(z)$ since both polynomials have degree $\leq k - 1$. □

Now we choose the numbers $z_0, z_1, \ldots, z_{k-1}$ to be the kth roots of unity and we obtain:

Theorem 8.3 *Given k complex numbers $w_0, w_1, \ldots, w_{k-1}$, there exist k uniquely determined complex numbers $a_0, a_1, \ldots, a_{k-1}$ such that*

$$w_m = \sum_{n=0}^{k-1} a_n e^{2\pi imn/k} \tag{1}$$

for $m = 0, 1, 2, \ldots, k - 1$. Moreover, the coefficients a_n are given by the formula

$$a_n = \frac{1}{k}\sum_{m=0}^{k-1} w_m e^{-2\pi imn/k} \quad \text{for } n = 0, 1, 2, \ldots, k - 1. \tag{2}$$

PROOF. Let $z_m = e^{2\pi im/k}$. The numbers $z_0, z_1, \ldots, z_{k-1}$ are distinct so there is a unique Lagrange polynomial

$$P(z) = \sum_{n=0}^{k-1} a_n z^n$$

such that $P(z_m) = w_m$ for each $m = 0, 1, 2, \ldots, k - 1$. This shows that there are uniquely determined numbers a_n satisfying (1). To deduce the formula (2) for a_n we multiply both sides of (1) by $e^{-2\pi imr/k}$, where m and r are nonnegative integers less than k, and sum on m to get

$$\sum_{m=0}^{k-1} w_m e^{-2\pi imr/k} = \sum_{n=0}^{k-1} a_n \sum_{m=0}^{k-1} e^{2\pi i(n-r)m/k}.$$

By Theorem 8.1, the sum on m is 0 unless $k \mid (n - r)$. But $|n - r| \leq k - 1$ so $k \mid (n - r)$ if, and only if, $n = r$. Therefore the only nonvanishing term on the right occurs when $n = r$ and we find

$$\sum_{m=0}^{k-1} w_m e^{-2\pi imr/k} = ka_r.$$

This equation gives us (2). □

Theorem 8.4 *Let f be an arithmetical function which is periodic* mod k. *Then there is a uniquely determined arithmetical function g, also periodic* mod k, *such that*

$$f(m) = \sum_{n=0}^{k-1} g(n)e^{2\pi imn/k}.$$

In fact, g is given by the formula

$$g(n) = \frac{1}{k}\sum_{m=0}^{k-1} f(m)e^{-2\pi imn/k}.$$

PROOF. Let $w_m = f(m)$ for $m = 0, 1, 2, \ldots, k - 1$ and apply Theorem 8.3 to determine the numbers $a_0, a_1, \ldots, a_{k-1}$. Define the function g by the relations $g(m) = a_m$ for $m = 0, 1, 2, \ldots, k - 1$ and extend the definition of $g(m)$ to all integers m by periodicity mod k. Then f is related to g by the equations in the theorem. □

Note. Since both f and g are periodic mod k we can rewrite the sums in Theorem 8.4 as follows:

$$f(m) = \sum_{n \bmod k} g(n)e^{2\pi imn/k} \tag{3}$$

and

$$g(n) = \frac{1}{k}\sum_{m \bmod k} f(m)e^{-2\pi imn/k}. \tag{4}$$

In each case the summation can be extended over any complete residue system modulo k. The sum in (3) is called the *finite Fourier expansion* of f and the numbers $g(n)$ defined by (4) are called the *Fourier coefficients* of f.

8.3 Ramanujan's sum and generalizations

In Exercise 2.14(b) it is shown that the Möbius function $\mu(k)$ is the sum of the primitive kth roots of unity. In this section we generalize this result. Specifically, let n be a fixed positive integer and consider the sum of the nth powers of the primitive kth roots of unity. This sum is known as *Ramanujan's sum* and is denoted by $c_k(n)$:

$$c_k(n) = \sum_{\substack{m \bmod k \\ (m,k)=1}} e^{2\pi imn/k}.$$

We have already noted that this sum reduces to the Möbius function when $n = 1$,

$$\mu(k) = c_k(1).$$

When $k|n$ the sum reduces to the Euler φ function since each term is 1 and the number of terms is $\varphi(k)$. Ramanujan showed that $c_k(n)$ is always an integer and that it has interesting multiplicative properties. He deduced these facts from the relation

$$c_k(n) = \sum_{d|(n,k)} d\mu\left(\frac{k}{d}\right). \tag{5}$$

This formula shows why $c_k(n)$ reduces to both $\mu(k)$ and $\varphi(k)$. In fact, when $n = 1$ there is only one term in the sum and we obtain $c_k(1) = \mu(k)$. And when $k|n$ we have $(n, k) = k$ and $c_k(n) = \sum_{d|k} d\mu(k/d) = \varphi(k)$. We shall deduce (5) as a special case of a more general result (Theorem 8.5).

Formula (5) for $c_k(n)$ suggests that we study general sums of the form

$$\sum_{d|(n,k)} f(d)g\left(\frac{k}{d}\right). \tag{6}$$

These resemble the sums for the Dirichlet convolution $f * g$ except that we sum over a *subset* of the divisors of k, namely those d which also divide n.

Denote the sum in (6) by $s_k(n)$. Since n occurs only in the gcd (n, k) we have

$$s_k(n + k) = s_k(n)$$

so $s_k(n)$ is a periodic function of n with period k. Hence this sum has a finite Fourier expansion. The next theorem tells us that its Fourier coefficients are given by a sum of the same type.

Theorem 8.5 *Let* $s_k(n) = \sum_{d|(n,k)} f(d)g(k/d)$. *Then* $s_k(n)$ *has the finite Fourier expansion*

$$s_k(n) = \sum_{m \bmod k} a_k(m)e^{2\pi imn/k} \tag{7}$$

where

$$a_k(m) = \sum_{d|(m,k)} g(d)f\left(\frac{k}{d}\right)\frac{d}{k}. \tag{8}$$

PROOF. By Theorem 8.4 the coefficients $a_k(m)$ are given by

$$\begin{aligned} a_k(m) &= \frac{1}{k}\sum_{n \bmod k} s_k(n)e^{-2\pi inm/k} \\ &= \frac{1}{k}\sum_{n=1}^{k}\sum_{\substack{d|n \\ d|k}} f(d)g\left(\frac{k}{d}\right)e^{-2\pi inm/k}. \end{aligned}$$

Now we write $n = cd$ and note that for each fixed d the index c runs from 1 to k/d and we obtain

$$a_k(m) = \frac{1}{k}\sum_{d|k} f(d)g\left(\frac{k}{d}\right)\sum_{c=1}^{k/d} e^{-2\pi icdm/k}.$$

Now we replace d by k/d in the sum on the right to get

$$a_k(m) = \frac{1}{k} \sum_{d|k} f\left(\frac{k}{d}\right) g(d) \sum_{c=1}^{d} e^{-2\pi i c m/d}.$$

But by Theorem 8.1 the sum on c is 0 unless $d|m$ in which case the sum has the value d. Hence

$$a_k(m) = \frac{1}{k} \sum_{\substack{d|k \\ d|m}} f\left(\frac{k}{d}\right) g(d) d$$

which proves (8). □

Now we specialize f and g to obtain the formula for Ramanujan's sum mentioned earlier.

Theorem 8.6 *We have*

$$c_k(n) = \sum_{d|(n,k)} d\mu\left(\frac{k}{d}\right).$$

PROOF. Taking $f(k) = k$ and $g(k) = \mu(k)$ in Theorem 8.5 we find

$$\sum_{d|(n,k)} d\mu\left(\frac{k}{d}\right) = \sum_{m \bmod k} a_k(m) e^{2\pi i m n/k}$$

where

$$a_k(m) = \sum_{d|(m,k)} \mu(d) = \left[\frac{1}{(m,k)}\right] = \begin{cases} 1 & \text{if } (m,k) = 1, \\ 0 & \text{if } (m,k) > 1. \end{cases}$$

Hence

$$\sum_{d|(n,k)} d\mu\left(\frac{k}{d}\right) = \sum_{\substack{m \bmod k \\ (m,k)=1}} e^{2\pi i m n/k} = c_k(n).$$ □

8.4 Multiplicative properties of the sums $s_k(n)$

Theorem 8.7 *Let*

$$s_k(n) = \sum_{d|(n,k)} f(d) g\left(\frac{k}{d}\right)$$

where f and g are multiplicative. Then we have

(9) $\quad s_{mk}(ab) = s_m(a)s_k(b)$ *whenever* $(a,k) = (b,m) = (m,k) = 1.$

In particular, we have

$$s_m(ab) = s_m(a) \quad \text{if } (b, m) = 1, \tag{10}$$

and

$$s_{mk}(a) = s_m(a)g(k) \quad \text{if } (a, k) = 1. \tag{11}$$

PROOF. The relations $(a, k) = (b, m) = 1$ imply (see Exercise 1.24)

$$(mk, ab) = (a, m)(k, b)$$

with (a, m) and (b, k) relatively prime. Therefore

$$s_{mk}(ab) = \sum_{d|(mk,\, ab)} f(d)g\left(\frac{mk}{d}\right) = \sum_{d|(a,\, m)(b,\, k)} f(d)g\left(\frac{mk}{d}\right).$$

Writing $d = d_1 d_2$ in the last sum we obtain

$$\begin{aligned} s_{mk}(ab) &= \sum_{d_1|(a,\, m)} \sum_{d_2|(b,\, k)} f(d_1 d_2)g\left(\frac{mk}{d_1 d_2}\right) \\ &= \sum_{d_1|(a,\, m)} f(d_1)g\left(\frac{m}{d_1}\right) \sum_{d_2|(b,\, k)} f(d_2)g\left(\frac{k}{d_2}\right) = s_m(a)s_k(b). \end{aligned}$$

This proves (9).

Taking $k = 1$ in (9) we get

$$s_m(ab) = s_m(a)s_1(b) = s_m(a)$$

since $s_1(b) = f(1)g(1) = 1$. This proves (10). Taking $b = 1$ in (9) we find

$$s_{mk}(a) = s_m(a)s_k(1) = s_m(a)g(k)$$

since $s_k(1) = f(1)g(k) = g(k)$. This proves (11). □

EXAMPLE For Ramanujan's sum we obtain the following multiplicative properties:

$$c_{mk}(ab) = c_m(a)c_k(b) \qquad \text{whenever } (a, k) = (b, m) = (m, k) = 1,$$
$$c_m(ab) = c_m(a) \qquad \text{whenever } (b, m) = 1,$$

and

$$c_{mk}(a) = c_m(a)\mu(k) \qquad \text{whenever } (a, k) = (m, k) = 1.$$

Sometimes the sums $s_k(n)$ can be evaluated in terms of the Dirichlet convolution $f * g$. In this connection we have:

Theorem 8.8 *Let f be completely multiplicative, and let $g(k) = \mu(k)h(k)$, where h is multiplicative. Assume that $f(p) \neq 0$ and $f(p) \neq h(p)$ for all primes p, and let*

$$s_k(n) = \sum_{d|(n,\, k)} f(d)g\left(\frac{k}{d}\right).$$

Then we have

$$s_k(n) = \frac{F(k)g(N)}{F(N)},$$

where $F = f * g$ *and* $N = k/(n, k)$.

Proof. First we note that

$$F(k) = \sum_{d|k} f(d)\mu\left(\frac{k}{d}\right)h\left(\frac{k}{d}\right) = \sum_{d|k} f\left(\frac{k}{d}\right)\mu(d)h(d) = f(k)\sum_{d|k}\mu(d)\frac{h(d)}{f(d)}$$

$$= f(k)\prod_{p|k}\left(1 - \frac{h(p)}{f(p)}\right).$$

Next, we write $a = (n, k)$, so that $k = aN$. Then we have

$$s_k(n) = \sum_{d|a} f(d)\mu\left(\frac{k}{d}\right)h\left(\frac{k}{d}\right) = \sum_{d|a} f(d)\mu\left(\frac{aN}{d}\right)h\left(\frac{aN}{d}\right)$$

$$= \sum_{d|a} f\left(\frac{a}{d}\right)\mu(Nd)h(Nd).$$

Now $\mu(Nd) = \mu(N)\mu(d)$ if $(N, d) = 1$, and $\mu(Nd) = 0$ if $(N, d) > 1$, so the last equation gives us

$$s_k(n) = \mu(N)h(N)\sum_{\substack{d|a \\ (N,d)=1}} f\left(\frac{a}{d}\right)\mu(d)h(d) = f(a)\mu(N)h(N)\sum_{\substack{d|a \\ (N,d)=1}}\mu(d)\frac{h(d)}{f(d)}$$

$$= f(a)\mu(N)h(N)\prod_{\substack{p|a \\ p\nmid N}}\left(1 - \frac{h(p)}{f(p)}\right) = f(a)\mu(N)h(N)\frac{\prod_{p|aN}\left(1 - \frac{h(p)}{f(p)}\right)}{\prod_{p|N}\left(1 - \frac{h(p)}{f(p)}\right)}$$

$$= f(a)\mu(N)h(N)\frac{F(k)}{f(k)}\frac{f(N)}{F(N)} = \frac{F(k)\mu(N)h(N)}{F(N)} = \frac{F(k)g(N)}{F(N)}. \qquad \square$$

Example For Ramanujan's sum we obtain the following simplification:

$$c_k(n) = \varphi(k)\mu(N)/\varphi(N) = \frac{\varphi(k)\mu\left(\frac{k}{(n, k)}\right)}{\varphi\left(\frac{k}{(n, k)}\right)}.$$

8.5 Gauss sums associated with Dirichlet characters

Definition For any Dirichlet character χ mod k the sum

$$G(n, \chi) = \sum_{m=1}^{k} \chi(m)e^{2\pi imn/k}$$

is called the Gauss sum associated with χ.

If $\chi = \chi_1$, the principal character mod k, we have $\chi_1(m) = 1$ if $(m, k) = 1$, and $\chi_1(m) = 0$ otherwise. In this case the Gauss sum reduces to Ramanujan's sum:

$$G(n, \chi_1) = \sum_{\substack{m=1 \\ (m,k)=1}}^{k} e^{2\pi imn/k} = c_k(n).$$

Thus, the Gauss sums $G(n, \chi)$ can be regarded as generalizations of Ramanujan's sum. We turn now to a detailed study of their properties.

The first result is a factorization property which plays an important role in the subsequent development.

Theorem 8.9 *If χ is any Dirichlet character* mod *k then*

$$G(n, \chi) = \bar{\chi}(n)G(1, \chi) \quad \textit{whenever } (n, k) = 1.$$

PROOF. When $(n, k) = 1$ the numbers nr run through a complete residue system mod k with r. Also, $|\chi(n)|^2 = \chi(n)\bar{\chi}(n) = 1$ so

$$\chi(r) = \bar{\chi}(n)\chi(n)\chi(r) = \bar{\chi}(n)\chi(nr).$$

Therefore the sum defining $G(n, \chi)$ can be written as follows:

$$G(n, \chi) = \sum_{r \bmod k} \chi(r)e^{2\pi inr/k} = \bar{\chi}(n) \sum_{r \bmod k} \chi(nr)e^{2\pi inr/k}$$

$$= \bar{\chi}(n) \sum_{m \bmod k} \chi(m)e^{2\pi im/k} = \bar{\chi}(n)G(1, \chi).$$

This proves the theorem. □

Definition The Gauss sum $G(n, \chi)$ is said to be separable if

$$G(n, \chi) = \bar{\chi}(n)G(1, \chi). \tag{12}$$

Theorem 8.9 tells us that $G(n, \chi)$ is separable whenever n is relatively prime to the modulus k. For those integers n not relatively prime to k we have the following theorem.

Theorem 8.10 *If χ is a character* mod k *the Gauss sum $G(n, \chi)$ is separable for every n if, and only if,*

$$G(n, \chi) = 0 \quad \textit{whenever } (n, k) > 1.$$

PROOF. Separability always holds if $(n, k) = 1$. But if $(n, k) > 1$ we have $\bar{\chi}(n) = 0$ so Equation (12) holds if and only if $G(n, \chi) = 0$. □

The next theorem gives an important consequence of separability.

Theorem 8.11 *If $G(n, \chi)$ is separable for every n then*

$$|G(1, \chi)|^2 = k. \tag{13}$$

PROOF. We have

$$\begin{aligned} |G(1, \chi)|^2 &= G(1, \chi)\overline{G(1, \chi)} = G(1, \chi) \sum_{m=1}^{k} \bar{\chi}(m)e^{-2\pi i m/k} \\ &= \sum_{m=1}^{k} G(m, \chi)e^{-2\pi i m/k} = \sum_{m=1}^{k} \sum_{r=1}^{k} \chi(r)e^{2\pi i m r/k}e^{-2\pi i m/k} \\ &= \sum_{r=1}^{k} \chi(r) \sum_{m=1}^{k} e^{2\pi i m(r-1)/k} = k\chi(1) = k, \end{aligned}$$

since the last sum over m is a geometric sum which vanishes unless $r = 1$.

8.6 Dirichlet characters with nonvanishing Gauss sums

For every character χ mod k we have seen that $G(n, \chi)$ is separable if $(n, k) = 1$, and that separability of $G(n, \chi)$ is equivalent to the vanishing of $G(n, \chi)$ for $(n, k) > 1$. Now we describe further properties of those characters such that $G(n, \chi) = 0$ whenever $(n, k) > 1$. Actually, it is simpler to study the complementary set. The next theorem gives a necessary condition for $G(n, \chi)$ to be nonzero for $(n, k) > 1$.

Theorem 8.12 *Let χ be a Dirichlet character* mod k *and assume that $G(n, \chi) \neq 0$ for some n satisfying $(n, k) > 1$. Then there exists a divisor d of k, $d < k$, such that*

$$\chi(a) = 1 \quad \textit{whenever } (a, k) = 1 \textit{ and } a \equiv 1 \pmod{d}. \tag{14}$$

PROOF. For the given n, let $q = (n, k)$ and let $d = k/q$. Then $d \mid k$ and, since $q > 1$, we have $d < k$. Choose any a satisfying $(a, k) = 1$ and $a \equiv 1 \pmod{d}$. We will prove that $\chi(a) = 1$.

Since $(a, k) = 1$, in the sum defining $G(n, \chi)$ we can replace the index of summation m by am and we find

$$G(n, \chi) = \sum_{m \bmod k} \chi(m)e^{2\pi inm/k} = \sum_{m \bmod k} \chi(am)e^{2\pi inam/k}$$

$$= \chi(a) \sum_{m \bmod k} \chi(m)e^{2\pi inam/k}.$$

Since $a \equiv 1 \pmod{d}$ and $d = k/q$ we can write $a = 1 + (bk/q)$ for some integer b, and we have

$$\frac{anm}{k} = \frac{nm}{k} + \frac{bknm}{qk} = \frac{nm}{k} + \frac{bnm}{q} \equiv \frac{nm}{k} \pmod{1}$$

since $q \mid n$. Hence $e^{2\pi inam/k} = e^{2\pi inm/k}$ and the sum for $G(n, \chi)$ becomes

$$G(n, \chi) = \chi(a) \sum_{m \bmod k} \chi(m)e^{2\pi inm/k} = \chi(a)G(n, \chi).$$

Since $G(n, \chi) \neq 0$ this implies $\chi(a) = 1$, as asserted. □

The foregoing theorem leads us to consider those characters χ mod k for which there is a divisor $d < k$ satisfying (14). These are treated next.

8.7 Induced moduli and primitive characters

Definition of induced modulus Let χ be a Dirichlet character mod k and let d be any positive divisor of k. The number d is called an induced modulus for χ if we have

(15) $\qquad \chi(a) = 1 \quad$ whenever $(a, k) = 1$ and $a \equiv 1 \pmod{d}$.

In other words, d is an induced modulus if the character χ mod k acts like a character mod d on the representatives of the residue class $\hat{1}$ mod d which are relatively prime to k. Note that k itself is always an induced modulus for χ.

Theorem 8.13 *Let χ be a Dirichlet character* mod k. *Then* 1 *is an induced modulus for χ if, and only if, $\chi = \chi_1$.*

PROOF. If $\chi = \chi_1$ then $\chi(a) = 1$ for all a relatively prime to k. But since every a satisfies $a \equiv 1 \pmod{1}$ the number 1 is an induced modulus.

Conversely, if 1 is an induced modulus, then $\chi(a) = 1$ whenever $(a, k) = 1$, so $\chi = \chi_1$ since χ vanishes on the numbers not prime to k. □

For any Dirichlet character mod k the modulus k itself is an induced modulus. If there are no others we call the character *primitive*. That is, we have:

Definition of primitive characters A Dirichlet character χ mod k is said to be primitive mod k if it has no induced modulus $d < k$. In other words, χ is primitive mod k if, and only if, for every divisor d of k, $0 < d < k$, there exists an integer $a \equiv 1 \pmod{d}$, $(a, k) = 1$, such that $\chi(a) \neq 1$.

If $k > 1$ the principal character χ_1 is not primitive since it has 1 as an induced modulus. Next we show that if the modulus is *prime* every nonprincipal character is primitive.

Theorem 8.14 *Every nonprincipal character χ modulo a prime p is a primitive character* mod p.

PROOF. The only divisors of p are 1 and p so these are the only candidates for induced moduli. But if $\chi \neq \chi_1$ the divisor 1 is not an induced modulus so χ has no induced modulus $< p$. Hence χ is primitive. □

Now we can restate the results of Theorems 8.10 through 8.12 in the terminology of primitive characters.

Theorem 8.15 *Let χ be a primitive Dirichlet character* mod k. *Then we have:*

(a) $G(n, \chi) = 0$ *for every n with* $(n, k) > 1$.
(b) $G(n, \chi)$ *is separable for every n.*
(c) $|G(1, \chi)|^2 = k$.

PROOF. If $G(n, \chi) \neq 0$ for some n with $(n, k) > 1$ then Theorem 8.12 shows that χ has an induced modulus $d < k$, so χ cannot be primitive. This proves (a).

Part (b) follows from (a) and Theorem 8.10. Part (c) follows from part (b) and Theorem 8.11. □

Note. Theorem 8.15(b) shows that the Gauss sum $G(n, \chi)$ is separable if χ is primitive. In a later section we prove the converse. That is, if $G(n, \chi)$ is separable for every n then χ is primitive. (See Theorem 8.19.)

8.8 Further properties of induced moduli

The next theorem refers to the action of χ on numbers which are congruent modulo an induced modulus.

Theorem 8.16 *Let χ be a Dirichlet character* mod k *and assume* $d \mid k$, $d > 0$. *Then d is an induced modulus for χ if, and only if,*

$$(16) \qquad \chi(a) = \chi(b) \quad \textit{whenever } (a, k) = (b, k) = 1 \textit{ and } a \equiv b \pmod{d}.$$

Proof. If (16) holds then d is an induced modulus since we may choose $b = 1$ and refer to Equation (15). Now we prove the converse.

Choose a and b so that $(a, k) = (b, k) = 1$ and $a \equiv b \pmod{d}$. We will show that $\chi(a) = \chi(b)$. Let a' be the reciprocal of a mod k, $aa' \equiv 1 \pmod{k}$. The reciprocal exists because $(a, k) = 1$. Now $aa' \equiv 1 \pmod{d}$ since $d \mid k$. Hence $\chi(aa') = 1$ since d is an induced modulus. But $aa' \equiv ba' \equiv 1 \pmod{d}$ because $a \equiv b \pmod{d}$, hence $\chi(aa') = \chi(ba')$, so

$$\chi(a)\chi(a') = \chi(b)\chi(a').$$

But $\chi(a') \neq 0$ since $\chi(a)\chi(a') = 1$. Canceling $\chi(a')$ we find $\chi(a) = \chi(b)$, and this completes the proof. □

Equation (16) tells us that χ is periodic mod d on those integers relatively prime to k. Thus χ acts very much like a character mod d. To further explore this relation it is worthwhile to consider a few examples.

Example 1 The following table describes one of the characters χ mod 9.

n	1	2	3	4	5	6	7	8	9
$\chi(n)$	1	−1	0	1	−1	0	1	−1	0

We note that this table is periodic modulo 3 so 3 is an induced modulus for χ. In fact, χ acts like the following character ψ modulo 3:

n	1	2	3
$\psi(n)$	1	−1	0

Since $\chi(n) = \psi(n)$ for all n we call χ an *extension* of ψ. It is clear that whenever χ is an extension of a character ψ modulo d then d will be an induced modulus for χ.

Example 2 Now we examine one of the characters χ modulo 6:

n	1	2	3	4	5	6
$\chi(n)$	1	0	0	0	−1	0

In this case the number 3 is an induced modulus because $\chi(n) = 1$ for all $n \equiv 1 \pmod{3}$ with $(n, 6) = 1$. (There is only one such n, namely, $n = 1$.)

However, χ is *not* an extension of any character ψ modulo 3, because the only characters modulo 3 are the principal character ψ_1, given by the table:

n	1	2	3
$\psi_1(n)$	1	1	0

and the character ψ shown in Example 1. Since $\chi(2) = 0$ it cannot be an extension of either ψ or ψ_1.

These examples shed some light on the next theorem.

Theorem 8.17 *Let χ be a Dirichlet character modulo k and assume $d \mid k$, $d > 0$. Then the following two statements are equivalent:*

(a) *d is an induced modulus for χ.*
(b) *There is a character ψ modulo d such that*

$$\chi(n) = \psi(n)\chi_1(n) \quad \textit{for all } n, \tag{17}$$

where χ_1 is the principal character modulo k.

Proof. Assume (b) holds. Choose n satisfying $(n, k) = 1$, $n \equiv 1 \pmod{d}$. Then $\chi_1(n) = \psi(n) = 1$ so $\chi(n) = 1$ and hence d is an induced modulus. Thus, (b) implies (a).

Now assume (a) holds. We will exhibit a character ψ modulo d for which (17) holds. We define $\psi(n)$ as follows: If $(n, d) > 1$, let $\psi(n) = 0$. In this case we also have $(n, k) > 1$ so (17) holds because both members are zero.

Now suppose $(n, d) = 1$. Then there exists an integer m such that $m \equiv n \pmod{d}$, $(m, k) = 1$. This can be proved immediately with Dirichlet's theorem. The arithmetic progression $xd + n$ contains infinitely many primes. We choose one that does not divide k and call this m. However, the result is not that deep; the existence of such an m can easily be established without using Dirichlet's theorem. (See Exercise 8.4 for an alternate proof.) Having chosen m, which is unique modulo d, we define

$$\psi(n) = \chi(m).$$

The number $\psi(n)$ is well-defined because χ takes equal values at numbers which are congruent modulo d and relatively prime to k.

The reader can easily verify that ψ is, indeed, a character mod d. We shall verify that Equation (17) holds for all n.

If $(n, k) = 1$ then $(n, d) = 1$ so $\psi(n) = \chi(m)$ for some $m \equiv n \pmod{d}$. Hence, by Theorem 8.16,

$$\chi(n) = \chi(m) = \psi(n) = \psi(n)\chi_1(n)$$

since $\chi_1(n) = 1$.

If $(n, k) > 1$, then $\chi(n) = \chi_1(n) = 0$ and both members of (17) are 0. Thus, (17) holds for all n. □

8.9 The conductor of a character

Definition Let χ be a Dirichlet character mod k. The smallest induced modulus d for χ is called the conductor of χ.

Theorem 8.18 *Every Dirichlet character* χ mod k *can be expressed as a product,*

$$\chi(n) = \psi(n)\chi_1(n) \quad \text{for all } n, \tag{18}$$

where χ_1 *is the principal character* mod k *and* ψ *is a primitive character modulo the conductor of* χ.

Proof. Let d be the conductor of χ. From Theorem 8.17 we know that χ can be expressed as a product of the form (18), where ψ is a character mod d. Now we shall prove that ψ is primitive mod d.

We assume that ψ is not primitive mod d and arrive at a contradiction. If ψ is not primitive mod d there is a divisor q of d, $q < d$, which is an induced modulus for ψ. We shall prove that this q, which divides k, is also an induced modulus for χ, contradicting the fact that d is the smallest induced modulus for χ.

Choose $n \equiv 1 \pmod q$, $(n, k) = 1$. Then

$$\chi(n) = \psi(n)\chi_1(n) = \psi(n) = 1$$

because q is an induced modulus for ψ. Hence q is also an induced modulus for χ and this is a contradiction. □

8.10 Primitive characters and separable Gauss sums

As an application of the foregoing theorems we give the following alternate description of primitive characters.

Theorem 8.19 *Let* χ *be a character* mod k. *Then* χ *is primitive* mod k *if, and only if, the Gauss sum*

$$G(n, \chi) = \sum_{m \bmod k} \chi(m)e^{2\pi imn/k}$$

is separable for every n.

Proof. If χ is primitive, then $G(n, \chi)$ is separable by Theorem 8.15(b). Now we prove the converse.

Because of Theorems 8.9 and 8.10 it suffices to prove that if χ is not primitive mod k then for some r satisfying $(r, k) > 1$ we have $G(r, \chi) \neq 0$. Suppose, then, that χ is not primitive mod k. This implies $k > 1$. Then χ has a conductor $d < k$. Let $r = k/d$. Then $(r, k) > 1$ and we shall prove that

$G(r, \chi) \neq 0$ for this r. By Theorem 8.18 there exists a primitive character ψ mod d such that $\chi(n) = \psi(n)\chi_1(n)$ for all n. Hence we can write

$$G(r, \chi) = \sum_{m \bmod k} \psi(m)\chi_1(m)e^{2\pi irm/k} = \sum_{\substack{m \bmod k \\ (m, k) = 1}} \psi(m)e^{2\pi irm/k}$$

$$= \sum_{\substack{m \bmod k \\ (m, k) = 1}} \psi(m)e^{2\pi im/d} = \frac{\varphi(k)}{\varphi(d)} \sum_{\substack{m \bmod d \\ (m, d) = 1}} \psi(m)e^{2\pi im/d},$$

where in the last step we used Theorem 5.33(a). Therefore we have

$$G(r, \chi) = \frac{\varphi(k)}{\varphi(d)} G(1, \psi).$$

But $|G(1, \psi)|^2 = d$ by Theorem 8.15 (since ψ is primitive mod d) and hence $G(r, \chi) \neq 0$. This completes the proof. □

8.11 The finite Fourier series of the Dirichlet characters

Since each Dirichlet character χ mod k is periodic mod k it has a finite Fourier expansion

$$\chi(m) = \sum_{n=1}^{k} a_k(n)e^{2\pi imn/k}, \tag{19}$$

and Theorem 8.4 tells us that its coefficients are given by the formula

$$a_k(n) = \frac{1}{k} \sum_{m=1}^{k} \chi(m)e^{-2\pi imn/k}.$$

The sum on the right is a Gauss sum $G(-n, \chi)$ so we have

$$a_k(n) = \frac{1}{k} G(-n, \chi). \tag{20}$$

When χ is primitive the Fourier expansion (19) can be expressed as follows:

Theorem 8.20 *The finite Fourier expansion of a primitive Dirichlet character χ mod k has the form*

$$\chi(m) = \frac{\tau_k(\chi)}{\sqrt{k}} \sum_{n=1}^{k} \bar{\chi}(n)e^{-2\pi imn/k} \tag{21}$$

where

$$\tau_k(\chi) = \frac{G(1, \chi)}{\sqrt{k}} = \frac{1}{\sqrt{k}} \sum_{m=1}^{k} \chi(m)e^{2\pi im/k}. \tag{22}$$

The numbers $\tau_k(\chi)$ have absolute value 1.

PROOF. Since χ is primitive we have $G(-n, \chi) = \bar{\chi}(-n)G(1, \chi)$ and (20) implies $a_k(n) = \bar{\chi}(-n)G(1, \chi)/k$. Therefore (19) can be written as

$$\chi(m) = \frac{G(1, \chi)}{k} \sum_{n=1}^{k} \bar{\chi}(-n)e^{2\pi imn/k} = \frac{G(1, \chi)}{k} \sum_{n=1}^{k} \bar{\chi}(n)e^{-2\pi imn/k},$$

which is the same as (21). Theorem 8.11 shows that the numbers $\tau_k(\chi)$ have absolute value 1. □

8.12 Pólya's inequality for the partial sums of primitive characters

The proof of Dirichlet's theorem given in Chapter 7 made use of the relation

$$\left| \sum_{m \le x} \chi(m) \right| \le \varphi(k)$$

which holds for any Dirichlet character χ mod k and every real $x \ge 1$. This cannot be improved because when $\chi = \chi_1$ we have $\sum_{m=1}^{k} \chi_1(m) = \varphi(k)$. However, Pólya showed that the inequality can be considerably improved when χ is a primitive character.

Theorem 8.21 Pólya's inequality. *If χ is any primitive character* mod k *then for all $x \ge 1$ we have*

$$\left| \sum_{m \le x} \chi(m) \right| < \sqrt{k} \log k. \tag{23}$$

PROOF. We express $\chi(m)$ by its finite Fourier expansion, as given in Theorem 8.20

$$\chi(m) = \frac{\tau_k(\chi)}{\sqrt{k}} \sum_{n=1}^{k} \bar{\chi}(n)e^{-2\pi imn/k},$$

and sum over all $m \le x$ to get

$$\sum_{m \le x} \chi(m) = \frac{\tau_k(\chi)}{\sqrt{k}} \sum_{n=1}^{k-1} \bar{\chi}(n) \sum_{m \le x} e^{-2\pi imn/k}$$

since $\chi(k) = 0$. Taking absolute values and multiplying by $\sqrt{k}$ we find

$$\sqrt{k} \left| \sum_{m \le x} \chi(m) \right| \le \sum_{n=1}^{k-1} \left| \sum_{m \le x} e^{-2\pi imn/k} \right| = \sum_{n=1}^{k-1} |f(n)|, \tag{24}$$

say, where

$$f(n) = \sum_{m \le x} e^{-2\pi imn/k}.$$

Now

$$f(k-n) = \sum_{m \le x} e^{-2\pi i m(k-n)/k} = \sum_{m \le x} e^{2\pi i m n/k} = \overline{f(n)}$$

so

$|f(k-n)| = |f(n)|$. Hence (24) can be written as

$$\sqrt{k}\left|\sum_{m \le x} \chi(m)\right| \le 2 \sum_{n < k/2} |f(n)| + \left|f\left(\frac{k}{2}\right)\right|, \tag{25}$$

the term $|f(k/2)|$ appearing only if k is even. But $f(n)$ is a geometric sum, of the form

$$f(n) = \sum_{m=1}^{r} y^m$$

where $r = [x]$ and $y = e^{-2\pi i n/k}$. Here $y \ne 1$ since $1 \le n \le k - 1$. Writing $z = e^{-\pi i n/k}$, we have $y = z^2$ and $z^2 \ne 1$ since $n \le k/2$. Hence we have

$$f(n) = y\,\frac{y^r - 1}{y - 1} = z^2\,\frac{z^{2r} - 1}{z^2 - 1} = z^{r+1}\,\frac{z^r - z^{-r}}{z - z^{-1}}$$

so

$$|f(n)| = \left|\frac{z^r - z^{-r}}{z - z^{-1}}\right| = \left|\frac{e^{-\pi i r n/k} - e^{\pi i r n/k}}{e^{-\pi i n/k} - e^{\pi i n/k}}\right| = \frac{\left|\sin\dfrac{\pi r n}{k}\right|}{\left|\sin\dfrac{\pi n}{k}\right|} \le \frac{1}{\sin\dfrac{\pi n}{k}}. \tag{26}$$

Now we use the inequality $\sin t \ge 2t/\pi$, valid for $0 \le t \le \pi/2$, with $t = \pi n/k$ to get

$$|f(n)| \le \frac{1}{\dfrac{2}{\pi}\dfrac{\pi n}{k}} = \frac{k}{2n}.$$

If k is odd, (25) becomes

$$\sqrt{k}\left|\sum_{m \le x} \chi(m)\right| \le k \sum_{n < k/2} \frac{1}{n} < k \log k.$$

But if k is even, $|f(k/2)| \le 1$, and (25) gives us

$$\sqrt{k}\left|\sum_{m \le x} \chi(m)\right| \le k\left\{\sum_{n < k/2} \frac{1}{n} + \frac{1}{k}\right\} < k \log k,$$

and this proves (23).

Note. Pólya's inequality can be extended to any nonprincipal character (see Theorem 13.15). For nonprimitive characters it takes the form

$$\sum_{m \le x} \chi(m) = O(\sqrt{k} \log k).$$

Exercises for Chapter 8

1. Let $x = e^{2\pi i/n}$ and prove that

$$\sum_{k=1}^{n-1} kx^k = \frac{n}{x-1}.$$

2. Let $((x)) = x - [x] - \frac{1}{2}$ if x is not an integer, and let $((x)) = 0$ otherwise. Note that $((x))$ is a periodic function of x with period 1. If k and n are integers, with $n > 0$, prove that

$$\left(\left(\frac{k}{n}\right)\right) = -\frac{1}{2n}\sum_{m=1}^{n-1} \cot\frac{\pi m}{n}\sin\frac{2\pi km}{n}.$$

3. Let $c_k(m)$ denote Ramanujan's sum and let $M(x) = \sum_{n \le x} \mu(n)$, the partial sums of the Möbius function.

(a) Prove that

$$\sum_{k=1}^{n} c_k(m) = \sum_{d|m} dM\left(\frac{n}{d}\right).$$

In particular, when $n = m$, we have

$$\sum_{k=1}^{m} c_k(m) = \sum_{d|m} dM\left(\frac{m}{d}\right).$$

(b) Use (a) to deduce that

$$M(m) = m \sum_{d|m} \frac{\mu(m/d)}{d} \sum_{k=1}^{d} c_k(d).$$

(c) Prove that

$$\sum_{m=1}^{n} c_k(m) = \sum_{d|k} d\mu\left(\frac{k}{d}\right)\left[\frac{n}{d}\right].$$

4. Let n, a, d be given integers with $(a, d) = 1$. Let $m = a + qd$ where q is the product (possibly empty) of all primes which divide n but not a. Prove that

$$m \equiv a \pmod{d} \qquad \text{and } (m, n) = 1.$$

5. Prove that there exists no real primitive character χ mod k if $k = 2m$, where m is odd.

6. Let χ be a character mod k. If k_1 and k_2 are induced moduli for χ prove that so too is (k_1, k_2), their gcd.

7. Prove that the conductor of χ divides every induced modulus for χ.

In Exercises 8 through 12, assume that $k = k_1 k_2 \cdots k_r$, where the positive integers k_i are relatively prime in pairs: $(k_i, k_j) = 1$ if $i \ne j$.

8. (a) Given any integer a, prove that there is an integer a_i such that

$$a_i \equiv a \pmod{k_i} \qquad \text{and } a_i \equiv 1 \pmod{k_j} \quad \text{for all } j \ne i.$$

(b) Let χ be a character mod k. Define χ_i by the equation

$$\chi_i(a) = \chi(a_i),$$

where a_i is the integer of part (a). Prove that χ_i is a character mod k_i.

9. Prove that every character χ mod k can be factored uniquely as a product of the form $\chi = \chi_1 \chi_2 \cdots \chi_r$, where χ_i is a character mod k_i.

10. Let $f(\chi)$ denote the conductor of χ. If χ has the factorization in Exercise 9, prove that $f(\chi) = f(\chi_1) \cdots f(\chi_r)$.

11. If χ has the factorization in Exercise 9, prove that for every integer a we have

$$G(a, \chi) = \prod_{i=1}^{r} \chi_i\left(\frac{k}{k_i}\right) G(a_i, \chi_i),$$

where a_i is the integer of Exercise 8.

12. If χ has the factorization in Exercise 9, prove that χ is primitive mod k if, and only if, each χ_i is primitive mod k_i. [*Hint*: Theorem 8.19 or Exercise 10.]

13. Let χ be a primitive character mod k. Prove that if $N < M$ we have

$$\left| \sum_{m=N+1}^{M} \frac{\chi(m)}{m} \right| < \frac{2}{N+1} \sqrt{k} \log k.$$

14. This exercise outlines a slight improvement in Pólya's inequality. Refer to the proof of Theorem 8.21. After inequality (26) write

$$\sum_{n \le k/2} |f(n)| \le \sum_{n \le k/2} \frac{1}{\sin \frac{\pi n}{k}} < \frac{1}{\sin \frac{\pi}{k}} + \int_1^{k/2} \frac{dt}{\sin \frac{\pi t}{k}}.$$

Show that the integral is less than $-(k/\pi)\log(\sin(\pi/2k))$ and deduce that

$$\left| \sum_{n \le x} \chi(n) \right| < \sqrt{k} + \frac{2}{\pi} \sqrt{k} \log k.$$

This improves Pólya's inequality by a factor $2/\pi$ in the principal term.

15. The Kloosterman sum $K(m, n; k)$ is defined as follows:

$$K(m, n; k) = \sum_{\substack{h \bmod k \\ (h, k) = 1}} e^{2\pi i(mh + nh')/k}$$

where h' is the reciprocal of h mod k. When $k \mid n$ this reduces to Ramanujan's sum $c_k(m)$. Derive the following properties of Kloosterman sums:

(a) $K(m, n; k) = K(n, m; k)$.

(b) $K(m, n; k) = K(1, mn; k)$ whenever $(m, k) = 1$.

(c) Given integers n, k_1, k_2 such that $(k_1, k_2) = 1$, show that there exist integers n_1 and n_2 such that

$$n \equiv n_1 k_2^2 + n_2 k_1^2 \pmod{k_1 k_2},$$

and that for these integers we have

$$K(m, n; k_1 k_2) = K(m, n_1; k_1) K(m, n_2; k_2).$$

This reduces the study of Kloosterman sums to the special case $K(m, n; p^\alpha)$, where p is prime.

16. If n and k are integers, $n > 0$, the sum

$$G(k; n) = \sum_{r=1}^{n} e^{2\pi i k r^2/n}$$

is called a quadratic Gauss sum. Derive the following properties of quadratic Gauss sums:

(a) $G(k; mn) = G(km; n)G(kn; m)$ whenever $(m, n) = 1$. This reduces the study of Gauss sums to the special case $G(k; p^\alpha)$, where p is prime.

(b) Let p be an odd prime, $p \nmid k$, $\alpha \geq 2$. Prove that $G(k; p^\alpha) = pG(k; p^{\alpha-2})$ and deduce that

$$G(k; p^\alpha) = \begin{cases} p^{\alpha/2} & \text{if } \alpha \text{ is even,} \\ p^{(\alpha-1)/2}G(k; p) & \text{if } \alpha \text{ is odd.} \end{cases}$$

Further properties of the Gauss sum $G(k; p)$ are developed in the next chapter where it is shown that $G(k; p)$ is the same as the Gauss sum $G(k, \chi)$ associated with a certain Dirichlet character χ mod p. (See Exercise 9.9.)

9 Quadratic Residues and the Quadratic Reciprocity Law

9.1 Quadratic residues

As shown in Chapter 5, the problem of solving a polynomial congruence

$$f(x) \equiv 0 \pmod{m}$$

can be reduced to polynomial congruences with prime moduli plus a set of linear congruences. This chapter is concerned with quadratic congruences of the form

$$x^2 \equiv n \pmod{p} \tag{1}$$

where p is an odd prime and $n \not\equiv 0 \pmod{p}$. Since the modulus is prime we know that (1) has at most two solutions. Moreover, if x is a solution so is $-x$, hence the number of solutions is either 0 or 2.

Definition If congruence (1) has a solution we say that n is a quadratic residue mod p and we write nRp. If (1) has no solution we say that n is a quadratic nonresidue mod p and we write $n\bar{R}p$.

Two basic problems dominate the theory of quadratic residues:

1. Given a prime p, determine which n are quadratic residues mod p and which are quadratic nonresidues mod p.
2. Given n, determine those primes p for which n is a quadratic residue mod p and those for which n is a quadratic nonresidue mod p.

We begin with some methods for solving problem 1.

EXAMPLE To find the quadratic residues modulo 11 we square the numbers 1, 2, ..., 10 and reduce mod 11. We obtain

$$1^2 \equiv 1, \quad 2^2 \equiv 4, \quad 3^2 \equiv 9, \quad 4^2 \equiv 5, \quad 5^2 \equiv 3 \pmod{11}.$$

It suffices to square only the first half of the numbers since

$$6^2 \equiv (-5)^2 \equiv 3, \quad 7^2 \equiv (-4)^2 \equiv 5, \ldots, 10^2 \equiv (-1)^2 \equiv 1 \pmod{11}.$$

Consequently, the quadratic residues mod 11 are 1, 3, 4, 5, 9, and the nonresidues are 2, 6, 7, 8, 10.

This example illustrates the following theorem.

Theorem 9.1 *Let p be an odd prime. Then every reduced residue system* mod p *contains exactly $(p-1)/2$ quadratic residues and exactly $(p-1)/2$ quadratic nonresidues* mod p. *The quadratic residues belong to the residue classes containing the numbers*

$$1^2, 2^2, 3^2, \ldots, \left(\frac{p-1}{2}\right)^2. \tag{2}$$

PROOF. First we note that the numbers in (2) are distinct mod p. In fact, if $x^2 \equiv y^2 \pmod{p}$ with $1 \le x \le (p-1)/2$ and $1 \le y \le (p-1)/2$, then

$$(x-y)(x+y) \equiv 0 \pmod{p}.$$

But $1 < x + y < p$ so $x - y \equiv 0 \pmod{p}$, hence $x = y$. Since

$$(p-k)^2 \equiv k^2 \pmod{p},$$

every quadratic residue is congruent mod p to exactly one of the numbers in (2). This completes the proof. □

The following brief table of quadratic residues R and nonresidues $\bar{R}$ was obtained with the help of Theorem 9.1.

	$p = 3$	$p = 5$	$p = 7$	$p = 11$	$p = 13$
R:	1	1, 4	1, 2, 4	1, 3, 4, 5, 9	1, 3, 4, 9, 10, 12
$\bar{R}$:	2	2, 3	3, 5, 6	2, 6, 7, 8, 10	2, 5, 6, 7, 8, 11

9.2 Legendre's symbol and its properties

Definition Let p be an odd prime. If $n \not\equiv 0 \pmod{p}$ we define Legendre's symbol $(n \mid p)$ as follows:

$$(n \mid p) = \begin{cases} +1 & \text{if } nRp, \\ -1 & \text{if } n\bar{R}p. \end{cases}$$

If $n \equiv 0 \pmod{p}$ we define $(n \mid p) = 0$.

EXAMPLES $(1|p) = 1$, $(m^2|p) = 1$, $(7|11) = -1$, $(22|11) = 0$.

Note. Some authors write $\left(\frac{n}{p}\right)$ instead of $(n|p)$.

It is clear that $(m|p) = (n|p)$ whenever $m \equiv n \pmod p$, so $(n|p)$ is a periodic function of n with period p.

The little Fermat theorem tells us that $n^{p-1} \equiv 1 \pmod p$ if $p \nmid n$. Since

$$n^{p-1} - 1 = (n^{(p-1)/2} - 1)(n^{(p-1)/2} + 1)$$

it follows that $n^{(p-1)/2} \equiv \pm 1 \pmod p$. The next theorem tells us that we get $+1$ if nRp and -1 if $n\bar{R}p$.

Theorem 9.2 Euler's criterion. *Let p be an odd prime. Then for all n we have*

$$(n|p) \equiv n^{(p-1)/2} \pmod p.$$

PROOF. If $n \equiv 0 \pmod p$ the result is trivial since both members are congruent to 0 mod p. Now suppose that $(n|p) = 1$. Then there is an x such that $x^2 \equiv n \pmod p$ and hence

$$n^{(p-1)/2} \equiv (x^2)^{(p-1)/2} = x^{p-1} \equiv 1 = (n|p) \pmod p.$$

This proves the theorem if $(n|p) = 1$.

Now suppose that $(n|p) = -1$ and consider the polynomial

$$f(x) = x^{(p-1)/2} - 1.$$

Since $f(x)$ has degree $(p-1)/2$ the congruence

$$f(x) \equiv 0 \pmod p$$

has at most $(p-1)/2$ solutions. But the $(p-1)/2$ quadratic residues mod p are solutions so the nonresidues are not. Hence

$$n^{(p-1)/2} \not\equiv 1 \pmod p \quad \text{if } (n|p) = -1.$$

But $n^{(p-1)/2} \equiv \pm 1 \pmod p$ so $n^{(p-1)/2} \equiv -1 \equiv (n|p) \pmod p$. This completes the proof. □

Theorem 9.3 *Legendre's symbol $(n|p)$ is a completely multiplicative function of n.*

PROOF. If $p|m$ or $p|n$ then $p|mn$ so $(mn|p) = 0$ and either $(m|p) = 0$ or $(n|p) = 0$. Therefore $(mn|p) = (m|p)(n|p)$ if $p|m$ or $p|n$.

If $p \nmid m$ and $p \nmid n$ then $p \nmid mn$ and we have

$$(mn|p) \equiv (mn)^{(p-1)/2} = m^{(p-1)/2}n^{(p-1)/2} \equiv (m|p)(n|p) \pmod p.$$

But each of $(mn|p)$, $(m|p)$ and $(n|p)$ is 1 or -1 so the difference

$$(mn|p) - (m|p)(n|p)$$

is either 0, 2, or -2. Since this difference is divisible by p it must be 0.

Note. Since $(n|p)$ is a completely multiplicative function of n which is periodic with period p and vanishes when $p|n$, it follows that $(n|p) = \chi(n)$, where χ is one of the Dirichlet characters modulo p. The Legendre symbol is called the *quadratic character* mod p.

9.3 Evaluation of $(-1|p)$ and $(2|p)$

Theorem 9.4 *For every odd prime p we have*

$$(-1|p) = (-1)^{(p-1)/2} = \begin{cases} 1 & \text{if } p \equiv 1 \pmod 4, \\ -1 & \text{if } p \equiv 3 \pmod 4. \end{cases}$$

PROOF. By Euler's criterion we have $(-1|p) \equiv (-1)^{(p-1)/2} \pmod p$. Since each member of this congruence is 1 or -1 the two members are equal. □

Theorem 9.5 *For every odd prime p we have*

$$(2|p) = (-1)^{(p^2-1)/8} = \begin{cases} 1 & \text{if } p \equiv \pm 1 \pmod 8, \\ -1 & \text{if } p \equiv \pm 3 \pmod 8. \end{cases}$$

PROOF. Consider the following $(p - 1)/2$ congruences:

$$\begin{aligned} p - 1 &\equiv 1(-1)^1 && \pmod p \\ 2 &\equiv 2(-1)^2 && \pmod p \\ p - 3 &\equiv 3(-1)^3 && \pmod p \\ 4 &\equiv 4(-1)^4 && \pmod p \\ &\vdots \\ r &\equiv \frac{p-1}{2}(-1)^{(p-1)/2} && \pmod p, \end{aligned}$$

where r is either $p - (p - 1)/2$ or $(p - 1)/2$. Multiply these together and note that each integer on the left is even. We obtain

$$2 \cdot 4 \cdot 6 \cdots (p-1) \equiv \left(\frac{p-1}{2}\right)!(-1)^{1+2+\cdots+(p-1)/2} \pmod p.$$

This gives us

$$2^{(p-1)/2}\left(\frac{p-1}{2}\right)! \equiv \left(\frac{p-1}{2}\right)!(-1)^{(p^2-1)/8} \pmod p.$$

Since $((p-1)/2)! \not\equiv 0 \pmod{p}$ this implies

$$2^{(p-1)/2} \equiv (-1)^{(p^2-1)/8} \pmod{p}.$$

By Euler's criterion we have $2^{(p-1)/2} \equiv (2|p) \pmod{p}$, and since each member is 1 or -1 the two members are equal. This completes the proof. □

9.4 Gauss' lemma

Although Euler's criterion gives a straightforward method for computing $(n|p)$, the calculation may become prohibitive for large n since it requires raising n to the power $(p-1)/2$. Gauss found another criterion which involves a simpler calculation.

Theorem 9.6 Gauss' lemma. *Assume $n \not\equiv 0 \pmod{p}$ and consider the least positive residues* mod p *of the following $(p-1)/2$ multiples of n:*

$$n, 2n, 3n, \ldots, \frac{p-1}{2}n. \tag{3}$$

If m denotes the number of these residues which exceed $p/2$, then

$$(n|p) = (-1)^m.$$

Proof. The numbers in (3) are incongruent mod p. We consider their least positive residues and distribute them into two disjoint sets A and B, according as the residues are $<p/2$ or $>p/2$. Thus

$$A = \{a_1, a_2, \ldots, a_k\}$$

where each $a_i \equiv tn \pmod{p}$ for some $t \le (p-1)/2$ and $0 < a_i < p/2$; and

$$B = \{b_1, b_2, \ldots, b_m\}$$

where each $b_i \equiv sn \pmod{p}$ for some $s \le (p-1)/2$ and $p/2 < b_i < p$. Note that $m + k = (p-1)/2$ since A and B are disjoint. The number m of elements in B is pertinent in this theorem. Form a new set C of m elements by subtracting each b_i from p. Thus

$$C = \{c_1, c_2, \ldots, c_m\}, \quad \text{where } c_i = p - b_i.$$

Now $0 < c_i < p/2$ so the elements of C lie in the same interval as the elements of A. We show next that the sets A and C are disjoint.

Assume that $c_i = a_j$ for some pair i and j. Then $p - b_i = a_j$, or $a_j + b_i \equiv 0 \pmod{p}$. Therefore

$$tn + sn = (t + s)n \equiv 0 \pmod{p}$$

for some s and t with $1 \le t < p/2$, $1 \le s < p/2$. But this is impossible since $p \nmid n$ and $0 < s + t < p$. Therefore A and C are disjoint, so their union

$A \cup C$ contains $m + k = (p - 1)/2$ integers in the interval $[1, (p - 1)/2]$. Hence

$$A \cup C = \{a_1, a_2, \ldots, a_k, c_1, c_2, \ldots, c_m\} = \left\{1, 2, \ldots, \frac{p-1}{2}\right\}.$$

Now form the product of all the elements in $A \cup C$ to obtain

$$a_1 a_2 \cdots a_k c_1 c_2 \cdots c_m = \left(\frac{p-1}{2}\right)!.$$

Since $c_i = p - b_i$ this gives us

$$\begin{aligned}
\left(\frac{p-1}{2}\right)! &= a_1 a_2 \cdots a_k (p - b_1)(p - b_2) \cdots (p - b_m) \\
&\equiv (-1)^m a_1 a_2 \cdots a_k b_1 b_2 \cdots b_m \pmod{p} \\
&\equiv (-1)^m n(2n)(3n) \cdots \left(\frac{p-1}{2} n\right) \pmod{p} \\
&\equiv (-1)^m n^{(p-1)/2} \left(\frac{p-1}{2}\right)! \pmod{p}.
\end{aligned}$$

Canceling the factorial we obtain

$$n^{(p-1)/2} \equiv (-1)^m \pmod{p}.$$

Euler's criterion shows that $(-1)^m \equiv (n|p) \pmod{p}$ hence $(-1)^m = (n|p)$ and the proof of Gauss' lemma is complete. □

To use Gauss' lemma in practice we need not know the exact value of m, but only its parity, that is, whether m is odd or even. The next theorem gives a relatively simple way to determine the parity of m.

Theorem 9.7 *Let m be the number defined in Gauss' lemma. Then*

$$m \equiv \sum_{t=1}^{(p-1)/2} \left[\frac{tn}{p}\right] + (n - 1)\frac{p^2 - 1}{8} \pmod{2}.$$

In particular, if n is odd we have

$$m \equiv \sum_{t=1}^{(p-1)/2} \left[\frac{tn}{p}\right] \pmod{2}.$$

PROOF. Recall that m is the number of least positive residues of the numbers

$$n, 2n, 3n, \ldots, \frac{p-1}{2} n$$

which exceed $p/2$. Take a typical number, say tn, divide it by p and examine the size of the remainder. We have

$$\frac{tn}{p} = \left[\frac{tn}{p}\right] + \left\{\frac{tn}{p}\right\}, \quad \text{where } 0 < \left\{\frac{tn}{p}\right\} < 1,$$

so

$$tn = p\left[\frac{tn}{p}\right] + p\left\{\frac{tn}{p}\right\} = p\left[\frac{tn}{p}\right] + r_t,$$

say, where $0 < r_t < p$. The number $r_t = tn - p[tn/p]$ is the least positive residue of tn modulo p. Referring again to the sets A and B used in the proof of Gauss' lemma we have

$$\{r_1, r_2, \ldots, r_{(p-1)/2}\} = \{a_1, a_2, \ldots, a_k, b_1, \ldots, b_m\}.$$

Recall also that

$$\left\{1, 2, \ldots, \frac{p-1}{2}\right\} = \{a_1, a_2, \ldots, a_k, c_1, \ldots, c_m\}$$

where each $c_i = p - b_i$. Now we compute the sums of the elements in these sets to obtain the two equations

$$\sum_{t=1}^{(p-1)/2} r_t = \sum_{i=1}^{k} a_i + \sum_{j=1}^{m} b_j$$

and

$$\sum_{t=1}^{(p-1)/2} t = \sum_{i=1}^{k} a_i + \sum_{j=1}^{m} c_j = \sum_{i=1}^{k} a_i + mp - \sum_{j=1}^{m} b_j.$$

In the first equation we replace r_t by its definition to obtain

$$\sum_{i=1}^{k} a_i + \sum_{j=1}^{m} b_j = n \sum_{t=1}^{(p-1)/2} t - p \sum_{t=1}^{(p-1)/2} \left[\frac{tn}{p}\right].$$

The second equation is

$$mp + \sum_{i=1}^{k} a_i - \sum_{j=1}^{n} b_j = \sum_{t=1}^{(p-1)/2} t.$$

Adding this to the previous equation we get

$$\begin{aligned} mp + 2\sum_{i=1}^{k} a_i &= (n+1) \sum_{t=1}^{(p-1)/2} t - p \sum_{t=1}^{(p-1)/2} \left[\frac{tn}{p}\right] \\ &= (n+1)\frac{p^2-1}{8} - p \sum_{t=1}^{(p-1)/2} \left[\frac{tn}{p}\right]. \end{aligned}$$

Now we reduce this modulo 2, noting that $n + 1 \equiv n - 1 \pmod 2$ and $p \equiv 1 \pmod 2$, and we obtain

$$m \equiv (n-1)\frac{p^2-1}{8} + \sum_{t=1}^{(p-1)/2}\left[\frac{tn}{p}\right] \pmod 2,$$

which completes the proof. □

9.5 The quadratic reciprocity law

Both Euler's criterion and Gauss' lemma give straightforward though sometimes lengthy procedures for solving the first basic problem of the theory of quadratic residues. The second problem is much more difficult. Its solution depends on a remarkable theorem known as the *quadratic reciprocity law*, first stated in a complicated form by Euler in the period 1744–1746, and rediscovered in 1785 by Legendre who gave a partial proof. Gauss discovered the reciprocity law independently at the age of eighteen and a year later in 1796 gave the first complete proof.

The quadratic reciprocity law states that if p and q are distinct odd primes, then $(p|q) = (q|p)$ unless $p \equiv q \equiv 3 \pmod 4$, in which case $(p|q) = -(q|p)$. The theorem is usually stated in the following symmetric form given by Legendre.

Theorem 9.8 Quadratic reciprocity law. *If p and q are distinct odd primes, then*

$$(p|q)(q|p) = (-1)^{(p-1)(q-1)/4}. \tag{4}$$

PROOF. By Gauss' lemma and Theorem 9.7 we have

$$(q|p) = (-1)^m$$

where

$$m \equiv \sum_{t=1}^{(p-1)/2}\left[\frac{tq}{p}\right] \pmod 2.$$

Similarly,

$$(p|q) = (-1)^n$$

where

$$n \equiv \sum_{s=1}^{(q-1)/2}\left[\frac{sp}{q}\right] \pmod 2.$$

Hence $(p|q)(q|p) = (-1)^{m+n}$, and (4) follows at once from the identity

$$\sum_{t=1}^{(p-1)/2}\left[\frac{tq}{p}\right] + \sum_{s=1}^{(q-1)/2}\left[\frac{sp}{q}\right] = \frac{p-1}{2}\frac{q-1}{2}. \tag{5}$$

To prove (5) consider the function

$$f(x, y) = qx - py \qquad \text{for } |x| < p/2 \text{ and } |y| < q/2.$$

If x and y are nonzero integers then $f(x, y)$ is a nonzero integer. Moreover, as x takes the values $1, 2, \ldots, (p - 1)/2$ and y takes the values $1, 2, \ldots, (q - 1)/2$ then $f(x, y)$ takes

$$\frac{p-1}{2}\frac{q-1}{2}$$

values, no two of which are equal since

$$f(x, y) - f(x', y') = f(x - x', y - y') \neq 0.$$

Now we count the number of values of $f(x, y)$ which are positive and the number which are negative.

For each fixed x we have $f(x, y) > 0$ if and only if $y < qx/p$, or $y \le [qx/p]$. Hence the total number of positive values is

$$\sum_{x=1}^{(p-1)/2} \left[\frac{qx}{p}\right].$$

Similarly, the number of negative values is

$$\sum_{y=1}^{(q-1)/2} \left[\frac{py}{q}\right].$$

Since the number of positive and negative values together is

$$\frac{p-1}{2}\frac{q-1}{2}$$

this proves (5) and hence (4). □

Note. The reader may find it instructive to interpret the foregoing proof of (5) geometrically, using lattice points in the plane.

At least 150 proofs of the quadratic reciprocity law have been published. Gauss himself supplied no less than eight, including a version of the one just given. A short proof of the quadratic reciprocity law is described in an article by M. Gerstenhaber [25].

9.6 Applications of the reciprocity law

The following examples show how the quadratic reciprocity law can be used to solve the two basic types of problems in the theory of quadratic residues.

EXAMPLE 1 Determine whether 219 is a quadratic residue or nonresidue mod 383.

Solution

We evaluate the Legendre symbol $(219|383)$ by using the multiplicative property, the reciprocity law, periodicity, and the special values $(-1|p)$ and $(2|p)$ calculated earlier.

Since $219 = 3 \cdot 73$ the multiplicative property implies

$$(219|383) = (3|383)(73|383).$$

Using the reciprocity law and periodicity we have

$$(3|383) = (383|3)(-1)^{(383-1)(3-1)/4} = -(-1|3) = -(-1)^{(3-1)/2} = 1,$$

and

$$\begin{aligned}(73|383) &= (383|73)(-1)^{(383-1)(73-1)/4} = (18|73) = (2|73)(9|73)\\ &= (-1)^{((73)^2-1)/8} = 1.\end{aligned}$$

Hence $(219|383) = 1$ so 219 is a quadratic residue mod 383.

Example 2 Determine those odd primes p for which 3 is a quadratic residue and those for which it is a nonresidue.

Solution

Again, by the reciprocity law we have

$$(3|p) = (p|3)(-1)^{(p-1)(3-1)/4} = (-1)^{(p-1)/2}(p|3).$$

To determine $(p|3)$ we need to know the value of p mod 3, and to determine $(-1)^{(p-1)/2}$ we need to know the value of $(p-1)/2$ mod 2, or the value of p mod 4. Hence we consider p mod 12. There are only four cases to consider, $p \equiv 1, 5, 7$, or $11 \pmod{12}$, the others being excluded since $\varphi(12) = 4$.

Case 1. $p \equiv 1 \pmod{12}$. In this case $p \equiv 1 \pmod 3$ so $(p|3) = (1|3) = 1$. Also $p \equiv 1 \pmod 4$ so $(p-1)/2$ is even, hence $(3|p) = 1$.

Case 2. $p \equiv 5 \pmod{12}$. In this case $p \equiv 2 \pmod 3$ so $(p|3) = (2|3) = (-1)^{(3^2-1)/8} = -1$. Again, $(p-1)/2$ is even since $p \equiv 1 \pmod 4$, so $(3|p) = -1$.

Case 3. $p \equiv 7 \pmod{12}$. In this case $p \equiv 1 \pmod 3$, so $(p|3) = (1|3) = 1$. Also $(p-1)/2$ is odd since $p \equiv 3 \pmod 4$, hence $(3|p) = -1$.

Case 4. $p \equiv 11 \pmod{12}$. In this case $p \equiv 2 \pmod 3$ so $(p|3) = (2|3) = -1$. Again $(p-1)/2$ is odd since $p \equiv 3 \pmod 4$, hence $(3|p) = 1$.

Summarizing the results of the four cases we find

$$3Rp \text{ if } p \equiv \pm 1 \pmod{12}$$
$$3\bar{R}p \text{ if } p \equiv \pm 5 \pmod{12}.$$

9.7 The Jacobi symbol

To determine if a composite number is a quadratic residue or nonresidue mod p it is necessary to consider several cases depending on the quadratic character of the factors. Some calculations can be simplified by using an extension of Legendre's symbol introduced by Jacobi.

Definition If P is a positive odd integer with prime factorization

$$P = \prod_{i=1}^{r} p_i^{a_i}$$

the Jacobi symbol $(n|P)$ is defined for all integers n by the equation

$$(n|P) = \prod_{i=1}^{r} (n|p_i)^{a_i}, \tag{6}$$

where $(n|p_i)$ is the Legendre symbol. We also define $(n|1) = 1$.

The possible values of $(n|P)$ are 1, -1, or 0, with $(n|P) = 0$ if and only if $(n, P) > 1$.

If the congruence

$$x^2 \equiv n \pmod{P}$$

has a solution then $(n|p_i) = 1$ for each prime p_i in (6), and hence $(n|P) = 1$. However, the converse is not true since $(n|P)$ can be 1 if an even number of factors -1 appears in (6).

The reader can verify that the following properties of the Jacobi symbol are easily deduced from properties of the Legendre symbol.

Theorem 9.9 *If P and Q are odd positive integers, we have*

(a) $(m|P)(n|P) = (mn|P)$,
(b) $(n|P)(n|Q) = (n|PQ)$,
(c) $(m|P) = (n|P)$ *whenever* $m \equiv n \pmod{P}$,
(d) $(a^2n|P) = (n|P)$ *whenever* $(a, P) = 1$.

The special formulas for evaluating the Legendre symbols $(-1|p)$ and $(2|p)$ also hold for the Jacobi symbol.

Theorem 9.10 *If P is an odd positive integer we have*

$$(-1|P) = (-1)^{(P-1)/2} \tag{7}$$

and

$$(2|P) = (-1)^{(P^2-1)/8}. \tag{8}$$

PROOF. Write $P = p_1 p_2 \cdots p_m$ where the prime factors p_i are not necessarily distinct. This can also be written as

$$P = \prod_{i=1}^{m}(1 + p_i - 1) = 1 + \sum_{i=1}^{m}(p_i - 1) + \sum_{i \neq j}(p_i - 1)(p_j - 1) + \cdots.$$

But each factor $p_i - 1$ is even so each sum after the first is divisible by 4. Hence

$$P \equiv 1 + \sum_{i=1}^{m}(p_i - 1) \pmod 4,$$

or

$$\frac{1}{2}(P - 1) \equiv \sum_{i=1}^{m} \frac{1}{2}(p_i - 1) \pmod 2.$$

Therefore

$$(-1|P) = \prod_{i=1}^{m}(-1|p_i) = \prod_{i=1}^{m}(-1)^{(p_i-1)/2} = (-1)^{(P-1)/2},$$

which proves (7).

To prove (8) we write

$$P^2 = \prod_{i=1}^{m}(1 + p_i^2 - 1) = 1 + \sum_{i=1}^{m}(p_i^2 - 1) + \sum_{i \neq j}(p_i^2 - 1)(p_j^2 - 1) + \cdots.$$

Since p_i is odd we have $p_i^2 - 1 \equiv 0 \pmod 8$ so

$$P^2 \equiv 1 + \sum_{i=1}^{m}(p_i^2 - 1) \pmod{64}$$

hence

$$\frac{1}{8}(P^2 - 1) = \sum_{i=1}^{m} \frac{1}{8}(p_i^2 - 1) \pmod 8.$$

This also holds mod 2, hence

$$(2|P) = \prod_{i=1}^{m}(2|p_i) = \prod_{i=1}^{m}(-1)^{(p_i^2-1)/8} = (-1)^{(P^2-1)/8},$$

which proves (8). □

Theorem 9.11 Reciprocity law for Jacobi symbols. *If P and Q are positive odd integers with $(P, Q) = 1$, then*

$$(P|Q)(Q|P) = (-1)^{(P-1)(Q-1)/4}.$$

Proof. Write $P = p_1 \cdots p_m$, $Q = q_1 \cdots q_n$, where the p_i and q_i are primes. Then

$$(P|Q)(Q|P) = \prod_{i=1}^{m}\prod_{j=1}^{n}(p_i|q_j)(q_j|p_i) = (-1)^r,$$

say. Applying the quadratic reciprocity law to each factor we find that

$$r = \sum_{i=1}^{m}\sum_{j=1}^{n}\frac{1}{2}(p_i - 1)\frac{1}{2}(q_j - 1) = \sum_{i=1}^{m}\frac{1}{2}(p_i - 1)\sum_{j=1}^{n}\frac{1}{2}(q_j - 1).$$

In the proof of Theorem 9.10 we showed that

$$\sum_{i=1}^{m} \frac{1}{2}(p_i - 1) \equiv \frac{1}{2}(P - 1) \pmod{2},$$

and a corresponding congruence holds for $\sum \frac{1}{2}(q_j - 1)$. Therefore

$$r \equiv \frac{P-1}{2}\frac{Q-1}{2} \pmod{2},$$

which completes the proof. □

Example 1 Determine whether 888 is a quadratic residue or nonresidue of the prime 1999.

Solution

We have

$$(888 \mid 1999) = (4 \mid 1999)(2 \mid 1999)(111 \mid 1999) = (111 \mid 1999).$$

To calculate $(111 \mid 1999)$ using Legendre symbols we would write

$$(111 \mid 1999) = (3 \mid 1999)(37 \mid 1999)$$

and apply the quadratic reciprocity law to each factor on the right. The calculation is simpler with Jacobi symbols since we have

$$(111 \mid 1999) = -(1999 \mid 111) = -(1 \mid 111) = -1.$$

Therefore 888 is a quadratic nonresidue of 1999.

Example 2 Determine whether -104 is a quadratic residue or nonresidue of the prime 997.

Solution

Since $104 = 2 \cdot 4 \cdot 13$ we have

$$\begin{aligned}(-104 \mid 997) &= (-1 \mid 997)(2 \mid 997)(13 \mid 997) = -(13 \mid 997)\\ &= -(997 \mid 13) = -(9 \mid 13) = -1.\end{aligned}$$

Therefore -104 is a quadratic nonresidue of 997.

9.8 Applications to Diophantine equations

Equations to be solved in integers are called *Diophantine equations* after Diophantus of Alexandria. An example is the equation

$$y^2 = x^3 + k \tag{9}$$

where k is a given integer. The problem is to decide, for a given k, whether or not the equation has integer solutions x, y and, if so, to exhibit all of them.

We discuss this equation here partly because it has a long history, going back to the seventeenth century, and partly because some cases can be treated with the help of quadratic residues. A general theorem states that the Diophantine equation

$$y^2 = f(x)$$

has at most a finite number of solutions if $f(x)$ is a polynomial of degree ≥ 3 with integer coefficients and with distinct zeros. (See Theorem 4-18 in LeVeque [44], Vol. 2.) However, no method is known for determining the solutions (or even the *number* of solutions) except for very special cases. The next theorem describes an infinite set of values of k for which (9) has *no* solutions.

Theorem 9.12 *The Diophantine equation*

$$y^2 = x^3 + k \tag{10}$$

has no solutions if k has the form

$$k = (4n - 1)^3 - 4m^2, \tag{11}$$

where m and n are integers such that no prime $p \equiv -1 \pmod 4$ divides m.

PROOF. We assume a solution x, y exists and obtain a contradiction by considering the equation modulo 4. Since $k \equiv -1 \pmod 4$ we have

$$y^2 \equiv x^3 - 1 \pmod 4. \tag{12}$$

Now $y^2 \equiv 0$ or $1 \pmod 4$ for every y, so (12) cannot be satisfied if x is even or if $x \equiv -1 \pmod 4$. Therefore we must have $x \equiv 1 \pmod 4$. Now let

$$a = 4n - 1$$

so that $k = a^3 - 4m^2$, and write (10) in the form

$$y^2 + 4m^2 = x^3 + a^3 = (x + a)(x^2 - ax + a^2). \tag{13}$$

Since $x \equiv 1 \pmod 4$ and $a \equiv -1 \pmod 4$ we have

$$x^2 - ax + a^2 \equiv 1 - a + a^2 \equiv -1 \pmod 4. \tag{14}$$

Hence $x^2 - ax + a^2$ is odd, and (14) shows that all its prime factors cannot be $\equiv 1 \pmod 4$. Therefore some prime $p \equiv -1 \pmod 4$ divides $x^2 - ax + a^2$, and (13) shows that this also divides $y^2 + 4m^2$. In other words,

$$y^2 \equiv -4m^2 \pmod p \quad \text{for some } p \equiv -1 \pmod 4. \tag{15}$$

But $p \nmid m$ by hypothesis, so $(-4m^2 | p) = (-1 | p) = -1$, contradicting (15). This proves that the Diophantine equation (10) has no solutions when k has the form (11). □

The following table gives some values of k covered by Theorem 9.12.

n	0	0	0	0	1	1	1	1	2	2	2	2
m	1	2	4	5	1	2	4	5	1	2	4	5
k	-5	-17	-65	-100	23	11	-37	-73	339	327	279	243

Note. All solutions of (10) have been calculated when k is in the interval $-100 \le k \le 100$. (See reference [32].) No solutions exist for the following positive values of $k \le 100$:

$$k = 6, 7, 11, 13, 14, 20, 21, 23, 29, 32, 34, 39, 42, 45, 46, 47, 51, 53, 58, 59, 60, 61, 62, 66, 67, 69, 70, 74, 75, 77, 78, 83, 84, 85, 86, 87, 88, 90, 93, 95, 96.$$

9.9 Gauss sums and the quadratic reciprocity law

This section gives another proof of the quadratic reciprocity law with the help of the Gauss sums

$$G(n, \chi) = \sum_{r \bmod p} \chi(r) e^{2\pi i n r/p}, \tag{16}$$

where $\chi(r) = (r|p)$ is the quadratic character mod p. Since the modulus is prime, χ is a primitive character and we have the separability property

$$G(n, \chi) = (n|p)G(1, \chi) \tag{17}$$

for every n. Also, Theorem 8.11 implies that $|G(1, \chi)|^2 = p$. The next theorem shows that $G(1, \chi)^2$ is $\pm p$.

Theorem 9.13 *If p is an odd prime and $\chi(r) = (r|p)$ we have*

$$G(1, \chi)^2 = (-1|p)p. \tag{18}$$

PROOF. We have

$$G(1, \chi)^2 = \sum_{r=1}^{p-1} \sum_{s=1}^{p-1} (r|p)(s|p) e^{2\pi i (r+s)/p}.$$

For each pair r, s there is a unique t mod p such that $s \equiv tr \pmod p$, and $(r|p)(s|p) = (r|p)(tr|p) = (r^2|p)(t|p) = (t|p)$. Hence

$$G(1, \chi)^2 = \sum_{t=1}^{p-1} \sum_{r=1}^{p-1} (t|p) e^{2\pi i r(1+t)/p} = \sum_{t=1}^{p-1} (t|p) \sum_{r=1}^{p-1} e^{2\pi i r(1+t)/p}.$$

The last sum on r is a geometric sum given by

$$\sum_{r=1}^{p-1} e^{2\pi i r(1+t)/p} = \begin{cases} -1 & \text{if } p \nmid (1+t), \\ p-1 & \text{if } p \mid (1+t). \end{cases}$$

Therefore

$$G(1, \chi)^2 = -\sum_{t=1}^{p-2}(t|p) + (p-1)(p-1|p) = -\sum_{t=1}^{p-1}(t|p) + p(-1|p)$$
$$= (-1|p)p$$

since $\sum_{t=1}^{p-1}(t|p) = 0$. This proves (18). □

Equation (18) shows that $G(1, \chi)^2$ is an integer, so $G(1, \chi)^{q-1}$ is also an integer for every odd q. The next theorem shows that the quadratic reciprocity law is connected to the value of this integer modulo q.

Theorem 9.14 *Let p and q be distinct odd primes and let χ be the quadratic character* mod p. *Then the quadratic reciprocity law*

$$(q|p) = (-1)^{(p-1)(q-1)/4}(p|q) \tag{19}$$

is equivalent to the congruence

$$G(1, \chi)^{q-1} \equiv (q|p) \pmod q. \tag{20}$$

Proof. From (18) we have

$$G(1, \chi)^{q-1} = (-1|p)^{(q-1)/2}p^{(q-1)/2} = (-1)^{(p-1)(q-1)/4}p^{(q-1)/2}. \tag{21}$$

By Euler's criterion we have $p^{(q-1)/2} \equiv (p|q) \pmod q$ so (21) implies

$$G(1, \chi)^{q-1} \equiv (-1)^{(p-1)(q-1)/4}(p|q) \pmod q. \tag{22}$$

If (20) holds we obtain

$$(q|p) \equiv (-1)^{(p-1)(q-1)/4}(p|q) \pmod q$$

which implies (19) since both members are ± 1. Conversely, if (19) holds then (22) implies (20). □

The next theorem gives an identity which we will use to deduce (20).

Theorem 9.15 *If p and q are distinct odd primes and if χ is the quadratic character* mod p *we have*

$$G(1, \chi)^{q-1} = (q|p) \sum_{\substack{r_1 \bmod p \\ r_1 + \cdots + r_q \equiv q \pmod p}} \cdots \sum_{r_q \bmod p} (r_1 \cdots r_q|p). \tag{23}$$

Proof. The Gauss sum $G(n, \chi)$ is a periodic function of n with period p. The same is true of $G(n, \chi)^q$ so we have a finite Fourier expansion

$$G(n, \chi)^q = \sum_{m \bmod p} a_q(m)e^{2\pi imn/p},$$

where the coefficients are given by

$$a_q(m) = \frac{1}{p} \sum_{n \bmod p} G(n, \chi)^q e^{-2\pi i m n/p}. \tag{24}$$

From the definition of $G(n, \chi)$ we have

$$\begin{aligned} G(n, \chi)^q &= \sum_{r_1 \bmod p} (r_1 | p) e^{2\pi i n r_1/p} \cdots \sum_{r_q \bmod p} (r_q | p) e^{2\pi i n r_q/p} \\ &= \sum_{r_1 \bmod p} \cdots \sum_{r_q \bmod p} (r_1 \cdots r_q | p) e^{2\pi i n(r_1 + \cdots + r_q)/p}, \end{aligned}$$

so (24) becomes

$$a_q(m) = \frac{1}{p} \sum_{r_1 \bmod p} \cdots \sum_{r_q \bmod p} (r_1 \cdots r_q | p) \sum_{n \bmod p} e^{2\pi i n(r_1 + \cdots + r_q - m)/p}.$$

The sum on n is a geometric sum which vanishes unless $r_1 + \cdots + r_q \equiv m \pmod p$, in which case the sum is equal to p. Hence

$$a_q(m) = \sum_{\substack{r_1 \bmod p \\ r_1 + \cdots + r_q \equiv m \pmod p}} \cdots \sum_{r_q \bmod p} (r_1 \cdots r_q | p). \tag{25}$$

Now we return to (24) and obtain an alternate expression for $a_q(m)$. Using the separability of $G(n, \chi)$ and the relation $(n|p)^q = (n|p)$ for odd q we find

$$\begin{aligned} a_q(m) &= \frac{1}{p} G(1, \chi)^q \sum_{n \bmod p} (n|p) e^{-2\pi i m n/p} = \frac{1}{p} G(1, \chi)^q G(-m, \chi) \\ &= \frac{1}{p} G(1, \chi)^q (m|p) G(-1, \chi) = (m|p) G(1, \chi)^{q-1} \end{aligned}$$

since

$$G(1, \chi) G(-1, \chi) = G(1, \chi)\overline{G(1, \chi)} = |G(1, \chi)|^2 = p.$$

In other words, $G(1, \chi)^{q-1} = (m|p) a_q(m)$. Taking $m = q$ and using (25) we obtain (23). □

PROOF OF THE RECIPROCITY LAW. To deduce the quadratic reciprocity law from (23) it suffices to show that

$$\sum_{r_1 \bmod p} \cdots \sum_{r_q \bmod p} (r_1 \cdots r_q | p) \equiv 1 \pmod q, \tag{26}$$

where the summation indices $r_1, \ldots, r_q$ are subject to the restriction

$$r_1 + \cdots + r_q \equiv q \pmod p. \tag{27}$$

If all the indices $r_1, \ldots, r_q$ are congruent to each other mod p, then their sum is congruent to qr_j for each $j = 1, 2, \ldots, q$, so (27) holds if, and only if,

$$qr_j \equiv q \pmod{p},$$

that is, if, and only if $r_j \equiv 1 \pmod{p}$ for each j. In this case the corresponding summand in (26) is $(1 | p) = 1$. For all other choices of indices satisfying (27) there must be at least two incongruent indices among $r_1, \ldots, r_q$. Therefore every cyclic permutation of $r_1, \ldots, r_q$ gives a new solution of (27) which contributes the same summand, $(r_1 \cdots r_q | p)$. Therefore each such summand appears q times and contributes 0 modulo q to the sum. Hence the only contribution to the sum in (26) which is nonzero modulo q is $(1 | p) = 1$. This completes the proof. □

9.10 The reciprocity law for quadratic Gauss sums

This section describes another proof of the quadratic reciprocity law based on the quadratic Gauss sums

$$G(n; m) = \sum_{r=1}^{m} e^{2\pi i n r^2/m}. \tag{28}$$

If p is an odd prime and $p \nmid n$ we have the formula

$$G(n; p) = (n | p) G(1; p) \tag{29}$$

which reduces the study of the sums $G(n; p)$ to the case $n = 1$. Equation (29) follows easily from (28) or by noting that $G(n; p) = G(n, \chi)$, where $\chi(n) = (n | p)$, and observing that $G(n, \chi)$ is separable.

Although each term of the sum $G(1; p)$ has absolute value 1, the sum itself has absolute value 0, $\sqrt{p}$ or $\sqrt{2p}$. In fact, Gauss proved the remarkable formula

$$G(1; m) = \frac{1}{2}\sqrt{m}(1 + i)(1 + e^{-\pi i m/2}) = \begin{cases} \sqrt{m} & \text{if } m \equiv 1 \pmod{4} \\ 0 & \text{if } m \equiv 2 \pmod{4} \\ i\sqrt{m} & \text{if } m \equiv 3 \pmod{4} \\ (1 + i)\sqrt{m} & \text{if } m \equiv 0 \pmod{4} \end{cases} \tag{30}$$

for every $m \geq 1$. A number of different proofs of (30) are known. We will deduce (30) by treating a related sum

$$S(a, m) = \sum_{r=0}^{m-1} e^{\pi i a r^2/m},$$

where a and m are positive integers. If $a = 2$, then $S(2, m) = G(1; m)$.

The sums $S(a, m)$ enjoy a reciprocity law (stated below in Theorem 9.16) which implies Gauss' formula (30) and also leads to another proof of the quadratic reciprocity law.

Theorem 9.16 *If the product ma is even, we have*

$$S(a, m) = \sqrt{\frac{m}{a}}\left(\frac{1+i}{\sqrt{2}}\right)\overline{S(m, a)}, \tag{31}$$

where the bar denotes the complex conjugate.

Note. To deduce Gauss' formula (30) we take $a = 2$ in (31) and observe that $\overline{S(m, 2)} = 1 + e^{-\pi i m/2}$.

PROOF. This proof is based on residue calculus. Let g be the function defined by the equation

$$g(z) = \sum_{r=0}^{m-1} e^{\pi i a(z+r)^2/m}. \tag{32}$$

Then g is analytic everywhere, and $g(0) = S(a, m)$. Since ma is even we find

$$g(z+1) - g(z) = e^{\pi i a z^2/m}(e^{2\pi i a z} - 1) = e^{\pi i a z^2/m}(e^{2\pi i z} - 1)\sum_{n=0}^{a-1} e^{2\pi i n z}.$$

Now define f by the equation

$$f(z) = \frac{g(z)}{e^{2\pi i z} - 1}.$$

Then f is analytic everywhere except for a first-order pole at each integer, and f satisfies the equation

$$f(z+1) = f(z) + \varphi(z), \tag{33}$$

where

$$\varphi(z) = e^{\pi i a z^2/m}\sum_{n=0}^{a-1} e^{2\pi i n z}. \tag{34}$$

The function φ is analytic everywhere.

At $z = 0$ the residue of f is $g(0)/(2\pi i)$ and hence

$$S(a, m) = g(0) = 2\pi i \operatorname*{Res}_{z=0} f(z) = \int_\gamma f(z)\,dz, \tag{35}$$

where γ is any positively oriented simple closed path whose graph contains only the pole $z = 0$ in its interior region. We will choose γ so that it describes a parallelogram with vertices A, $A + 1$, $B + 1$, B where

$$A = -\frac{1}{2} - Re^{\pi i/4} \text{ and } B = -\frac{1}{2} + Re^{\pi i/4},$$

as shown in Figure 9.1. Integrating f along γ we have

$$\int_\gamma f = \int_A^{A+1} f + \int_{A+1}^{B+1} f + \int_{B+1}^{B} f + \int_B^A f.$$

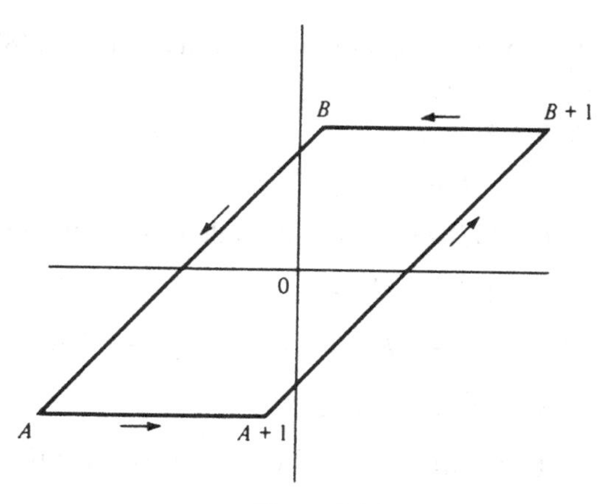

Figure 9.1

In the integral $\int_{A+1}^{B+1} f$ we make the change of variable $w = z + 1$ and then use (33) to get

$$\int_{A+1}^{B+1} f(w)\,dw = \int_{A}^{B} f(z+1)\,dz = \int_{A}^{B} f(z)\,dz + \int_{A}^{B} \varphi(z)\,dz.$$

Therefore (35) becomes

$$S(a, m) = \int_{A}^{B} \varphi(z)\,dz + \int_{A}^{A+1} f(z)\,dz - \int_{B}^{B+1} f(z)\,dz. \tag{36}$$

Now we show that the integrals along the horizontal segments from A to $A + 1$ and from B to $B + 1$ tend to 0 as $R \to +\infty$. To do this we estimate the integrand on these segments. We write

$$|f(z)| = \frac{|g(z)|}{|e^{2\pi i z} - 1|}, \tag{37}$$

and estimate the numerator and denominator separately.

On the segment joining B to $B + 1$ we let

$$\gamma(t) = t + Re^{\pi i/4}, \quad \text{where } -\frac{1}{2} \le t \le \frac{1}{2}.$$

From (32) we find

$$|g[\gamma(t)]| \le \sum_{r=0}^{m-1} \left| \exp\left\{ \frac{\pi i a (t + Re^{\pi i/4} + r)^2}{m} \right\} \right|, \tag{38}$$

where $\exp z = e^z$. The expression in braces has real part

$$\frac{-\pi a(\sqrt{2}tR + R^2 + \sqrt{2}rR)}{m}.$$

Since $|e^{x+iy}| = e^x$ and $\exp\{-\pi a\sqrt{2}rR/m\} \le 1$, each term in (38) has absolute value not exceeding $\exp\{-\pi aR^2/m\}\exp\{-\sqrt{2}\pi atR/m\}$. But $-1/2 \le t \le 1/2$, so we obtain the estimate

$$|g[\gamma(t)]| \le me^{\pi\sqrt{2}aR/(2m)}e^{-\pi aR^2/m}.$$

For the denominator in (37) we use the triangle inequality in the form

$$|e^{2\pi iz} - 1| \ge ||e^{2\pi iz}| - 1|.$$

Since $|\exp\{2\pi i\gamma(t)\}| = \exp\{-2\pi R\sin(\pi/4)\} = \exp\{-\sqrt{2}\pi R\}$, we find

$$|e^{2\pi i\gamma(t)} - 1| \ge 1 - e^{-\sqrt{2}\pi R}.$$

Therefore on the line segment joining B to $B + 1$ we have the estimate

$$|f(z)| \le \frac{me^{\pi\sqrt{2}aR/(2m)}e^{-\pi aR^2/m}}{1 - e^{-\sqrt{2}\pi R}} = o(1) \quad \text{as } R \to +\infty.$$

A similar argument shows that the integrand tends to 0 on the segment joining A to $A + 1$ as $R \to +\infty$. Since the length of the path of integration is 1 in each case, this shows that the second and third integrals on the right of (36) tend to 0 as $R \to +\infty$. Therefore we can write (36) in the form

$$S(a, m) = \int_A^B \varphi(z)\,dz + o(1) \quad \text{as } R \to +\infty. \tag{39}$$

To deal with the integral $\int_A^B \varphi$ we apply Cauchy's theorem, integrating φ around the parallelogram with vertices A, B, α, $-\alpha$, where $\alpha = B + \frac{1}{2} = Re^{\pi i/4}$. (See Figure 9.2.) Since φ is analytic everywhere, its integral around this parallelogram is 0, so

$$\int_A^B \varphi + \int_B^\alpha \varphi + \int_\alpha^{-\alpha} \varphi + \int_{-\alpha}^A \varphi = 0. \tag{40}$$

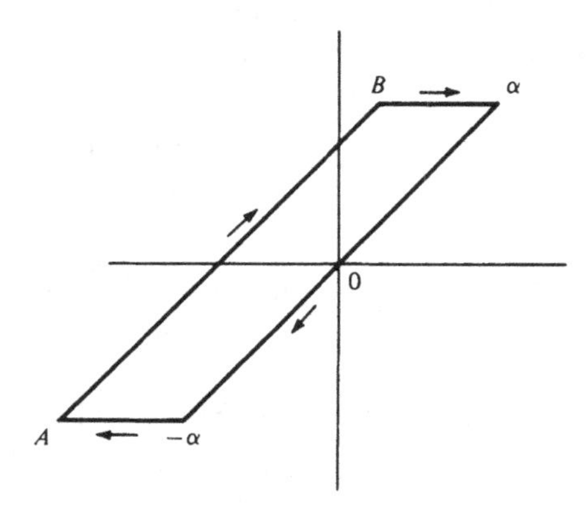

Figure 9.2

Because of the exponential factor $e^{\pi i a z^2/m}$ in (34), an argument similar to that given above shows that the integral of φ along each horizontal segment $\to 0$ as $R \to +\infty$. Therefore (40) gives us

$$\int_A^B \varphi = \int_{-\alpha}^{\alpha} \varphi + o(1) \quad \text{as } R \to +\infty,$$

and (39) becomes

$$S(a, m) = \int_{-\alpha}^{\alpha} \varphi(z)\, dz + o(1) \quad \text{as } R \to +\infty, \tag{41}$$

where $\alpha = Re^{\pi i/4}$. Using (34) we find

$$\int_{-\alpha}^{\alpha} \varphi(z)\, dz = \sum_{n=0}^{a-1} \int_{-\alpha}^{\alpha} e^{\pi i a z^2/m} e^{2\pi i n z}\, dz = \sum_{n=0}^{a-1} e^{-\pi i m n^2/a} I(a, m, n, R),$$

where

$$I(a, m, n, R) = \int_{-\alpha}^{\alpha} \exp\left\{\frac{\pi i a}{m}\left(z + \frac{nm}{a}\right)^2\right\} dz.$$

Applying Cauchy's theorem again to the parallelogram with vertices $-\alpha$, α, $\alpha - (nm/a)$, and $-\alpha - (nm/a)$, we find as before that the integrals along the horizontal segments $\to 0$ as $R \to +\infty$, so

$$I(a, m, n, R) = \int_{-\alpha - nm/a}^{\alpha - mn/a} \exp\left\{\frac{\pi i a}{m}\left(z + \frac{nm}{a}\right)^2\right\} dz + o(1) \quad \text{as } R \to +\infty.$$

The change of variable $w = \sqrt{a/m}(z + (nm/a))$ puts this into the form

$$I(a, m, n, R) = \sqrt{\frac{m}{a}} \int_{-\alpha\sqrt{a/m}}^{\alpha\sqrt{a/m}} e^{\pi i w^2}\, dw + o(1) \quad \text{as } R \to +\infty.$$

Letting $R \to +\infty$ in (41), we find

$$S(a, m) = \sum_{n=0}^{a-1} e^{-\pi i m n^2/a} \sqrt{\frac{m}{a}} \lim_{R \to +\infty} \int_{-R\sqrt{a/m}e^{\pi i/4}}^{R\sqrt{a/m}e^{\pi i/4}} e^{\pi i w^2}\, dw. \tag{42}$$

By writing $T = \sqrt{a/m}R$, we see that the last limit is equal to

$$\lim_{T \to +\infty} \int_{-Te^{\pi i/4}}^{Te^{\pi i/4}} e^{\pi i w^2}\, dw = I$$

say, where I is a number independent of a and m. Therefore (42) gives us

$$S(a, m) = \sqrt{\frac{m}{a}}\, I\, \overline{S(m, a)}. \tag{43}$$

To evaluate I we take $a = 1$ and $m = 2$ in (43). Then $S(1, 2) = 1 + i$ and $S(2, 1) = 1$, so (43) implies $I = (1 + i)/\sqrt{2}$, and (43) reduces to (31). □

Theorem 9.16 implies a reciprocity law for quadratic Gauss sums.

Theorem 9.17 *If $h > 0$, $k > 0$, h odd, then*

$$G(h; k) = \sqrt{\frac{k}{h}}\,\frac{1+i}{2}\,(1 + e^{-\pi ihk/2})\overline{G(k; h)}. \tag{44}$$

PROOF. Take $a = 2h$, $m = k$ in Theorem 9.16 to obtain

$$G(h; k) = S(2h, k) = \sqrt{\frac{k}{2h}}\,\frac{1+i}{\sqrt{2}}\,\overline{S(k, 2h)} = \sqrt{\frac{k}{h}}\,\frac{1+i}{2}\sum_{r=0}^{2h-1} e^{-\pi ikr^2/(2h)}. \tag{45}$$

We split the sum on r into two parts corresponding to even and odd r. For even r we write $r = 2s$ where $s = 0, 1, 2, \ldots, h - 1$. For odd r we note that $(r + 2h)^2 \equiv r^2 \pmod{4h}$ so the sum can be extended over the odd numbers in any complete residue system mod $2h$. We sum over the odd numbers in the interval $h \le r < 3h$, writing $r = 2s + h$, where $s = 0, 1, 2, \ldots, h - 1$. (The numbers $2s + h$ are odd and distinct mod $2h$.) This gives us

$$\begin{aligned}\sum_{r=0}^{2h-1} e^{-\pi ikr^2/(2h)} &= \sum_{s=0}^{h-1} e^{-\pi ik(2s)^2/(2h)} + \sum_{s=0}^{h-1} e^{-\pi ik(2s+h)^2/(2h)} \\ &= \sum_{s=0}^{h-1} e^{-2\pi iks^2/h}(1 + e^{-\pi ihk/2}) \\ &= (1 + e^{-\pi ihk/2})\overline{G(k; h)}.\end{aligned}$$

Using this in (45) we obtain (44). □

9.11 Another proof of the quadratic reciprocity law

Gauss' formula (30) leads to a quick proof of the quadratic reciprocity law. First we note that (30) implies

$$G(1; k) = i^{(k-1)^2/4}\sqrt{k}$$

if k is odd. Also, we have the multiplicative property (see Exercise 8.16(a))

$$G(m; n)G(n; m) = G(1; mn) \quad \text{if } (m, n) = 1.$$

Therefore, if p and q are distinct odd primes we have

$$\begin{aligned}G(p; q) &= (p \mid q)G(1; q) = (p \mid q)i^{(q-1)^2/4}\sqrt{q} \\ G(q; p) &= (q \mid p)G(1; p) = (q \mid p)i^{(p-1)^2/4}\sqrt{p}\end{aligned}$$

and

$$G(p; q)G(q; p) = G(1; pq) = i^{(pq-1)^2/4}\sqrt{pq}.$$

Comparing the last equation with the previous two we find

$$(p|q)(q|p)i^{\{(q-1)^2+(p-1)^2\}/4} = i^{(pq-1)^2/4},$$

and the quadratic reciprocity law follows by observing that

$$i^{\{(pq-1)^2-(q-1)^2-(p-1)^2\}/4} = (-1)^{(p-1)(q-1)/4}. \qquad \square$$

Exercises for Chapter 9

1. Determine those odd primes p for which $(-3|p) = 1$ and those for which $(-3|p) = -1$.

2. Prove that 5 is a quadratic residue of an odd prime p if $p \equiv \pm 1 \pmod{10}$, and that 5 is a nonresidue if $p \equiv \pm 3 \pmod{10}$.

3. Let p be an odd prime. Assume that the set $\{1, 2, \ldots, p-1\}$ can be expressed as the union of two nonempty subsets S and T, $S \neq T$, such that the product (mod p) of any two elements in the same subset lies in S, whereas the product (mod p) of any element in S with any element in T lies in T. Prove that S consists of the quadratic residues and T of the nonresidues mod p.

4. Let $f(x)$ be a polynomial which takes integer values when x is an integer.
   (*a*) If a and b are integers, prove that

   $$\sum_{x \bmod p} (f(ax+b)|p) = \sum_{x \bmod p} (f(x)|p) \quad \text{if } (a, p) = 1,$$

   and that

   $$\sum_{x \bmod p} (af(x)|p) = (a|p) \sum_{x \bmod p} (f(x)|p) \quad \text{for all } a.$$

   (b) Prove that

   $$\sum_{x \bmod p} (ax + b|p) = 0 \quad \text{if } (a, p) = 1.$$

   (c) Let $f(x) = x(ax + b)$, where $(a, p) = (b, p) = 1$. Prove that

   $$\sum_{x=1}^{p-1} (f(x)|p) = \sum_{x=1}^{p-1} (a + bx|p) = -(a|p).$$

   [*Hint*: As x runs through a reduced residue system mod p, so does x', the reciprocal of x mod p.]

5. Let α and β be integers whose possible values are ± 1. Let $N(\alpha, \beta)$ denote the number of integers x among $1, 2, \ldots, p-2$ such that

   $$(x|p) = \alpha \qquad \text{and } (x+1|p) = \beta,$$

   where p is an odd prime. Prove that

   $$4N(\alpha, \beta) = \sum_{x=1}^{p-2} \{1 + \alpha(x|p)\}\{1 + \beta(x+1|p)\},$$

and use Exercise 4 to deduce that

$$4N(\alpha, \beta) = p - 2 - \beta - \alpha\beta - \alpha(-1|p).$$

In particular this gives

$$N(1, 1) = \frac{p - 4 - (-1|p)}{4},$$

$$N(-1, -1) = N(-1, 1) = \frac{p - 2 + (-1|p)}{4},$$

$$N(1, -1) = 1 + N(1, 1).$$

6. Use Exercise 5 to show that for every prime p there exist integers x and y such that $x^2 + y^2 + 1 \equiv 0 \pmod{p}$.

7. Let p be an odd prime. Prove each of the following statements:

(a) $\sum_{r=1}^{p-1} r(r|p) = 0 \quad \text{if } p \equiv 1 \pmod{4}.$

(b) $\sum_{\substack{r=1 \\ (r|p)=1}}^{p-1} r = \frac{p(p-1)}{4} \quad \text{if } p \equiv 1 \pmod{4}.$

(c) $\sum_{r=1}^{p-1} r^2(r|p) = p\sum_{r=1}^{p-1} r(r|p) \quad \text{if } p \equiv 3 \pmod{4}.$

(d) $\sum_{r=1}^{p-1} r^3(r|p) = \frac{3}{2}p\sum_{r=1}^{p-1} r^2(r|p) \quad \text{if } p \equiv 1 \pmod{4}.$

(e) $\sum_{r=1}^{p-1} r^4(r|p) = 2p\sum_{r=1}^{p-1} r^3(r|p) - p^2\sum_{r=1}^{p-1} r^2(r|p) \quad \text{if } p \equiv 3 \pmod{4}.$

[*Hint:* $p - r$ runs through the numbers $1, 2, \ldots, p - 1$ with r.]

8. Let p be an odd prime, $p \equiv 3 \pmod{4}$, and let $q = (p - 1)/2$.

(a) Prove that

$$\{1 - 2(2|p)\}\sum_{r=1}^{q} r(r|p) = p\frac{1 - (2|p)}{2}\sum_{r=1}^{q}(r|p).$$

[*Hint:* As r runs through the numbers $1, 2, \ldots, q$ then r and $p - r$ together run through the numbers $1, 2, \ldots, p - 1$, as do $2r$ and $p - 2r$.]

(b) Prove that

$$\{(2|p) - 2\}\sum_{r=1}^{p-1} r(r|p) = p\sum_{r=1}^{q}(r|p).$$

9. If p is an odd prime, let $\chi(n) = (n|p)$. Prove that the Gauss sum $G(n, \chi)$ associated with χ is the same as the quadratic Gauss sum $G(n; p)$ introduced in Exercise 8.16 if $(n, p) = 1$. In other words, if $p \nmid n$ we have

$$G(n, \chi) = \sum_{m \bmod p} \chi(m)e^{2\pi imn/p} = \sum_{r=1}^{p} e^{2\pi inr^2/p} = G(n; p).$$

It should be noted that $G(n, \chi) \neq G(n; p)$ if $p|n$ because $G(p, \chi) = 0$ but $G(p; p) = p$.

10. Evaluate the quadratic Gauss sum $G(2; p)$ using one of the reciprocity laws. Compare the result with the formula $G(2; p) = (2|p)G(1; p)$ and deduce that $(2|p) = (-1)^{(p^2-1)/8}$ if p is an odd prime.

10 Primitive Roots

10.1 The exponent of a number mod m. Primitive roots

Let a and m be relatively prime integers, with $m \geq 1$, and consider all the positive powers of a:

$$a, a^2, a^3, \ldots$$

We know, from the Euler–Fermat theorem, that $a^{\varphi(m)} \equiv 1 \pmod{m}$. However, there may be an earlier power a^f such that $a^f \equiv 1 \pmod{m}$. We are interested in the smallest positive f with this property.

Definition The smallest positive integer f such that

$$a^f \equiv 1 \pmod{m}$$

is called the exponent of a modulo m, and is denoted by writing

$$f = \exp_m(a).$$

If $\exp_m(a) = \varphi(m)$ then a is called a primitive root mod m.

The Euler–Fermat theorem tells us that $\exp_m(a) \leq \varphi(m)$. The next theorem shows that $\exp_m(a)$ divides $\varphi(m)$.

Theorem 10.1 *Given $m \geq 1$, $(a, m) = 1$, let $f = \exp_m(a)$. Then we have:*

(a) $a^k \equiv a^h \pmod{m}$ *if, and only if,* $k \equiv h \pmod{f}$.
(b) $a^k \equiv 1 \pmod{m}$ *if, and only if,* $k \equiv 0 \pmod{f}$. *In particular,* $f \mid \varphi(m)$.
(c) *The numbers* $1, a, a^2, \ldots, a^{f-1}$ *are incongruent* mod m.

PROOF. Parts (b) and (c) follow at once from (a), so we need only prove (a). If $a^k \equiv a^h \pmod{m}$ then $a^{k-h} \equiv 1 \pmod{m}$. Write

$$k - h = qf + r, \quad \text{where } 0 \le r < f.$$

Then $1 \equiv a^{k-h} = a^{qf+r} \equiv a^r \pmod{m}$, so $r = 0$ and $k \equiv h \pmod{f}$.

Conversely, if $k \equiv h \pmod{f}$ then $k - h = qf$ so $a^{k-h} \equiv 1 \pmod{m}$ and hence $a^k \equiv a^h \pmod{m}$. □

10.2 Primitive roots and reduced residue systems

Theorem 10.2 *Let* $(a, m) = 1$. *Then a is a primitive root* mod *m if, and only if, the numbers*

$$a, a^2, \ldots, a^{\varphi(m)} \tag{1}$$

form a reduced residue system mod *m*.

PROOF. If a is a primitive root the numbers in (1) are incongruent mod m, by Theorem 10.1(c). Since there are $\varphi(m)$ such numbers they form a reduced residue system mod m.

Conversely, if the numbers in (1) form a reduced residue system, then $a^{\varphi(m)} \equiv 1 \pmod{m}$ but no smaller power is congruent to 1, so a is a primitive root. □

Note. In Chapter 6 we found that the reduced residue classes mod m form a group. If m has a primitive root a, Theorem 10.2 shows that this group is the cyclic group generated by the residue class $\hat{a}$.

The importance of primitive roots is explained by Theorem 10.2. If m has a primitive root then each reduced residue system mod m can be expressed as a geometric progression. This gives a powerful tool that can be used in problems involving reduced residue systems. Unfortunately, not all moduli have primitive roots. In the next few sections we will prove that primitive roots exist only for the following moduli:

$$m = 1, 2, 4, p^\alpha, \text{ and } 2p^\alpha,$$

where p is an odd prime and $\alpha \ge 1$.

The first three cases are easily settled. The case $m = 1$ is trivial. For $m = 2$ the number 1 is a primitive root. For $m = 4$ we have $\varphi(4) = 2$ and $3^2 \equiv 1 \pmod{4}$, so 3 is a primitive root. Next we show that there are no primitive roots mod 2^α if $\alpha \ge 3$.

10.3 The nonexistence of primitive roots mod 2^α for $\alpha \geq 3$

Theorem 10.3 *Let x be an odd integer. If* $\alpha \geq 3$ *we have*

$$x^{\varphi(2^\alpha)/2} \equiv 1 \pmod{2^\alpha}, \tag{2}$$

so there are no primitive roots mod 2^α.

PROOF. If $\alpha = 3$ congruence (2) states that $x^2 \equiv 1 \pmod 8$ for x odd. This is easily verified by testing $x = 1, 3, 5, 7$ or by noting that

$$(2k + 1)^2 = 4k^2 + 4k + 1 = 4k(k + 1) + 1$$

and observing that $k(k + 1)$ is even.

Now we prove the theorem by induction on α. We assume (2) holds for α and prove that it also holds for $\alpha + 1$. The induction hypothesis is that

$$x^{\varphi(2^\alpha)/2} = 1 + 2^\alpha t,$$

where t is an integer. Squaring both sides we obtain

$$x^{\varphi(2^\alpha)} = 1 + 2^{\alpha+1}t + 2^{2\alpha}t^2 \equiv 1 \pmod{2^{\alpha+1}}$$

because $2\alpha \geq \alpha + 1$. This completes the proof since $\varphi(2^\alpha) = 2^{\alpha-1} = \varphi(2^{\alpha+1})/2$. □

10.4 The existence of primitive roots mod p for odd primes p

First we prove the following lemma.

Lemma 1 *Given* $(a, m) = 1$, *let* $f = \exp_m(a)$. *Then*

$$\exp_m(a^k) = \frac{\exp_m(a)}{(k, f)}.$$

In particular, $\exp_m(a^k) = \exp_m(a)$ *if, and only if,* $(k, f) = 1$.

PROOF. The exponent of a^k is the smallest positive x such that

$$a^{xk} \equiv 1 \pmod m.$$

This is also the smallest $x > 0$ such that $kx \equiv 0 \pmod f$. But this latter congruence is equivalent to the congruence

$$x \equiv 0 \left(\bmod \frac{f}{d}\right),$$

where $d = (k, f)$. The smallest positive solution of this congruence is f/d, so $\exp_m(a^k) = f/d$, as asserted. □

Lemma 1 will be used to prove the existence of primitive roots for prime moduli. In fact, we shall determine the exact number of primitive roots mod p.

Theorem 10.4 *Let p be an odd prime and let d be any positive divisor of $p - 1$. Then in every reduced residue system* mod p *there are exactly $\varphi(d)$ numbers a such that*

$$\exp_p(a) = d.$$

In particular, when $d = \varphi(p) = p - 1$ there are exactly $\varphi(p - 1)$ primitive roots mod p.

PROOF. We use the method employed in Chapter 2 to prove the relation

$$\sum_{d|n} \varphi(d) = n.$$

The numbers $1, 2, \ldots, p - 1$ are distributed into disjoint sets $A(d)$, each set corresponding to a divisor d of $p - 1$. Here we define

$$A(d) = \{x : 1 \leq x \leq p - 1 \text{ and } \exp_p(x) = d\}.$$

Let $f(d)$ be the number of elements in $A(d)$. Then $f(d) \geq 0$ for each d. Our goal is to prove that $f(d) = \varphi(d)$.

Since the sets $A(d)$ are disjoint and since each $x = 1, 2, \ldots, p - 1$ falls into some $A(d)$, we have

$$\sum_{d|p-1} f(d) = p - 1.$$

But we also have

$$\sum_{d|p-1} \varphi(d) = p - 1$$

so

$$\sum_{d|p-1} \{\varphi(d) - f(d)\} = 0.$$

To show each term in this sum is zero it suffices to prove that $f(d) \leq \varphi(d)$. We do this by showing that either $f(d) = 0$ or $f(d) = \varphi(d)$; or, in other words, that $f(d) \neq 0$ implies $f(d) = \varphi(d)$.

Suppose that $f(d) \neq 0$. Then $A(d)$ is nonempty so $a \in A(d)$ for some a. Therefore

$$\exp_p(a) = d, \quad \text{hence } a^d \equiv 1 \pmod{p}.$$

But every power of a satisfies the same congruence, so the d numbers

$$a, a^2, \ldots, a^d \tag{3}$$

are solutions of the polynomial congruence

$$x^d - 1 \equiv 0 \pmod{p}, \tag{4}$$

these solutions being incongruent mod p since $d = \exp_p(a)$. But (4) has at most d solutions since the modulus is prime, so the d numbers in (3) must be *all* the solutions of (4). Hence each number in $A(d)$ must be of the form a^k for some $k = 1, 2, \ldots, d$. When is $\exp_p(a^k) = d$? According to Lemma 1 this occurs if, and only if, $(k, d) = 1$. In other words, among the d numbers in (3) there are $\varphi(d)$ which have exponent d modulo p. Thus we have shown that $f(d) = \varphi(d)$ if $f(d) \neq 0$. As noted earlier, this completes the proof. □

10.5 Primitive roots and quadratic residues

Theorem 10.5 *Let g be a primitive root* mod p, *where p is an odd prime. Then the even powers*

$$g^2, g^4, \ldots, g^{p-1}$$

are the quadratic residues mod p, *and the odd powers*

$$g, g^3, \ldots, g^{p-2}$$

are the quadratic nonresidues mod p.

PROOF. If n is even, say $n = 2m$, then $g^n = (g^m)^2$ so

$$g^n \equiv x^2 \pmod{p}, \quad \text{where } x = g^m.$$

Hence $g^n R p$. But there are $(p - 1)/2$ distinct even powers $g^2, \ldots, g^{p-1}$ modulo p and the same number of quadratic residues mod p. Therefore the even powers are the quadratic residues and the odd powers are the nonresidues. □

10.6 The existence of primitive roots mod p^α

We turn next to the case $m = p^\alpha$, where p is an odd prime and $\alpha \geq 2$. In seeking primitive roots mod p^α it is natural to consider as candidates the primitive roots mod p. Let g be such a primitive root and let us ask whether g might also be a primitive root mod p^2. Now $g^{p-1} \equiv 1 \pmod{p}$ and, since $\varphi(p^2) = p(p - 1) > p - 1$, this g will certainly not be a primitive root mod p^2 if $g^{p-1} \equiv 1 \pmod{p^2}$. Therefore the relation

$$g^{p-1} \not\equiv 1 \pmod{p^2}$$

is a *necessary* condition for a primitive root g mod p to also be a primitive root mod p^2. Remarkably enough, this condition is also *sufficient* for g to be a primitive root mod p^2 and, more generally, mod p^α for all powers $\alpha \geq 2$. In fact, we have the following theorem.

Theorem 10.6 *Let p be an odd prime. Then we have:*

(a) *If g is a primitive root* mod p *then g is also a primitive root* mod p^α *for all $\alpha \geq 1$ if, and only if,*

(5) $$g^{p-1} \not\equiv 1 \pmod{p^2}.$$

(b) *There is at least one primitive root g* mod p *which satisfies* (5), *hence there exists at least one primitive root* mod p^α *if $\alpha \geq 2$.*

PROOF. We prove (b) first. Let g be a primitive root mod p. If $g^{p-1} \not\equiv 1 \pmod{p^2}$ there is nothing to prove. However, if $g^{p-1} \equiv 1 \pmod{p^2}$ we can show that $g_1 = g + p$, which is another primitive root modulo p, satisfies the condition

$$g_1{}^{p-1} \not\equiv 1 \pmod{p^2}.$$

In fact, we have

$$\begin{aligned} g_1{}^{p-1} &= (g + p)^{p-1} = g^{p-1} + (p-1)g^{p-2}p + tp^2 \\ &\equiv g^{p-1} + (p^2 - p)g^{p-2} \pmod{p^2} \\ &\equiv 1 - pg^{p-2} \pmod{p^2}. \end{aligned}$$

But we cannot have $pg^{p-2} \equiv 0 \pmod{p^2}$ for this would imply $g^{p-2} \equiv 0 \pmod{p}$, contradicting the fact that g is a primitive root mod p. Hence $g_1{}^{p-1} \not\equiv 1 \pmod{p^2}$, so (b) is proved.

Now we prove (a). Let g be a primitive root modulo p. If this g is a primitive root mod p^α for all $\alpha \geq 1$ then, in particular, it is a primitive root mod p^2 and, as we have already noted, this implies (5).

Now we prove the converse statement. Suppose that g is a primitive root mod p which satisfies (5). We must show that g is also a primitive root mod p^α for all $\alpha \geq 2$. Let t be the exponent of g modulo p^α. We wish to show that $t = \varphi(p^\alpha)$. Since $g^t \equiv 1 \pmod{p^\alpha}$ we also have $g^t \equiv 1 \pmod{p}$ so $\varphi(p) \mid t$ and we can write

(6) $$t = q\varphi(p).$$

Now $t \mid \varphi(p^\alpha)$ so $q\varphi(p) \mid \varphi(p^\alpha)$. But $\varphi(p^\alpha) = p^{\alpha-1}(p-1)$ hence

$$q(p-1) \mid p^{\alpha-1}(p-1)$$

which means $q \mid p^{\alpha-1}$. Therefore $q = p^\beta$ where $\beta \leq \alpha - 1$, and (6) becomes

$$t = p^\beta(p-1).$$

If we prove that $\beta = \alpha - 1$ then $t = \varphi(p^\alpha)$ and the proof will be complete.

Suppose, on the contrary, that $\beta < \alpha - 1$. Then $\beta \leq \alpha - 2$ and we have

$$t = p^\beta(p-1) \mid p^{\alpha-2}(p-1) = \varphi(p^{\alpha-1}).$$

Thus, since $\varphi(p^{\alpha-1})$ is a multiple of t, this implies,

(7) $$g^{\varphi(p^{\alpha-1})} \equiv 1 \pmod{p^\alpha}.$$

But now we make use of the following Lemma which shows that (7) is a contradiction. This contradiction will complete the proof of Theorem 10.6. □

Lemma 2 *Let g be a primitive root modulo p such that*

$$g^{p-1} \not\equiv 1 \pmod{p^2}. \tag{8}$$

Then for every $\alpha \geq 2$ we have

$$g^{\varphi(p^{\alpha-1})} \not\equiv 1 \pmod{p^\alpha}. \tag{9}$$

Proof of Lemma 2. We use induction on α. For $\alpha = 2$, relation (9) reduces to (8). Suppose then, that (9) holds for α. By the Euler–Fermat theorem we have

$$g^{\varphi(p^{\alpha-1})} \equiv 1 \pmod{p^{\alpha-1}}$$

so

$$g^{\varphi(p^{\alpha-1})} = 1 + kp^{\alpha-1}$$

where $p \nmid k$ because of (9). Raising both sides of this last relation to the pth power we find

$$g^{\varphi(p^\alpha)} = (1 + kp^{\alpha-1})^p = 1 + kp^\alpha + k^2 \frac{p(p-1)}{2} p^{2(\alpha-1)} + rp^{3(\alpha-1)}.$$

Now $2\alpha - 1 \geq \alpha + 1$ and $3\alpha - 3 \geq \alpha + 1$ since $\alpha \geq 2$. Hence, the last equation gives us the congruence

$$g^{\varphi(p^\alpha)} \equiv 1 + kp^\alpha \pmod{p^{\alpha+1}}$$

where $p \nmid k$. In other words, $g^{\varphi(p^\alpha)} \not\equiv 1 \pmod{p^{\alpha+1}}$ so (9) holds for $\alpha + 1$ if it holds for α. This completes the proof of Lemma 2 and also of Theorem 10.6. □

10.7 The existence of primitive roots mod $2p^\alpha$

Theorem 10.7 *If p is an odd prime and $\alpha \geq 1$ there exist odd primitive roots g modulo p^α. Each such g is also a primitive root modulo $2p^\alpha$.*

Proof. If g is a primitive root modulo p^α so is $g + p^\alpha$. But one of g or $g + p^\alpha$ is odd so *odd* primitive roots mod p^α always exist. Let g be an odd primitive root mod p^α and let f be the exponent of g mod $2p^\alpha$. We wish to show that $f = \varphi(2p^\alpha)$. Now $f \mid \varphi(2p^\alpha)$, and $\varphi(2p^\alpha) = \varphi(2)\varphi(p^\alpha) = \varphi(p^\alpha)$ so $f \mid \varphi(p^\alpha)$. On the other hand, $g^f \equiv 1 \pmod{2p^\alpha}$ so $g^f \equiv 1 \pmod{p^\alpha}$, hence $\varphi(p^\alpha) \mid f$ since g is a primitive root mod p^α. Therefore $f = \varphi(p^\alpha) = \varphi(2p^\alpha)$, so g is a primitive root mod $2p^\alpha$. □

10.8 The nonexistence of primitive roots in the remaining cases

Theorem 10.8 *Given $m \geq 1$ where m is not of the form $m = 1, 2, 4, p^\alpha$, or $2p^\alpha$, where p is an odd prime. Then for any a with $(a, m) = 1$ we have*

$$a^{\varphi(m)/2} \equiv 1 \pmod{m},$$

so there are no primitive roots mod m.

PROOF. We have already shown that there are no primitive roots mod 2^α if $\alpha \geq 3$. Therefore we can suppose that m has the factorization

$$m = 2^\alpha p_1^{\alpha_1} \cdots p_s^{\alpha_s}$$

where the p_i are odd primes, $s \geq 1$, and $\alpha \geq 0$. Since m is not of the form $1, 2, 4, p^\alpha$ or $2p^\alpha$ we have $\alpha \geq 2$ if $s = 1$ and $s \geq 2$ if $\alpha = 0$ or 1. Note that

$$\varphi(m) = \varphi(2^\alpha)\varphi(p_1^{\alpha_1}) \cdots \varphi(p_s^{\alpha_s}).$$

Now let a be any integer relatively prime to m. We wish to prove that

$$a^{\varphi(m)/2} \equiv 1 \pmod{m}.$$

Let g be a primitive root mod $p_1^{\alpha_1}$ and choose k so that

$$a \equiv g^k \pmod{p_1^{\alpha_1}}.$$

Then we have

$$a^{\varphi(m)/2} \equiv g^{k\varphi(m)/2} \equiv g^{t\varphi(p_1^{\alpha_1})} \pmod{p_1^{\alpha_1}} \tag{10}$$

where

$$t = k\varphi(2^\alpha)\varphi(p_2^{\alpha_2}) \cdots \varphi(p_s^{\alpha_s})/2.$$

We will show that t is an integer. If $\alpha \geq 2$ the factor $\varphi(2^\alpha)$ is even and hence t is an integer. If $\alpha = 0$ or 1 then $s \geq 2$ and the factor $\varphi(p_2^{\alpha_2})$ is even, so t is an integer in this case as well. Hence congruence (10) gives us

$$a^{\varphi(m)/2} \equiv 1 \pmod{p_1^{\alpha_1}}.$$

In the same way we find

$$a^{\varphi(m)/2} \equiv 1 \pmod{p_i^{\alpha_i}} \tag{11}$$

for each $i = 1, 2, \ldots, s$. Now we show that this congruence also holds mod 2^α. If $\alpha \geq 3$ the condition $(a, m) = 1$ requires a to be odd and we may apply Theorem 10.3 to write

$$a^{\varphi(2^\alpha)/2} \equiv 1 \pmod{2^\alpha}.$$

Since $\varphi(2^\alpha) \mid \varphi(m)$ this gives us

$$a^{\varphi(m)/2} \equiv 1 \pmod{2^\alpha} \tag{12}$$

for $\alpha \geq 3$.

If $\alpha \leq 2$ we have

$$a^{\varphi(2^\alpha)} \equiv 1 \pmod{2^\alpha}. \tag{13}$$

But $s \geq 1$ so $\varphi(m) = \varphi(2^\alpha)\varphi(p_1^{\alpha_1}) \cdots \varphi(p_s^{\alpha_s}) = 2r\varphi(2^\alpha)$ where r is an integer. Hence $\varphi(2^\alpha) \mid \varphi(m)/2$ and (13) implies (12) for $\alpha \leq 2$. Hence (12) holds for all α. Multiplying together the congruences (11) and (12) we obtain

$$a^{\varphi(m)/2} \equiv 1 \pmod{m},$$

and this shows that a cannot be a primitive root mod m. □

10.9 The number of primitive roots mod m

We have shown that an integer $m \geq 1$ has a primitive root if and only if

$$m = 1, 2, 4, p^\alpha \text{ or } 2p^\alpha,$$

where p is an odd prime and $\alpha \geq 1$. The next theorem tells us how many primitive roots exist for each such m.

Theorem 10.9 *If m has a primitive root g then m has exactly $\varphi(\varphi(m))$ incongruent primitive roots and they are given by the numbers in the set*

$$S = \{g^n : 1 \leq n \leq \varphi(m), \text{ and } (n, \varphi(m)) = 1\}.$$

Proof. We have $\exp_m(g) = \varphi(m)$, and Lemma 1 shows that $\exp_m(g^n) = \exp_m(g)$ if and only if $(n, \varphi(m)) = 1$. Therefore each element of S is a primitive root mod m.

Conversely, if a is a primitive root mod m, then $a \equiv g^k \pmod{m}$ for some $k = 1, 2, \ldots, \varphi(m)$. Hence $\exp_m(g^k) = \exp_m(a) = \varphi(m)$, and Lemma 1 implies $(k, \varphi(m)) = 1$. Therefore every primitive root is a member of S. Since S contains $\varphi(\varphi(m))$ incongruent members mod m the proof is complete. □

Although we have shown the existence of primitive roots for certain moduli, no direct method is known for calculating these roots in general without a great deal of computation, especially for large moduli. Let $g(p)$ denote the smallest primitive root mod p. Table 10.1 lists $g(p)$ for all odd primes $p < 1000$.

Table 10.1 $g(p)$ is the smallest primitive root of the prime p

p	$g(p)$	p	$g(p)$	p	$g(p)$	p	$g(p)$	p	$g(p)$	p	$g(p)$
2	1	109	6	269	2	439	15	617	3	811	3
3	2	113	3	271	6	443	2	619	2	821	2
5	2	127	3	277	5	449	3	631	3	823	3
7	3	131	2	281	3	457	13	641	3	827	2
11	2	137	3	283	3	461	2	643	11	829	2
13	2	139	2	293	2	463	3	647	5	839	11
17	3	149	2	307	5	467	2	653	2	853	2
19	2	151	6	311	17	479	13	659	2	857	3
23	5	157	5	313	10	487	3	661	2	859	2
29	2	163	2	317	2	491	2	673	5	863	5
31	3	167	5	331	3	499	7	677	2	877	2
37	2	173	2	337	10	503	5	683	5	881	3
41	6	179	2	347	2	509	2	691	3	883	2
43	3	181	2	349	2	521	3	701	2	887	5
47	5	191	19	353	3	523	2	709	2	907	2
53	2	193	5	359	7	541	2	719	11	911	17
59	2	197	2	367	6	547	2	727	5	919	7
61	2	199	3	373	2	557	2	733	6	929	3
67	2	211	2	379	2	563	2	739	3	937	5
71	7	223	3	383	5	569	3	743	5	941	2
73	5	227	2	389	2	571	3	751	3	947	2
79	3	229	6	397	5	577	5	757	2	953	3
83	2	233	3	401	3	587	2	761	6	967	5
89	3	239	7	409	21	593	3	769	11	971	6
97	5	241	7	419	2	599	7	773	2	977	3
101	2	251	6	421	2	601	7	787	2	983	5
103	5	257	3	431	7	607	3	797	2	997	7
107	2	263	5	433	5	613	2	809	3		

10.10 The index calculus

If m has a primitive root g the numbers $1, g, g^2, \ldots, g^{\varphi(m)-1}$ form a reduced residue system mod m. If $(a, m) = 1$ there is a unique integer k in the interval $0 \le k \le \varphi(m) - 1$ such that

$$a \equiv g^k \pmod{m}.$$

This integer is called the *index* of a to the base g (mod m), and we write

$$k = \operatorname{ind}_g a$$

or simply $k = \operatorname{ind} a$ if the base g is understood.

The following theorem shows that indices have properties analogous to those of logarithms. The proof is left as an exercise for the reader.

Theorem 10.10 *Let g be a primitive root* mod m. *If* $(a, m) = (b, m) = 1$ *we have:*

(a) $\operatorname{ind}(ab) \equiv \operatorname{ind} a + \operatorname{ind} b \pmod{\varphi(m)}$.
(b) $\operatorname{ind} a^n \equiv n \operatorname{ind} a \pmod{\varphi(m)}$ *if* $n \geq 1$.
(c) $\operatorname{ind} 1 = 0$ *and* $\operatorname{ind} g = 1$.
(d) $\operatorname{ind}(-1) = \varphi(m)/2$ *if* $m > 2$.
(e) *If g' is also a primitive root* mod m *then*

$$\operatorname{ind}_g a \equiv \operatorname{ind}_{g'} a \cdot \operatorname{ind}_g g' \pmod{\varphi(m)}.$$

Table 10.2 on pp. 216–217 lists indices for all numbers $a \not\equiv 0 \pmod{p}$ and all odd primes $p < 50$. The base g is the smallest primitive root of p.

The following examples illustrate the use of indices in solving congruences.

Example 1 Linear congruences. Assume m has a primitive root and let $(a, m) = (b, m) = 1$. Then the linear congruence

$$ax \equiv b \pmod{m} \tag{14}$$

is equivalent to the congruence

$$\operatorname{ind} a + \operatorname{ind} x \equiv \operatorname{ind} b \pmod{\varphi(m)},$$

so the unique solution of (14) satisfies the congruence

$$\operatorname{ind} x \equiv \operatorname{ind} b - \operatorname{ind} a \pmod{\varphi(m)}.$$

To treat a numerical example, consider the linear congruence

$$9x \equiv 13 \pmod{47}.$$

The corresponding index relation is

$$\operatorname{ind} x \equiv \operatorname{ind} 13 - \operatorname{ind} 9 \pmod{46}.$$

From Table 10.2 we find $\operatorname{ind} 13 = 11$ and $\operatorname{ind} 9 = 40$ (for $p = 47$), so

$$\operatorname{ind} x \equiv 11 - 40 \equiv -29 \equiv 17 \pmod{46}.$$

Again from Table 10.2 we find $x \equiv 38 \pmod{47}$.

Example 2 Binomial congruences. A congruence of the form

$$x^n \equiv a \pmod{m}$$

is called a *binomial congruence*. If m has a primitive root and if $(a, m) = 1$ this is equivalent to the congruence

$$n \operatorname{ind} x \equiv \operatorname{ind} a \pmod{\varphi(m)},$$

which is linear in the unknown ind x. As such, it has a solution if, and only if, ind a is divisible by $d = (n, \varphi(m))$, in which case it has exactly d solutions.

To illustrate with a numerical example, consider the binomial congruence

$$x^8 \equiv a \pmod{17}. \tag{15}$$

The corresponding index relation is

$$8 \text{ ind } x \equiv \text{ind } a \pmod{16}. \tag{16}$$

In this example $d = (8, 16) = 8$. Table 10.2 shows that 1 and 16 are the only numbers mod 17 whose index is divisible by 8. In fact, ind $1 = 0$ and ind $16 = 8$. Hence (15) has no solutions if $a \not\equiv 1$ or $a \not\equiv 16 \pmod{17}$.

For $a = 1$ congruence (16) becomes

$$8 \text{ ind } x \equiv 0 \pmod{16}, \tag{17}$$

and for $a = 16$ it becomes

$$8 \text{ ind } x \equiv 8 \pmod{16}. \tag{18}$$

Each of these has exactly eight solutions mod 16. The solutions of (17) are those x whose index is even,

$$x \equiv 1, 2, 4, 8, 9, 13, 15, 16 \pmod{17}.$$

These, of course, are the quadratic residues of 17. The solutions of (18) are those x whose index is odd, the quadratic nonresidues of 17,

$$x \equiv 3, 5, 6, 7, 10, 11, 12, 14 \pmod{17}.$$

Example 3 Exponential congruences. An exponential congruence is one of the form

$$a^x \equiv b \pmod{m}.$$

If m has a primitive root and if $(a, m) = (b, m) = 1$ this is equivalent to the linear congruence

$$x \text{ ind } a \equiv \text{ind } b \pmod{\varphi(m)}. \tag{19}$$

Let $d = (\text{ind } a, \varphi(m))$. Then (19) has a solution if, and only if, $d \mid \text{ind } b$, in which case there are exactly d solutions. In the numerical example

$$25^x \equiv 17 \pmod{47} \tag{20}$$

we have ind $25 = 2$, ind $17 = 16$, and $d = (2, 46) = 2$. Therefore (19) becomes

$$2x \equiv 16 \pmod{46},$$

with two solutions, $x \equiv 8$ and $31 \pmod{46}$. These are also the solutions of (20) mod 47.

Table 10.2 Indices of all numbers $a \not\equiv 0 \pmod{p}$ for odd primes $p < 50$. The base g is the smallest primitive root of p.

a	Primes 3	5	7	11	13	17	19	23	29	31	37	41	43	47
1	0	0	0	0	0	0	0	0	0	0	0	0	0	0
2	1	1	2	1	1	14	1	2	1	24	1	26	27	18
3		3	1	8	4	1	13	16	5	1	26	15	1	20
4		2	4	2	2	12	2	4	2	18	2	12	12	36
5			5	4	9	5	16	1	22	20	23	22	25	1
6			3	9	5	15	14	18	6	25	27	1	28	38
7				7	11	11	6	19	12	28	32	39	35	32
8				3	3	10	3	6	3	12	3	38	39	8
9				6	8	2	8	10	10	2	16	30	2	40
10				5	10	3	17	3	23	14	24	8	10	19
11					7	7	12	9	25	23	30	3	30	7
12					6	13	15	20	7	19	28	27	13	10
13						4	5	14	18	11	11	31	32	11
14						9	7	21	13	22	33	25	20	4
15						6	11	17	27	21	13	37	26	21
16						8	4	8	4	6	4	24	24	26
17							10	7	21	7	7	33	38	16
18							9	12	11	26	17	16	29	12
19								15	9	4	35	9	19	45
20								5	24	8	25	34	37	37

21	13	17	29	22	14	36	6
22	11	26	17	31	29	15	25
23		20	27	15	36	16	5
24		8	13	29	13	40	28
25		16	10	10	4	8	2
26		19	5	12	17	17	29
27		15	3	6	5	3	14
28		14	16	34	11	5	22
29			9	21	7	41	35
30			15	14	23	11	39
31				9	28	34	3
32				5	10	9	44
33				20	18	31	27
34				8	19	23	34
35				19	21	18	33
36				18	2	14	30
37					32	7	42
38					35	4	17
39					6	33	31
40					20	22	9
41						6	15
42						21	24
43							13
44							43
45							41
46							23

10.11 Primitive roots and Dirichlet characters

Primitive roots and indices can be used to construct explicitly all the Dirichlet characters mod m. First we consider a prime power modulus p^α, where p is an odd prime and $\alpha \geq 1$.

Let g be a primitive root mod p which is also a primitive root mod p^β for all $\beta \geq 1$. Such a g exists by Theorem 10.6. If $(n, p) = 1$ let $b(n) = \operatorname{ind}_g n$ (mod p^α), so that $b(n)$ is the unique integer satisfying the conditions

$$n \equiv g^{b(n)} \pmod{p^\alpha}, \quad 0 \leq b(n) < \varphi(p^\alpha).$$

For $h = 0, 1, 2, \ldots, \varphi(p^\alpha) - 1$, define χ_h by the relations

$$\chi_h(n) = \begin{cases} e^{2\pi i h b(n)/\varphi(p^\alpha)} & \text{if } p \nmid n, \\ 0 & \text{if } p \mid n. \end{cases} \tag{21}$$

Using the properties of indices it is easy to verify that χ_h is completely multiplicative and periodic with period p^α, so χ_h is a Dirichlet character mod p^α, with χ_0 being the principal character. This verification is left as an exercise for the reader.

Since

$$\chi_h(g) = e^{2\pi i h/\varphi(p^\alpha)}$$

the characters $\chi_0, \chi_1, \ldots, \chi_{\varphi(p^\alpha)-1}$ are distinct because they take distinct values at g. Therefore, since there are $\varphi(p^\alpha)$ such functions they represent all the Dirichlet characters mod p^α. The same construction works for the modulus 2^α if $\alpha = 1$ or $\alpha = 2$, using $g = 3$ as the primitive root.

Now if $m = p_1^{\alpha_1} \cdots p_r^{\alpha_r}$, where the p_i are distinct odd primes, and if χ_i is a Dirichlet character mod $p_i^{\alpha_i}$, then the product $\chi = \chi_1 \cdots \chi_r$ is a Dirichlet character mod m. Since $\varphi(m) = \varphi(p_1^{\alpha_1}) \cdots \varphi(p_r^{\alpha_r})$ we get $\varphi(m)$ such characters as each χ_i runs through the $\varphi(p_i^{\alpha_i})$ characters mod $p_i^{\alpha_i}$. Thus we have explicitly constructed all characters mod m for every odd modulus m.

If $\alpha \geq 3$ the modulus 2^α has no primitive root and a slightly different construction is needed to obtain the characters mod 2^α. The following theorem shows that 5 is a good substitute for a primitive root mod 2^α.

Theorem 10.11 *Assume $\alpha \geq 3$. Then for every odd integer n there is a uniquely determined integer $b(n)$ such that*

$$n \equiv (-1)^{(n-1)/2} 5^{b(n)} \pmod{2^\alpha}, \quad \textit{with } 1 \leq b(n) \leq \varphi(2^\alpha)/2.$$

PROOF. Let $f = \exp_{2^\alpha}(5)$ so that $5^f \equiv 1 \pmod{2^\alpha}$. We will show that $f = \varphi(2^\alpha)/2$. Now $f \mid \varphi(2^\alpha) = 2^{\alpha-1}$, so $f = 2^\beta$ for some $\beta \leq \alpha - 1$. From Theorem 10.8 we know that

$$5^{\varphi(2^\alpha)/2} \equiv 1 \pmod{2^\alpha},$$

hence $f \le \varphi(2^\alpha)/2 = 2^{\alpha-2}$. Therefore $\beta \le \alpha - 2$. We will show that $\beta = \alpha - 2$.

Raise both members of the equation $5 = 1 + 2^2$ to the $f = 2^\beta$ power to obtain

$$5^f = (1 + 2^2)^{2^\beta} = 1 + 2^{\beta+2} + r2^{\beta+3} = 1 + 2^{\beta+2}(1 + 2r)$$

where r is an integer. Hence $5^f - 1 = 2^{\beta+2}t$ where t is odd. But $2^\alpha \mid (5^f - 1)$ so $\alpha \le \beta + 2$, or $\beta \ge \alpha - 2$. Hence $\beta = \alpha - 2$ and $f = 2^{\alpha-2} = \varphi(2^\alpha)/2$. Therefore the numbers

$$5, 5^2, \ldots, 5^f \tag{22}$$

are incongruent mod 2^α. Also each is $\equiv 1 \pmod 4$ since $5 \equiv 1 \pmod 4$. Similarly, the numbers

$$-5, -5^2, \ldots, -5^f \tag{23}$$

are incongruent mod 2^α and each is $\equiv 3 \pmod 4$ since $-5 \equiv 3 \pmod 4$. There are $2f = \varphi(2^\alpha)$ numbers in (22) and (23) together. Moreover, we cannot have $5^a \equiv -5^b \pmod{2^\alpha}$ because this would imply $1 \equiv -1 \pmod 4$. Hence the numbers in (22) together with those in (23) represent $\varphi(2^\alpha)$ incongruent odd numbers mod 2^α. Each odd $n \equiv 1 \pmod 4$ is congruent mod 2^α to one of the numbers in (22), and each odd $n \equiv 3 \pmod 4$ is congruent to one in (23). This proves the theorem. □

With the help of Theorem 10.11 we can construct all the characters mod 2^α if $\alpha \ge 3$. Let

$$f(n) = \begin{cases} (-1)^{(n-1)/2} & \text{if } n \text{ is odd,} \\ 0 & \text{if } n \text{ is even,} \end{cases} \tag{24}$$

and let

$$g(n) = \begin{cases} e^{2\pi i b(n)/2^{\alpha-2}} & \text{if } n \text{ is odd,} \\ 0 & \text{if } n \text{ is even,} \end{cases}$$

where $b(n)$ is the integer given by Theorem 10.11. Then it is easy to verify that each of f and g is a character mod 2^α. So is each product

$$\chi_{a,c}(n) = f(n)^a g(n)^c \tag{25}$$

where $a = 1, 2$ and $c = 1, 2, \ldots, \varphi(2^\alpha)/2$. Moreover these $\varphi(2^\alpha)$ characters are distinct so they represent all the characters mod 2^α.

Now if $m = 2^\alpha Q$ where Q is odd, we form the products $\chi = \chi_1\chi_2$ where χ_1 runs through the $\varphi(2^\alpha)$ characters mod 2^α and χ_2 runs through the $\varphi(Q)$ characters mod Q to obtain all the characters mod m.

10.12 Real-valued Dirichlet characters mod p^α

If χ is a real-valued Dirichlet character mod m and $(n, m) = 1$, the number $\chi(n)$ is both a root of unity and real, so $\chi(n) = \pm 1$. From the construction in the foregoing section we can determine all real Dirichlet characters mod p^α.

Theorem 10.12 *For an odd prime p and $\alpha \geq 1$, consider the $\varphi(p^\alpha)$ Dirichlet characters χ_h* mod p^α *given by* (21). *Then χ_h is real if, and only if, $h = 0$ or $h = \varphi(p^\alpha)/2$. Hence there are exactly two real characters* mod p^α.

PROOF. We have $e^{\pi i z} = \pm 1$ if, and only if, z is an integer. If $p \nmid n$ we have

$$\chi_h(n) = e^{2\pi i h b(n)/\varphi(p^\alpha)}$$

so $\chi_h(n) = \pm 1$ if, and only if, $\varphi(p^\alpha) \mid 2hb(n)$. This condition is satisfied for all n if $h = 0$ or if $h = \varphi(p^\alpha)/2$. Conversely, if $\varphi(p^\alpha) \mid 2hb(n)$ for all n then when $b(n) = 1$ we have $\varphi(p^\alpha) \mid 2h$ or $\varphi(p^\alpha)/2 \mid h$. Hence $h = 0$ or $h = \varphi(p^\alpha)/2$ since these are the only multiples of $\varphi(p^\alpha)/2$ less than $\varphi(p^\alpha)$. □

Note. The character corresponding to $h = 0$ is the principal character. When $\alpha = 1$ the quadratic character $\chi(n) = (n \mid p)$ is the only other real character mod p.

For the moduli $m = 1$, 2 and 4, all the Dirichlet characters are real. The next theorem describes the real characters mod 2^α when $\alpha \geq 3$.

Theorem 10.13 *If $\alpha \geq 3$, consider the $\varphi(2^\alpha)$ Dirichlet characters $\chi_{a,c}$* mod 2^α *given by* (25). *Then $\chi_{a,c}$ is real if, and only if, $c = \varphi(2^\alpha)/2$ or $c = \varphi(2^\alpha)/4$. Hence there are exactly four real characters* mod 2^α *if $\alpha \geq 3$.*

PROOF. If $\alpha \geq 3$ and n is odd we have, by (25),

$$\chi_{a,c}(n) = f(n)^a g(n)^c$$

where $f(n) = \pm 1$ and

$$g(n)^c = e^{2\pi i c b(n)/2^{\alpha-2}},$$

with $1 \leq c \leq 2^{\alpha-2}$. This is ± 1 if, and only if, $2^{\alpha-2} \mid 2cb(n)$, or $2^{\alpha-3} \mid cb(n)$. Since $\varphi(2^\alpha) = 2^{\alpha-1}$ this condition is satisfied if $c = \varphi(2^\alpha)/2 = 2^{\alpha-2}$ or if $c = \varphi(2^\alpha)/4 = 2^{\alpha-3}$. Conversely, if $2^{\alpha-3} \mid cb(n)$ for all n then $b(n) = 1$ requires $2^{\alpha-3} \mid c$ so $c = 2^{\alpha-3}$ or $2^{\alpha-2}$ since $1 \leq c \leq 2^{\alpha-2}$. □

10.13 Primitive Dirichlet characters mod p^α

In Theorem 8.14 we proved that every nonprincipal character χ mod p is primitive if p is prime. Now we determine all the primitive Dirichlet characters mod p^α.

We recall (Section 8.7) that χ is primitive mod k if, and only if, χ has no induced modulus $d < k$. An induced modulus is a divisor d of k such that

$$\chi(n) = 1 \quad \text{whenever } (n, k) = 1 \text{ and } n \equiv 1 \pmod{d}.$$

If $k = p^\alpha$ and χ is imprimitive mod p^α then one of the divisors $1, p, \ldots, p^{\alpha-1}$ is an induced modulus, and hence $p^{\alpha-1}$ is an induced modulus. Therefore, χ is primitive mod p^α if, and only if, $p^{\alpha-1}$ is not an induced modulus for χ.

Theorem 10.14 *For an odd prime p and $\alpha \geq 2$, consider the $\varphi(p^\alpha)$ Dirichlet characters χ_h* mod *p^α given by* (21). *Then χ_h is primitive* mod *p^α if, and only if, $p \nmid h$.*

PROOF. We will show that $p^{\alpha-1}$ is an induced modulus if, and only if, $p \mid h$. If $p \nmid n$ we have, by (21),

$$\chi_h(n) = e^{2\pi i h b(n)/\varphi(p^\alpha)},$$

where $n \equiv g^{b(n)} \pmod{p^\alpha}$ and g is a primitive root mod p^β for all $\beta \geq 1$. Therefore

$$g^{b(n)} \equiv n \pmod{p^{\alpha-1}}.$$

Now if $n \equiv 1 \pmod{p^{\alpha-1}}$ then $g^{b(n)} \equiv 1 \pmod{p^{\alpha-1}}$ and, since g is a primitive root of $p^{\alpha-1}$, we have $\varphi(p^{\alpha-1}) \mid b(n)$, or

$$b(n) = t\varphi(p^{\alpha-1}) = t\varphi(p^\alpha)/p$$

for some integer t. Therefore

$$\chi_h(n) = e^{2\pi i h t/p}.$$

If $p \mid h$ this equals 1 and hence χ_h is imprimitive mod p^α. If $p \nmid h$ take $n = 1 + p^{\alpha-1}$. Then $n \equiv 1 \pmod{p^{\alpha-1}}$ but $n \not\equiv 1 \pmod{p^\alpha}$ so $0 < b(n) < \varphi(p^\alpha)$. Therefore $p \nmid t$, $p \nmid ht$ and $\chi_h(n) \neq 1$. This shows that χ_h is primitive if $p \nmid h$. □

When $m = 1$ or 2, there is only one character χ mod m, the principal character. If $m = 4$ there are two characters mod 4, the principal character and the primitive character f given by (24). The next theorem describes all the primitive characters mod 2^α for $\alpha \geq 3$. The proof is similar to that of Theorem 10.14 and is left to the reader.

Theorem 10.15 *If $\alpha \geq 3$, consider the $\varphi(2^\alpha)$ Dirichlet characters $\chi_{a,c}$* mod 2^α *given by* (25). *Then $\chi_{a,c}$ is primitive* mod 2^α *if, and only if, c is odd.*

The foregoing results describe all primitive characters mod p^α for all prime powers. To determine the primitive characters for a composite modulus k we write

$$k = p_1^{\alpha_1} \cdots p_r^{\alpha_r}.$$

Then every character χ mod k can be factored in the form

$$\chi = \chi_1 \cdots \chi_r$$

where each χ_i is a character mod $p_i^{\alpha_i}$. Moreover, by Exercise 8.12, χ is primitive mod k if, and only if, each χ_i is primitive mod $p_i^{\alpha_i}$. Therefore we have a complete description of all primitive characters mod k.

Exercises for Chapter 10

1. Prove that m is prime if and only if $\exp_m(a) = m - 1$ for some a.

2. If $(a, m) = (b, m) = 1$ and if $(\exp_m(a), \exp_m(b)) = 1$, prove that

$$\exp_m(ab) = \exp_m(a)\exp_m(b).$$

3. Let g be a primitive root of an odd prime p. Prove that $-g$ is also a primitive root of p if $p \equiv 1 \pmod 4$, but that $\exp_p(-g) = (p - 1)/2$ if $p \equiv 3 \pmod 4$.

4. (a) Prove that 3 is a primitive root mod p if p is a prime of the form $2^n + 1$, $n > 1$.
(b) Prove that 2 is a primitive root mod p if p is a prime of the form $4q + 1$, where q is an odd prime.

5. Let $m > 2$ be an integer having a primitive root, and let $(a, m) = 1$. We write aRm if there exists an x such that $a \equiv x^2 \pmod m$. Prove that:

(a) aRm if, and only if, $a^{\varphi(m)/2} \equiv 1 \pmod m$.
(b) If aRm the congruence $x^2 \equiv a \pmod m$ has exactly two solutions.
(c) There are exactly $\varphi(m)/2$ integers a, incongruent mod m, such that $(a, m) = 1$ and aRm.

6. Assume $m > 2$, $(a, m) = 1$, aRm. Prove that the congruence $x^2 \equiv a \pmod m$ has exactly two solutions if, and only if, m has a primitive root.

7. Let $S_n(p) = \sum_{k=1}^{p-1} k^n$, where p is an odd prime and $n > 1$. Prove that

$$S_n(p) \equiv \begin{cases} 0 \pmod p & \text{if } n \not\equiv 0 \pmod{p-1}, \\ -1 \pmod p & \text{if } n \equiv 0 \pmod{p-1}. \end{cases}$$

8. Prove that the sum of the primitive roots mod p is congruent to $\mu(p - 1)$ mod p.

9. If p is an odd prime > 3 prove that the product of the primitive roots mod p is congruent to 1 mod p.

10. Let p be an odd prime of the form $2^{2^k} + 1$. Prove that the set of primitive roots mod p is equal to the set of quadratic nonresidues mod p. Use this result to prove that 7 is a primitive root of every such prime.

11. Assume $d \mid \varphi(m)$. If $d = \exp_m(a)$ we say that a is a primitive root of the congruence

$$x^d \equiv 1 \pmod{m}.$$

Prove that if the congruence

$$x^{\varphi(m)} \equiv 1 \pmod{m}$$

has a primitive root then it has $\varphi(\varphi(m))$ primitive roots, incongruent mod m.

12. Prove the properties of indices described in Theorem 10.10.

13. Let p be an odd prime. If $(h, p) = 1$ let

$$S(h) = \{h^n : 1 \le n \le p - 1, (n, p - 1) = 1\}.$$

If h is a primitive root of p the numbers in the set $S(h)$ are distinct mod p (they are, in fact, the primitive roots of p). Prove that there is an integer h, not a primitive root of p, such that the numbers in $S(h)$ are distinct mod p if, and only if, $p \equiv 3 \pmod{4}$.

14. If $m > 1$ let $p_1, \ldots, p_k$ be the distinct prime divisors of $\varphi(m)$. If $(g, m) = 1$ prove that g is a primitive root of m if, and only if, g does not satisfy any of the congruences $g^{\varphi(m)/p_i} \equiv 1 \pmod{m}$ for $i = 1, 2, \ldots, k$.

15. The prime $p = 71$ has 7 as a primitive root. Find all primitive roots of 71 and also find a primitive root for p^2 and for $2p^2$.

16. Solve each of the following congruences:

(a) $8x \equiv 7 \pmod{43}$.
(b) $x^8 \equiv 17 \pmod{43}$.
(c) $8^x \equiv 3 \pmod{43}$.

17. Let q be an odd prime and suppose that $p = 4q + 1$ is also prime.

(a) Prove that the congruence $x^2 \equiv -1 \pmod{p}$ has exactly two solutions, each of which is quadratic nonresidue of p.
(b) Prove that every quadratic nonresidue of p is a primitive root of p, with the exception of the two nonresidues in (a).
(c) Find all the primitive roots of 29.

18. (Extension of Exercise 17.) Let q be an odd prime and suppose that $p = 2^n q + 1$ is prime. Prove that every quadratic nonresidue a of p is a primitive root of p if $a^{2^n} \not\equiv 1 \pmod{p}$.

19. Prove that there are only two real primitive characters mod 8 and make a table showing their values.

20. Let χ be a real primitive character mod m. If m is not a power of 2 prove that m has the form

$$m = 2^\alpha p_1 \cdots p_r$$

where the p_i are distinct odd primes and $\alpha = 0, 2,$ or 3. If $\alpha = 0$ show that

$$\chi(-1) = \prod_{p \mid m} (-1)^{(p-1)/2}$$

and find a corresponding formula for $\chi(-1)$ when $\alpha = 2$.

11 Dirichlet Series and Euler Products

11.1 Introduction

In 1737 Euler proved Euclid's theorem on the existence of infinitely many primes by showing that the series $\sum p^{-1}$, extended over all primes, diverges. He deduced this from the fact that the zeta function $\zeta(s)$, given by

$$\zeta(s) = \sum_{n=1}^{\infty} \frac{1}{n^s} \tag{1}$$

for real $s > 1$, tends to ∞ as $s \to 1$. In 1837 Dirichlet proved his celebrated theorem on primes in arithmetical progressions by studying the series

$$L(s, \chi) = \sum_{n=1}^{\infty} \frac{\chi(n)}{n^s} \tag{2}$$

where χ is a Dirichlet character and $s > 1$.

The series in (1) and (2) are examples of series of the form

$$\sum_{n=1}^{\infty} \frac{f(n)}{n^s} \tag{3}$$

where $f(n)$ is an arithmetical function. These are called Dirichlet series with coefficients $f(n)$. They constitute one of the most useful tools in analytic number theory.

This chapter studies general properties of Dirichlet series. The next chapter makes a more detailed study of the Riemann zeta function $\zeta(s)$ and the Dirichlet L-functions $L(s, \chi)$.

Notation Following Riemann, we let s be a complex variable and write

$$s = \sigma + it,$$

where σ and t are real. Then $n^s = e^{s \log n} = e^{(\sigma + it) \log n} = n^\sigma e^{it \log n}$. This shows that $|n^s| = n^\sigma$ since $|e^{i\theta}| = 1$ for real θ.

The set of points $s = \sigma + it$ such that $\sigma > a$ is called a *half-plane*. We will show that for each Dirichlet series there is a half-plane $\sigma > \sigma_c$ in which the series converges, and another half-plane $\sigma > \sigma_a$ in which it converges absolutely. We will also show that in the half-plane of convergence the series represents an analytic function of the complex variable s.

11.2 The half-plane of absolute convergence of a Dirichlet series

First we note that if $\sigma \geq a$ we have $|n^s| = n^\sigma \geq n^a$ hence

$$\left|\frac{f(n)}{n^s}\right| \leq \frac{|f(n)|}{n^a}.$$

Therefore, if a Dirichlet series $\sum f(n)n^{-s}$ converges absolutely for $s = a + ib$, then by the comparison test it also converges absolutely for all s with $\sigma \geq a$. This observation implies the following theorem.

Theorem 11.1 *Suppose the series $\sum |f(n)n^{-s}|$ does not converge for all s or diverge for all s. Then there exists a real number σ_a, called the abscissa of absolute convergence, such that the series $\sum f(n)n^{-s}$ converges absolutely if $\sigma > \sigma_a$ but does not converge absolutely if $\sigma < \sigma_a$.*

PROOF. Let D be the set of all real σ such that $\sum |f(n)n^{-s}|$ diverges. D is not empty because the series does not converge for all s, and D is bounded above because the series does not diverge for all s. Therefore D has a least upper bound which we call σ_a. If $\sigma < \sigma_a$ then $\sigma \in D$, otherwise σ would be an upper bound for D smaller than the least upper bound. If $\sigma > \sigma_a$ then $\sigma \notin D$ since σ_a is an upper bound for D. This proves the theorem. □

Note. If $\sum |f(n)n^{-s}|$ converges everywhere we define $\sigma_a = -\infty$. If the series $\sum |f(n)n^{-s}|$ converges nowhere we define $\sigma_a = +\infty$.

EXAMPLE 1 Riemann zeta function. The Dirichlet series $\sum_{n=1}^\infty n^{-s}$ converges absolutely for $\sigma > 1$. When $s = 1$ the series diverges, so $\sigma_a = 1$. The sum of this series is denoted by $\zeta(s)$ and is called the Riemann zeta function.

EXAMPLE 2 If f is bounded, say $|f(n)| \leq M$ for all $n \geq 1$, then $\sum f(n)n^{-s}$ converges absolutely for $\sigma > 1$, so $\sigma_a \leq 1$. In particular if χ is a Dirichlet character the L-series $L(s, \chi) = \sum \chi(n)n^{-s}$ converges absolutely for $\sigma > 1$.

EXAMPLE 3 The series $\sum n^n n^{-s}$ diverges for every s so $\sigma_a = +\infty$.

EXAMPLE 4 The series $\sum n^{-n} n^{-s}$ converges absolutely for every s so $\sigma_a = -\infty$.

11.3 The function defined by a Dirichlet series

Assume that $\sum f(n)n^{-s}$ converges absolutely for $\sigma > \sigma_a$ and let $F(s)$ denote the sum function

$$F(s) = \sum_{n=1}^{\infty} \frac{f(n)}{n^s} \quad \text{for } \sigma > \sigma_a. \tag{4}$$

This section derives some properties of $F(s)$. First we prove the following lemma.

Lemma 1 *If* $N \geq 1$ *and* $\sigma \geq c > \sigma_a$ *we have*

$$\left| \sum_{n=N}^{\infty} f(n)n^{-s} \right| \leq N^{-(\sigma-c)} \sum_{n=N}^{\infty} |f(n)|n^{-c}.$$

PROOF. We have

$$\left| \sum_{n=N}^{\infty} f(n)n^{-s} \right| \leq \sum_{n=N}^{\infty} |f(n)|n^{-\sigma} = \sum_{n=N}^{\infty} |f(n)|n^{-c}n^{-(\sigma-c)}$$

$$\leq N^{-(\sigma-c)} \sum_{n=N}^{\infty} |f(n)|n^{-c}.$$ □

The next theorem describes the behavior of $F(s)$ as $\sigma \to +\infty$.

Theorem 11.2 *If* $F(s)$ *is given by* (4), *then*

$$\lim_{\sigma \to +\infty} F(\sigma + it) = f(1)$$

uniformly for $-\infty < t < +\infty$.

PROOF. Since $F(s) = f(1) + \sum_{n=2}^{\infty} f(n)n^{-s}$ we need only prove that the second term tends to 0 as $\sigma \to +\infty$. Choose $c > \sigma_a$. Then for $\sigma \geq c$ the lemma implies

$$\left| \sum_{n=2}^{\infty} \frac{f(n)}{n^s} \right| \leq 2^{-(\sigma-c)} \sum_{n=2}^{\infty} |f(n)|n^{-c} = \frac{A}{2^\sigma}$$

where A is independent of σ and t. Since $A/2^\sigma \to 0$ as $\sigma \to +\infty$ this proves the theorem. □

EXAMPLES $\zeta(\sigma + it) \to 1$ and $L(\sigma + it, \chi) \to 1$ as $\sigma \to +\infty$.

We prove next that all the coefficients are uniquely determined by the sum function.

Theorem 11.3 Uniqueness theorem. *Given two Dirichlet series*

$$F(s) = \sum_{n=1}^{\infty} \frac{f(n)}{n^s} \quad \text{and } G(s) = \sum_{n=1}^{\infty} \frac{g(n)}{n^s},$$

both absolutely convergent for $\sigma > \sigma_a$. If $F(s) = G(s)$ for each s in an infinite sequence $\{s_k\}$ such that $\sigma_k \to +\infty$ as $k \to \infty$, then $f(n) = g(n)$ for every n.

PROOF. Let $h(n) = f(n) - g(n)$ and let $H(s) = F(s) - G(s)$. Then $H(s_k) = 0$ for each k. To prove that $h(n) = 0$ for all n we assume that $h(n) \neq 0$ for some n and obtain a contradiction.

Let N be the smallest integer for which $h(n) \neq 0$. Then

$$H(s) = \sum_{n=N}^{\infty} \frac{h(n)}{n^s} = \frac{h(N)}{N^s} + \sum_{n=N+1}^{\infty} \frac{h(n)}{n^s}.$$

Hence

$$h(N) = N^s H(s) - N^s \sum_{n=N+1}^{\infty} \frac{h(n)}{n^s}.$$

Putting $s = s_k$ we have $H(s_k) = 0$ hence

$$h(N) = -N^{s_k} \sum_{n=N+1}^{\infty} h(n) n^{-s_k}.$$

Choose k so that $\sigma_k > c$ where $c > \sigma_a$. Then Lemma 1 implies

$$|h(N)| \leq N^{\sigma_k}(N+1)^{-(\sigma_k - c)} \sum_{n=N+1}^{\infty} |h(n)| n^{-c} = \left(\frac{N}{N+1}\right)^{\sigma_k} A$$

where A is independent of k. Letting $k \to \infty$ we find $(N/(N+1))^{\sigma_k} \to 0$ so $h(N) = 0$, a contradiction. □

The uniqueness theorem implies the existence of a half-plane in which a Dirichlet series does not vanish (unless, of course, the series vanishes identically).

Theorem 11.4 *Let $F(s) = \sum f(n) n^{-s}$ and assume that $F(s) \neq 0$ for some s with $\sigma > \sigma_a$. Then there is a half-plane $\sigma > c \geq \sigma_a$ in which $F(s)$ is never zero.*

PROOF. Assume no such half-plane exists. Then for every $k = 1, 2, \ldots$ there is a point s_k with $\sigma_k > k$ such that $F(s_k) = 0$. Since $\sigma_k \to +\infty$ as $k \to \infty$ the uniqueness theorem shows that $f(n) = 0$ for all n, contradicting the hypothesis that $F(s) \neq 0$ for some s. □

11.4 Multiplication of Dirichlet series

The next theorem relates products of Dirichlet series with the Dirichlet convolution of their coefficients.

Theorem 11.5 *Given two functions $F(s)$ and $G(s)$ represented by Dirichlet series,*

$$F(s) = \sum_{n=1}^{\infty} \frac{f(n)}{n^s} \quad \text{for } \sigma > a,$$

and

$$G(s) = \sum_{n=1}^{\infty} \frac{g(n)}{n^s} \quad \text{for } \sigma > b.$$

Then in the half-plane where both series converge absolutely we have

$$F(s)G(s) = \sum_{n=1}^{\infty} \frac{h(n)}{n^s}, \tag{5}$$

where $h = f * g$, the Dirichlet convolution of f and g:

$$h(n) = \sum_{d|n} f(d)g\left(\frac{n}{d}\right).$$

Conversely, if $F(s)G(s) = \sum \alpha(n)n^{-s}$ for all s in a sequence $\{s_k\}$ with $\sigma_k \to +\infty$ as $k \to \infty$ then $\alpha = f * g$.

PROOF. For any s for which both series converge absolutely we have

$$F(s)G(s) = \sum_{n=1}^{\infty} f(n)n^{-s} \sum_{m=1}^{\infty} g(m)m^{-s} = \sum_{n=1}^{\infty} \sum_{m=1}^{\infty} f(n)g(m)(mn)^{-s}.$$

Because of absolute convergence we can multiply these series together and rearrange the terms in any way we please without altering the sum. Collect together those terms for which mn is constant, say $mn = k$. The possible values of k are $1, 2, \ldots$, hence

$$F(s)G(s) = \sum_{k=1}^{\infty} \left(\sum_{mn=k} f(n)g(m) \right) k^{-s} = \sum_{k=1}^{\infty} h(k)k^{-s}$$

where $h(k) = \sum_{mn=k} f(n)g(m) = (f * g)(k)$. This proves the first assertion, and the second follows from the uniqueness theorem. □

EXAMPLE 1 Both series $\sum n^{-s}$ and $\sum \mu(n)n^{-s}$ converge absolutely for $\sigma > 1$. Taking $f(n) = 1$ and $g(n) = \mu(n)$ in (5) we find $h(n) = [1/n]$, so

$$\zeta(s) \sum_{n=1}^{\infty} \frac{\mu(n)}{n^s} = 1 \quad \text{if } \sigma > 1.$$

In particular, this shows that $\zeta(s) \neq 0$ for $\sigma > 1$ and that

$$\sum_{n=1}^{\infty} \frac{\mu(n)}{n^s} = \frac{1}{\zeta(s)} \quad \text{if } \sigma > 1.$$

Example 2 More generally, assume $f(1) \neq 0$ and let $g = f^{-1}$, the Dirichlet inverse of f. Then in any half-plane where both series $F(s) = \sum f(n)n^{-s}$ and $G(s) = \sum g(n)n^{-s}$ converge absolutely we have $F(s) \neq 0$ and $G(s) = 1/F(s)$.

Example 3 Assume $F(s) = \sum f(n)n^{-s}$ converges absolutely for $\sigma > \sigma_a$. If f is completely multiplicative we have $f^{-1}(n) = \mu(n)f(n)$. Since $|f^{-1}(n)| \leq |f(n)|$ the series $\sum \mu(n)f(n)n^{-s}$ also converges absolutely for $\sigma > \sigma_a$ and we have

$$\sum_{n=1}^{\infty} \frac{\mu(n)f(n)}{n^s} = \frac{1}{F(s)} \quad \text{if } \sigma > \sigma_a.$$

In particular for every Dirichlet character χ we have

$$\sum_{n=1}^{\infty} \frac{\mu(n)\chi(n)}{n^s} = \frac{1}{L(s, \chi)} \quad \text{if } \sigma > 1.$$

Example 4 Take $f(n) = 1$ and $g(n) = \varphi(n)$, Euler's totient. Since $\varphi(n) \leq n$ the series $\sum \varphi(n)n^{-s}$ converges absolutely for $\sigma > 2$. Also, $h(n) = \sum_{d|n} \varphi(d) = n$ so (5) gives us

$$\zeta(s) \sum_{n=1}^{\infty} \frac{\varphi(n)}{n^s} = \sum_{n=1}^{\infty} \frac{n}{n^s} = \zeta(s - 1) \quad \text{if } \sigma > 2.$$

Therefore

$$\sum_{n=1}^{\infty} \frac{\varphi(n)}{n^s} = \frac{\zeta(s - 1)}{\zeta(s)} \quad \text{if } \sigma > 2.$$

Example 5 Take $f(n) = 1$ and $g(n) = n^\alpha$. Then $h(n) = \sum_{d|n} d^\alpha = \sigma_\alpha(n)$, and (5) gives us

$$\zeta(s)\zeta(s - \alpha) = \sum_{n=1}^{\infty} \frac{\sigma_\alpha(n)}{n^s} \quad \text{if } \sigma > \max\{1, 1 + \text{Re}(\alpha)\}.$$

Example 6 Take $f(n) = 1$ and $g(n) = \lambda(n)$, Liouville's function. Then

$$h(n) = \sum_{d|n} \lambda(d) = \begin{cases} 1 & \text{if } n = m^2 \text{ for some } m, \\ 0 & \text{otherwise,} \end{cases}$$

so (5) gives us

$$\zeta(s) \sum_{n=1}^{\infty} \frac{\lambda(n)}{n^s} = \sum_{\substack{n=1 \\ n=\text{square}}}^{\infty} \frac{1}{n^s} = \sum_{m=1}^{\infty} \frac{1}{m^{2s}} = \zeta(2s).$$

Hence

$$\sum_{n=1}^{\infty} \frac{\lambda(n)}{n^s} = \frac{\zeta(2s)}{\zeta(s)} \quad \text{if } \sigma > 1.$$

11.5 Euler products

The next theorem, discovered by Euler in 1737, is sometimes called the analytic version of the fundamental theorem of arithmetic.

Theorem 11.6 *Let f be a multiplicative arithmetical function such that the series $\sum f(n)$ is absolutely convergent. Then the sum of the series can be expressed as an absolutely convergent infinite product,*

$$\sum_{n=1}^{\infty} f(n) = \prod_p \{1 + f(p) + f(p^2) + \cdots\} \tag{6}$$

extended over all primes. If f is completely multiplicative, the product simplifies and we have

$$\sum_{n=1}^{\infty} f(n) = \prod_p \frac{1}{1 - f(p)}. \tag{7}$$

Note. In each case the product is called the *Euler product* of the series.

PROOF. Consider the finite product

$$P(x) = \prod_{p \le x} \{1 + f(p) + f(p^2) + \cdots\}$$

extended over all primes $p \le x$. Since this is the product of a finite number of absolutely convergent series we can multiply the series and rearrange the terms in any fashion without altering the sum. A typical term is of the form

$$f(p_1^{a_1})f(p_2^{a_2}) \cdots f(p_r^{a_r}) = f(p_1^{a_1}p_2^{a_2} \cdots p_r^{a_r})$$

since f is multiplicative. By the fundamental theorem of arithmetic we can write

$$P(x) = \sum_{n \in A} f(n)$$

where A consists of those n having all their prime factors $\le x$. Therefore

$$\sum_{n=1}^{\infty} f(n) - P(x) = \sum_{n \in B} f(n),$$

where B is the set of n having at least one prime factor $> x$. Therefore

$$\left| \sum_{n=1}^{\infty} f(n) - P(x) \right| \le \sum_{n \in B} |f(n)| \le \sum_{n > x} |f(n)|.$$

As $x \to \infty$ the last sum on the right $\to 0$ since $\sum |f(n)|$ is convergent. Hence $P(x) \to \sum f(n)$ as $x \to \infty$.

Now an infinite product of the form $\prod(1 + a_n)$ converges absolutely whenever the corresponding series $\sum a_n$ converges absolutely. In this case we have

$$\sum_{p\le x} |f(p) + f(p^2) + \cdots| \le \sum_{p\le x} (|f(p)| + |f(p^2)| + \cdots) \le \sum_{n=2}^{\infty} |f(n)|.$$

Since all the partial sums are bounded, the series of positive terms

$$\sum_p |f(p) + f(p^2) + \cdots|$$

converges, and this implies absolute convergence of the product in (6).

Finally, when f is completely multiplicative we have $f(p^n) = f(p)^n$ and each series on the right of (6) is a convergent geometric series with sum $(1 - f(p))^{-1}$. □

Applying Theorem 11.6 to absolutely convergent Dirichlet series we immediately obtain:

Theorem 11.7 *Assume $\sum f(n)n^{-s}$ converges absolutely for $\sigma > \sigma_a$. If f is multiplicative we have*

$$\sum_{n=1}^{\infty} \frac{f(n)}{n^s} = \prod_p \left\{1 + \frac{f(p)}{p^s} + \frac{f(p^2)}{p^{2s}} + \cdots\right\} \quad \text{if } \sigma > \sigma_a, \tag{8}$$

and if f is completely multiplicative we have

$$\sum_{n=1}^{\infty} \frac{f(n)}{n^s} = \prod_p \frac{1}{1 - f(p)p^{-s}} \quad \text{if } \sigma > \sigma_a.$$

It should be noted that the general term of the product in (8) is the Bell series $f_p(x)$ of the function f with $x = p^{-s}$. (See Section 2.16.)

EXAMPLES Taking $f(n) = 1$, $\mu(n)$, $\varphi(n)$, $\sigma_\alpha(n)$, $\lambda(n)$ and $\chi(n)$, respectively, we obtain the following Euler products:

$$\zeta(s) = \sum_{n=1}^{\infty} \frac{1}{n^s} = \prod_p \frac{1}{1 - p^{-s}} \quad \text{if } \sigma > 1.$$

$$\frac{1}{\zeta(s)} = \sum_{n=1}^{\infty} \frac{\mu(n)}{n^s} = \prod_p (1 - p^{-s}) \quad \text{if } \sigma > 1.$$

$$\frac{\zeta(s-1)}{\zeta(s)} = \sum_{n=1}^{\infty} \frac{\varphi(n)}{n^s} = \prod_p \frac{1 - p^{-s}}{1 - p^{1-s}} \quad \text{if } \sigma > 2.$$

$$\zeta(s)\zeta(s-\alpha) = \sum_{n=1}^{\infty} \frac{\sigma_\alpha(n)}{n^s} = \prod_p \frac{1}{(1 - p^{-s})(1 - p^{\alpha - s})} \quad \text{if } \sigma > \max\{1, 1 + \text{Re}(\alpha)\},$$

$$\frac{\zeta(2s)}{\zeta(s)} = \sum_{n=1}^{\infty} \frac{\lambda(n)}{n^s} = \prod_p \frac{1}{1 + p^{-s}} \quad \text{if } \sigma > 1,$$

$$L(s, \chi) = \sum_{n=1}^{\infty} \frac{\chi(n)}{n^s} = \prod_p \frac{1}{1 - \chi(p)p^{-s}} \quad \text{if } \sigma > 1.$$

Note. If $\chi = \chi_1$, the principal character mod k, then $\chi_1(p) = 0$ if $p \mid k$ and $\chi_1(p) = 1$ if $p \nmid k$, so the Euler product for $L(s, \chi_1)$ becomes

$$L(s, \chi_1) = \prod_{p \nmid k} \frac{1}{1 - p^{-s}} = \prod_{p} \frac{1}{1 - p^{-s}} \cdot \prod_{p \mid k} (1 - p^{-s}) = \zeta(s) \prod_{p \mid k} (1 - p^{-s}).$$

Thus the L-function $L(s, \chi_1)$ is equal to the zeta function $\zeta(s)$ multiplied by a finite number of factors.

11.6 The half-plane of convergence of a Dirichlet series

To prove the existence of a half-plane of convergence we use the following lemma.

Lemma 2 *Let $s_0 = \sigma_0 + it_0$ and assume that the Dirichlet series $\sum f(n)n^{-s_0}$ has bounded partial sums, say*

$$\left| \sum_{n \le x} f(n)n^{-s_0} \right| \le M$$

for all $x \ge 1$. Then for each s with $\sigma > \sigma_0$ we have

$$\left| \sum_{a < n \le b} f(n)n^{-s} \right| \le 2Ma^{\sigma_0 - \sigma}\left(1 + \frac{|s - s_0|}{\sigma - \sigma_0}\right). \tag{9}$$

PROOF. Let $a(n) = f(n)n^{-s_0}$ and let $A(x) = \sum_{n \le x} a(n)$. Then $f(n)n^{-s} = a(n)n^{s_0 - s}$ so we can apply Theorem 4.2 (with $f(x) = x^{s_0 - s}$) to obtain

$$\sum_{a < n \le b} f(n)n^{-s} = A(b)b^{s_0 - s} - A(a)a^{s_0 - s} + (s - s_0)\int_a^b A(t)t^{s_0 - s - 1}\, dt.$$

Since $|A(x)| \le M$ this gives us

$$\begin{aligned} \left| \sum_{a < n \le b} f(n)n^{-s} \right| &\le Mb^{\sigma_0 - \sigma} + Ma^{\sigma_0 - \sigma} + |s - s_0|M \int_a^b t^{\sigma_0 - \sigma - 1}\, dt \\ &\le 2Ma^{\sigma_0 - \sigma} + |s - s_0|M \left| \frac{b^{\sigma_0 - \sigma} - a^{\sigma_0 - \sigma}}{\sigma_0 - \sigma} \right| \\ &\le 2Ma^{\sigma_0 - \sigma}\left(1 + \frac{|s - s_0|}{\sigma - \sigma_0}\right). \end{aligned}$$

□

EXAMPLES If the partial sums $\sum_{n \le x} f(n)$ are bounded, Lemma 2 implies that $\sum f(n)n^{-s}$ converges for $\sigma > 0$. In fact, if we take $s_0 = \sigma_0 = 0$ in (9) we obtain, for $\sigma > 0$,

$$\left| \sum_{a < n \le b} f(n)n^{-s} \right| \le Ka^{-\sigma}$$

where K is independent of a. Letting $a \to +\infty$ we find that $\sum f(n)n^{-s}$ converges if $\sigma > 0$. In particular, this shows that the Dirichlet series

$$\sum_{n=1}^{\infty} \frac{(-1)^n}{n^s}$$

converges for $\sigma > 0$ since $|\sum_{n \le x} (-1)^n| \le 1$. Similarly, if χ is any nonprincipal Dirichlet character mod k we have $|\sum_{n \le x} \chi(n)| \le \varphi(k)$ so

$$\sum_{n=1}^{\infty} \frac{\chi(n)}{n^s}$$

converges for $\sigma > 0$. The same type of reasoning gives the following theorem.

Theorem 11.8 *If the series $\sum f(n)n^{-s}$ converges for $s = \sigma_0 + it_0$ then it also converges for all s with $\sigma > \sigma_0$. If it diverges for $s = \sigma_0 + it_0$ then it diverges for all s with $\sigma < \sigma_0$.*

PROOF. The second statement follows from the first. To prove the first statement, choose any s with $\sigma > \sigma_0$. Lemma 2 shows that

$$\left| \sum_{a<n\le b} f(n)n^{-s} \right| \le K a^{\sigma_0 - \sigma}$$

where K is independent of a. Since $a^{\sigma_0 - \sigma} \to 0$ as $a \to +\infty$, the Cauchy condition shows that $\sum f(n)n^{-s}$ converges. □

Theorem 11.9 *If the series $\sum f(n)n^{-s}$ does not converge everywhere or diverge everywhere, then there exists a real number σ_c, called the abscissa of convergence, such that the series converges for all s in the half-plane $\sigma > \sigma_c$ and diverges for all s in the half-plane $\sigma < \sigma_c$.*

PROOF. We argue as in the proof of Theorem 11.1, taking σ_c to be the least upper bound of all σ for which $\sum f(n)n^{-s}$ diverges. □

Note. If the series converges everywhere we define $\sigma_c = -\infty$, and if it converges nowhere we define $\sigma_c = +\infty$.

Since absolute converge implies convergence, we always have $\sigma_a \ge \sigma_c$. If $\sigma_a > \sigma_c$ there is an infinite strip $\sigma_c < \sigma < \sigma_a$ in which the series converges conditionally (see Figure 11.1.) The next theorem shows that the width of this strip does not exceed 1.

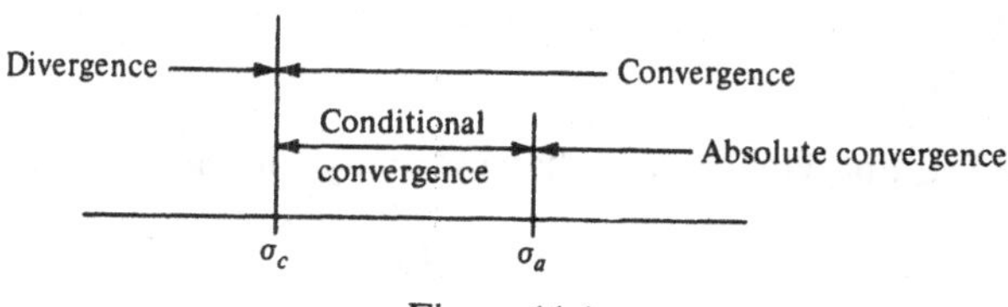

Figure 11.1

Theorem 11.10 *For any Dirichlet series with σ_c finite we have*

$$0 \le \sigma_a - \sigma_c \le 1.$$

PROOF. It suffices to show that if $\sum f(n)n^{-s_0}$ converges for some s_0 then it converges absolutely for all s with $\sigma > \sigma_0 + 1$. Let A be an upper bound for the numbers $|f(n)n^{-s_0}|$. Then

$$\left|\frac{f(n)}{n^s}\right| = \left|\frac{f(n)}{n^{s_0}}\right| \left|\frac{1}{n^{s-s_0}}\right| \le \frac{A}{n^{\sigma-\sigma_0}}$$

so $\sum |f(n)n^{-s}|$ converges by comparison with $\sum n^{\sigma_0-\sigma}$. □

EXAMPLE The series

$$\sum_{n=1}^{\infty} \frac{(-1)^n}{n^s}$$

converges if $\sigma > 0$, but the convergence is absolute only if $\sigma > 1$. Therefore in this example $\sigma_c = 0$ and $\sigma_a = 1$.

Convergence properties of Dirichlet series can be compared with those of power series. Every power series has a disk of convergence, whereas every Dirichlet series has a half-plane of convergence. For power series the interior of the disk of convergence is also the domain of absolute convergence. For Dirichlet series the domain of absolute convergence may be a proper subset of the domain of convergence. A power series represents an analytic function inside its disk of convergence. We show next that a Dirichlet series represents an analytic function inside its half-plane of convergence.

11.7 Analytic properties of Dirichlet series

Analytic properties of Dirichlet series will be deduced from the following general theorem of complex function theory which we state as a lemma.

Lemma 3 *Let $\{f_n\}$ be a sequence of functions analytic on an open subset S of the complex plane, and assume that $\{f_n\}$ converges uniformly on every compact subset of S to a limit function f. Then f is analytic on S and the sequence of derivatives $\{f_n'\}$ converges uniformly on every compact subset of S to the derivative f'.*

PROOF. Since f_n is analytic on S we have Cauchy's integral formula

$$f_n(a) = \frac{1}{2\pi i} \int_{\partial D} \frac{f_n(z)}{z - a}\, dz$$

where D is any compact disk in S, ∂D is its positively oriented boundary, and a is any interior point of D. Because of uniform convergence we can pass to the limit under the integral sign and obtain

$$f(a) = \frac{1}{2\pi i} \int_{\partial D} \frac{f(z)}{z - a}\, dz$$

which implies that f is analytic inside D. For the derivatives we have

$$f_n'(a) = \frac{1}{2\pi i} \int_{\partial D} \frac{f_n(z)}{(z - a)^2}\, dz \qquad \text{and } f'(a) = \frac{1}{2\pi i} \int_{\partial D} \frac{f(z)}{(z - a)^2}\, dz$$

from which it follows easily that $f_n'(a) \to f'(a)$ uniformly on every compact subset of S as $n \to \infty$. □

To apply the lemma to Dirichlet series we show first that we have uniform convergence on compact subsets of the half-plane of convergence.

Theorem 11.11 *A Dirichlet series $\sum f(n)n^{-s}$ converges uniformly on every compact subset lying interior to the half-plane of convergence $\sigma > \sigma_c$.*

PROOF. It suffices to show that $\sum f(n)n^{-s}$ converges uniformly on every compact rectangle $R = [\alpha, \beta] \times [c, d]$ with $\alpha > \sigma_c$. To do this we use the estimate obtained in Lemma 2,

$$\left| \sum_{a<n\le b} f(n)n^{-s} \right| \le 2Ma^{\sigma_0 - \sigma}\left(1 + \frac{|s - s_0|}{\sigma - \sigma_0}\right) \tag{10}$$

where $s_0 = \sigma_0 + it_0$ is any point in the half-plane $\sigma > \sigma_c$ and s is any point with $\sigma > \sigma_0$. We choose $s_0 = \sigma_0$ where $\sigma_c < \sigma_0 < \alpha$. (See Figure 11.2.)

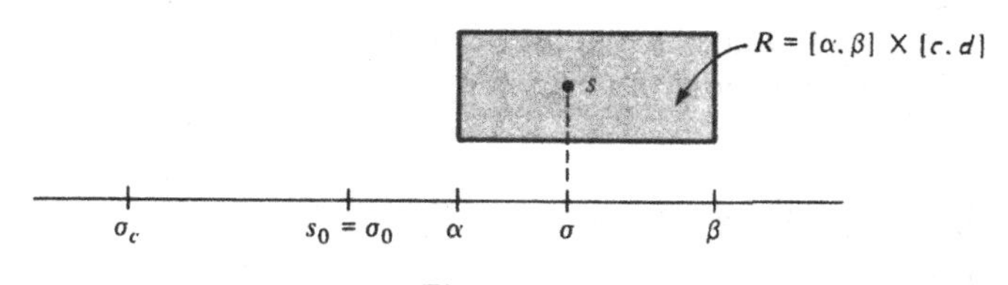

Figure 11.2

Then if $s \in R$ we have $\sigma - \sigma_0 \ge \alpha - \sigma_0$ and $|s_0 - s| < C$, where C is a constant depending on s_0 and R but not on s. Then (10) implies

$$\left| \sum_{a<n\le b} f(n)n^{-s} \right| \le 2Ma^{\sigma_0 - \alpha}\left(1 + \frac{C}{\alpha - \sigma_0}\right) = Ba^{\sigma_0 - \alpha}$$

where B is independent of s. Since $a^{\sigma_0 - \alpha} \to 0$ as $a \to +\infty$ the Cauchy condition for uniform convergence is satisfied. □

Theorem 11.12 *The sum function $F(s) = \sum f(n)n^{-s}$ of a Dirichlet series is analytic in its half-plane of convergence $\sigma > \sigma_c$, and its derivative $F'(s)$ is represented in this half-plane by the Dirichlet series*

$$F'(s) = -\sum_{n=1}^{\infty} \frac{f(n)\log n}{n^s}, \tag{11}$$

obtained by differentiating term by term.

PROOF. We apply Theorem 11.11 and Lemma 3 to the sequence of partial sums. □

Notes. The derived series in (11) has the same abscissa of convergence and the same abscissa of absolute convergence as the series for $F(s)$.

Applying Theorem 11.12 repeatedly we find that the kth derivative is given by

$$F^{(k)}(s) = (-1)^k \sum_{n=1}^{\infty} \frac{f(n)(\log n)^k}{n^s} \quad \text{for } \sigma > \sigma_c.$$

EXAMPLES For $\sigma > 1$ we have

$$\zeta'(s) = -\sum_{n=1}^{\infty} \frac{\log n}{n^s} \tag{12}$$

and

$$-\frac{\zeta'(s)}{\zeta(s)} = \sum_{n=1}^{\infty} \frac{\Lambda(n)}{n^s}. \tag{13}$$

Equation (12) follows by differentiating the series for the zeta function term by term, and (13) is obtained by multiplying the two Dirichlet series $\sum \Lambda(n)n^{-s}$ and $\sum n^{-s}$ and using the identity $\sum_{d|n} \Lambda(d) = \log n$.

11.8 Dirichlet series with nonnegative coefficients

Some functions which are defined by Dirichlet series in their half-plane of convergence $\sigma > \sigma_c$ can be continued analytically beyond the line $\sigma = \sigma_c$. For example, in the next chapter we will show that the Riemann zeta function $\zeta(s)$ can be continued analytically beyond the line $\sigma = 1$ to a function which is analytic for all s except for a simple pole at $s = 1$. Similarly, if χ is a nonprincipal Dirichlet character, the L-function $L(s, \chi)$ can be continued analytically beyond the line $\sigma = 1$ to an entire function (analytic for all s). The singularity for the zeta function is explained by the following theorem of Landau which deals with Dirichlet series having nonnegative coefficients.

Theorem 11.13 *Let $F(s)$ be represented in the half-plane $\sigma > c$ by the Dirichlet series*

$$F(s) = \sum_{n=1}^{\infty} \frac{f(n)}{n^s}, \tag{14}$$

where c is finite, and assume that $f(n) \geq 0$ for all $n \geq n_0$. If $F(s)$ is analytic in some disk about the point $s = c$, then the Dirichlet series converges in the half-plane $\sigma > c - \varepsilon$ for some $\varepsilon > 0$. Consequently, if the Dirichlet series has a finite abscissa of convergence σ_c, then $F(s)$ has a singularity on the real axis at the point $s = \sigma_c$.

PROOF. Let $a = 1 + c$. Since F is analytic at a it can be represented by an absolutely convergent power series expansion about a,

$$F(s) = \sum_{k=0}^{\infty} \frac{F^{(k)}(a)}{k!}(s - a)^k, \tag{15}$$

and the radius of convergence of this power series exceeds 1 since F is analytic at c. (See Figure 11.3.) By Theorem 11.12 the derivatives $F^{(k)}(a)$ can be determined by repeated differentiation of (14). This gives us

$$F^{(k)}(a) = (-1)^k \sum_{n=1}^{\infty} f(n)(\log n)^k n^{-a},$$

so (15) can be rewritten as

$$F(s) = \sum_{k=0}^{\infty} \sum_{n=1}^{\infty} \frac{(a - s)^k}{k!} f(n)(\log n)^k n^{-a}. \tag{16}$$

Since the radius of convergence exceeds 1, this formula is valid for some real $s = c - \varepsilon$ where $\varepsilon > 0$ (see Figure 11.3.) Then $a - s = 1 + \varepsilon$ for this s and the double series in (16) has nonnegative terms for $n \geq n_0$. Therefore we can interchange the order of summation to obtain

$$F(c - \varepsilon) = \sum_{n=1}^{\infty} \frac{f(n)}{n^a} \sum_{k=0}^{\infty} \frac{\{(1 + \varepsilon)\log n\}^k}{k!} = \sum_{n=1}^{\infty} \frac{f(n)}{n^a} e^{(1+\varepsilon)\log n} = \sum_{n=1}^{\infty} \frac{f(n)}{n^{c-\varepsilon}}.$$

In other words, the Dirichlet series $\sum f(n)n^{-s}$ converges for $s = c - \varepsilon$, hence it also converges in the half-plane $\sigma > c - \varepsilon$. □

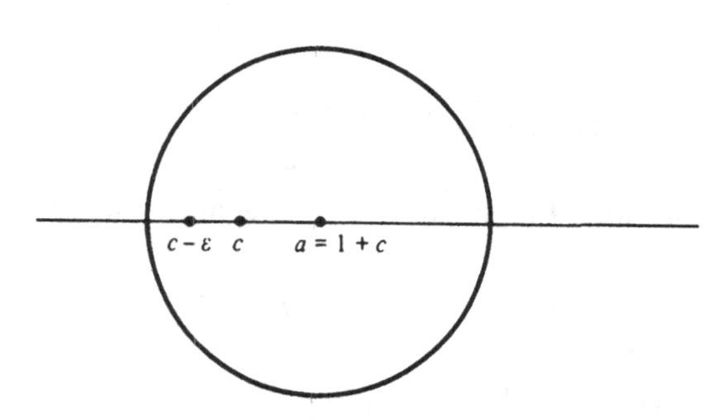

Figure 11.3

11.9 Dirichlet series expressed as exponentials of Dirichlet series

A Dirichlet series $F(s) = \sum f(n)n^{-s}$ which does not vanish identically has a half-plane in which it never vanishes. The next theorem shows that in this half-plane $F(s)$ is the exponential of another Dirichlet series if $f(1) \neq 0$.

Theorem 11.14 *Let $F(s) = \sum f(n)n^{-s}$ be absolutely convergent for $\sigma > \sigma_a$ and assume that $f(1) \neq 0$. If $F(s) \neq 0$ for $\sigma > \sigma_0 \geq \sigma_a$, then for $\sigma > \sigma_0$ we have*

$$F(s) = e^{G(s)}$$

with

$$G(s) = \log f(1) + \sum_{n=2}^{\infty} \frac{(f' * f^{-1})(n)}{\log n} n^{-s},$$

where f^{-1} is the Dirichlet inverse of f and $f'(n) = f(n)\log n$.

Note. For complex $z \neq 0$, $\log z$ denotes that branch of the logarithm which is real when $z > 0$.

Proof. Since $F(s) \neq 0$ we can write $F(s) = e^{G(s)}$ for some function $G(s)$ which is analytic for $\sigma > \sigma_0$. Differentiation gives us

$$F'(s) = e^{G(s)}G'(s) = F(s)G'(s),$$

so $G'(s) = F'(s)/F(s)$. But

$$F'(s) = -\sum_{n=1}^{\infty} \frac{f(n)\log n}{n^s} = -\sum_{n=1}^{\infty} \frac{f'(n)}{n^s} \quad \text{and} \quad \frac{1}{F(s)} = \sum_{n=1}^{\infty} \frac{f^{-1}(n)}{n^s},$$

hence

$$G'(s) = F'(s) \cdot \frac{1}{F(s)} = -\sum_{n=2}^{\infty} \frac{(f' * f^{-1})(n)}{n^s}.$$

Integration gives

$$G(s) = C + \sum_{n=2}^{\infty} \frac{(f' * f^{-1})(n)}{\log n} n^{-s}$$

where C is a constant. Letting $\sigma \to +\infty$ we find $\lim_{\sigma \to \infty} G(\sigma + it) = C$, hence

$$f(1) = \lim_{\sigma \to \infty} F(\sigma + it) = e^C$$

so $C = \log f(1)$. This completes the proof. The proof also shows that the series for $G(s)$ converges absolutely if $\sigma > \sigma_0$. □

EXAMPLE 1 When $f(n) = 1$ we have $f'(n) = \log n$ and $f^{-1}(n) = \mu(n)$ so

$$(f' * f^{-1})(n) = \sum_{d|n} \log d\mu\left(\frac{n}{d}\right) = \Lambda(n).$$

Therefore if $\sigma > 1$ we have

$$\zeta(s) = e^{G(s)} \tag{17}$$

where

$$G(s) = \sum_{n=2}^{\infty} \frac{\Lambda(n)}{\log n} n^{-s}.$$

EXAMPLE 2 A similar argument shows that if f is completely multiplicative and $F(s) = \sum f(n)n^{-s}$ then in the half-plane of absolute convergence $\sigma > \sigma_a$ we have

$$F(s) = e^{G(s)}$$

where

$$G(s) = \sum_{n=2}^{\infty} \frac{f(n)\Lambda(n)}{\log n} n^{-s}$$

since $(f' * f^{-1})(n) = \sum_{d|n} f(d)\log d\mu(n/d)f(n/d) = f(n)\Lambda(n)$.

The formulas in the foregoing examples can also be deduced with the help of Euler products. For example, for the Riemann zeta function we have

$$\zeta(s) = \prod_p \frac{1}{1 - p^{-s}}.$$

Keep s real, $s > 1$, so that $\zeta(s)$ is positive. Taking logarithms and using the power series $-\log(1 - x) = \sum x^m/m$ we find

$$\log \zeta(s) = -\sum_p \log(1 - p^{-s}) = \sum_p \sum_{m=1}^{\infty} \frac{p^{-ms}}{m} = \sum_{n=1}^{\infty} \Lambda_1(n)n^{-s}$$

where

$$\Lambda_1(n) = \begin{cases} \dfrac{1}{m} & \text{if } n = p^m \text{ for some prime } p, \\ 0 & \text{otherwise.} \end{cases}$$

But if $n = p^m$ then $\log n = m \log p = m\Lambda(n)$ so $1/m = \Lambda(n)/\log n$. Therefore

$$\log \zeta(s) = \sum_{n=2}^{\infty} \frac{\Lambda(n)}{\log n} n^{-s}$$

which implies (17) for real $s > 1$. But each member of (17) is analytic in the half-plane $\sigma > 1$ so, by analytic continuation, (17) also holds for $\sigma > 1$.

11.10 Mean value formulas for Dirichlet series

Theorem 11.15 *Given two Dirichlet series $F(s) = \sum f(n)n^{-s}$ and $G(s) = \sum g(n)n^{-s}$ with abscissae of absolute convergence σ_1 and σ_2, respectively. Then for $a > \sigma_1$ and $b > \sigma_2$ we have*

$$\lim_{T\to\infty} \frac{1}{2T}\int_{-T}^{T} F(a + it)G(b - it)\,dt = \sum_{n=1}^{\infty} \frac{f(n)g(n)}{n^{a+b}}.$$

PROOF. We have

$$F(a + it)G(b - it) = \left(\sum_{m=1}^{\infty} \frac{f(m)}{m^{a+it}}\right)\left(\sum_{n=1}^{\infty} \frac{g(n)}{n^{b-it}}\right) = \sum_{m=1}^{\infty}\sum_{n=1}^{\infty} \frac{f(m)g(n)}{m^a n^b}\left(\frac{n}{m}\right)^{it}$$

$$= \sum_{n=1}^{\infty} \frac{f(n)g(n)}{n^{a+b}} + \sum_{\substack{m=1\\ m\neq n}}^{\infty}\sum_{n=1}^{\infty} \frac{f(m)g(n)}{m^a n^b}\left(\frac{n}{m}\right)^{it}.$$

Now

$$\sum_{m=1}^{\infty}\sum_{n=1}^{\infty} \left|\frac{f(m)g(n)}{m^a n^b}\left(\frac{n}{m}\right)^{it}\right| \leq \sum_{m=1}^{\infty} \frac{|f(m)|}{m^a} \sum_{n=1}^{\infty} \frac{|g(n)|}{n^b}$$

so the series is absolutely convergent, and this convergence is also *uniform* for all t. Hence we can integrate term by term and divide by $2T$ to obtain

$$\frac{1}{2T}\int_{-T}^{T} F(a + it)G(b - it)\,dt$$

$$= \sum_{n=1}^{\infty} \frac{f(n)g(n)}{n^{a+b}} + \sum_{\substack{m,n=1\\ m\neq n}}^{\infty} \frac{f(m)g(n)}{m^a n^b}\frac{1}{2T}\int_{-T}^{T} e^{it\log(n/m)}\,dt.$$

But for $m \neq n$ we have

$$\int_{-T}^{T} e^{it\log(n/m)}\,dt = \frac{e^{it\log(n/m)}}{i\log(n/m)}\bigg|_{-T}^{T} = \frac{2\sin\left[T\log\left(\frac{n}{m}\right)\right]}{\log\left(\frac{n}{m}\right)}$$

so we obtain

$$\frac{1}{2T}\int_{-T}^{T} F(a + it)G(b - it)\,dt$$

$$= \sum_{n=1}^{\infty} \frac{f(n)g(n)}{n^{a+b}} + \sum_{\substack{m,n=1\\ m\neq n}}^{\infty} \frac{f(m)g(n)}{m^a n^b}\frac{\sin\left[T\log\left(\frac{n}{m}\right)\right]}{T\log\left(\frac{n}{m}\right)}.$$

Again, the double series converges uniformly with respect to T since $(\sin x)/x$ is bounded for every x. Hence, we can pass to the limit term by term to obtain the statement of the theorem. □

Theorem 11.16 *If $F(s) = \sum_{n=1}^{\infty} f(n)n^{-s}$ converges absolutely for $\sigma > \sigma_a$ then for $\sigma > \sigma_a$ we have*

$$\lim_{T\to\infty} \frac{1}{2T}\int_{-T}^{T} |F(\sigma + it)|^2\, dt = \sum_{n=1}^{\infty} \frac{|f(n)|^2}{n^{2\sigma}}. \tag{18}$$

In particular, if $\sigma > 1$ we have

(a) $$\lim_{T\to\infty} \frac{1}{2T}\int_{-T}^{T} |\zeta(\sigma + it)|^2\, dt = \sum_{n=1}^{\infty} \frac{1}{n^{2\sigma}} = \zeta(2\sigma).$$

(b) $$\lim_{T\to\infty} \frac{1}{2T}\int_{-T}^{T} |\zeta^{(k)}(\sigma + it)|^2\, dt = \sum_{n=1}^{\infty} \frac{\log^{2k} n}{n^{2\sigma}} = \zeta^{(2k)}(2\sigma).$$

(c) $$\lim_{T\to\infty} \frac{1}{2T}\int_{-T}^{T} |\zeta(\sigma + it)|^{-2}\, dt = \sum_{n=1}^{\infty} \frac{\mu^2(n)}{n^{2\sigma}} = \frac{\zeta(2\sigma)}{\zeta(4\sigma)}.$$

(d) $$\lim_{T\to\infty} \frac{1}{2T}\int_{-T}^{T} |\zeta(\sigma + it)|^{4}\, dt = \sum_{n=1}^{\infty} \frac{\sigma_0{}^2(n)}{n^{2\sigma}} = \frac{\zeta^4(2\sigma)}{\zeta(4\sigma)}.$$

Proof. Formula (18) follows by taking $g(n) = \overline{f(n)}$ in Theorem 11.15. To deduce the special formulas (a) through (d) we need only evaluate the Dirichlet series $\sum |f(n)|^2 n^{-2\sigma}$ for the following choices of $f(n)$: (a) $f(n) = 1$; (b) $f(n) = (-1)^k \log^k n$; (c) $f(n) = \mu(n)$; (d) $f(n) = \sigma_0(n)$. The formula (a) is clear, and formula (b) follows from the relation

$$\zeta^{(k)}(s) = (-1)^k \sum_{n=1}^{\infty} \frac{\log^k n}{n^s}.$$

To prove (c) and (d) we use Euler products. For (c) we have

$$\sum_{n=1}^{\infty} \frac{\mu^2(n)}{n^s} = \prod_p (1 + p^{-s}) = \prod_p \frac{1 - p^{-2s}}{1 - p^{-s}} = \frac{\zeta(s)}{\zeta(2s)}.$$

Replacing s by 2σ we get (c). For (d) we write

$$\begin{aligned}\sum_{n=1}^{\infty} \frac{\sigma_0{}^2(n)}{n^s} &= \prod_p \{1 + \sigma_0{}^2(p)p^{-s} + \sigma_0{}^2(p^2)p^{-2s} + \cdots\}\\ &= \prod_p \{1 + 2^2p^{-s} + 3^2p^{-2s} + \cdots\}\\ &= \prod_p \left\{\sum_{n=0}^{\infty} (n+1)^2 p^{-ns}\right\} = \prod_p \frac{1 - p^{-2s}}{(1 - p^{-s})^4} = \frac{\zeta^4(s)}{\zeta(2s)}\end{aligned}$$

since $\sum_{n=0}^{\infty} (n + 1)^2 x^n = \dfrac{x + 1}{(1 - x)^3} = \dfrac{1 - x^2}{(1 - x)^4}$. Now replace s by 2σ to get (d).

□

11.11 An integral formula for the coefficients of a Dirichlet series

Theorem 11.17 *Assume the series* $F(s) = \sum_{n=1}^{\infty} f(n)n^{-s}$ *converges absolutely for* $\sigma > \sigma_a$. *Then for* $\sigma > \sigma_a$ *and* $x > 0$ *we have*

$$\lim_{T\to\infty} \frac{1}{2T}\int_{-T}^{T} F(\sigma + it)x^{\sigma+it}\,dt = \begin{cases} f(n) & \text{if } x = n,\\ 0 & \text{otherwise.}\end{cases}$$

PROOF. For $\sigma > \sigma_a$ we have

$$\frac{1}{2T}\int_{-T}^{T} F(\sigma + it)x^{\sigma+it}\,dt = \frac{x^\sigma}{2T}\int_{-T}^{T}\sum_{n=1}^{\infty}\frac{f(n)}{n^\sigma}\left(\frac{x}{n}\right)^{it} dt \tag{19}$$

$$= \frac{x^\sigma}{2T}\sum_{n=1}^{\infty}\frac{f(n)}{n^\sigma}\int_{-T}^{T} e^{it\log(x/n)}\,dt,$$

since the series is uniformly convergent for all t in any interval $[-T, T]$. If x is not an integer then $x/n \neq 1$ for all n and we have

$$\int_{-T}^{T} e^{it\log(x/n)}\,dt = \frac{2\sin\left[T\log\left(\dfrac{x}{n}\right)\right]}{\log\left(\dfrac{x}{n}\right)}$$

and the series becomes

$$\frac{x^\sigma}{T}\sum_{n=1}^{\infty}\frac{f(n)}{n^\sigma}\frac{\sin\left[T\log\left(\dfrac{x}{n}\right)\right]}{\log\left(\dfrac{x}{n}\right)}$$

which tends to 0 as $T \to \infty$. However, if x is an integer, say $x = k$, then the term in (19) with $n = k$ contributes

$$\int_{-T}^{T}\left(\frac{x}{n}\right)^{it} dt = \int_{-T}^{T}\left(\frac{k}{k}\right)^{it} dt = \int_{-T}^{T} dt = 2T,$$

and hence

$$\frac{x^\sigma}{2T}\sum_{n=1}^{\infty}\frac{f(n)}{n^\sigma}\int_{-T}^{T}\left(\frac{x}{n}\right)^{it} dt = f(k) + \frac{k^\sigma}{2T}\sum_{\substack{n=1\\ n\neq k}}^{\infty}\frac{f(n)}{n^\sigma}\int_{-T}^{T}\left(\frac{k}{n}\right)^{it} dt.$$

The second term tends to 0 as $T \to \infty$ as was shown in first part of the argument. □

11.12 An integral formula for the partial sums of a Dirichlet series

In this section we derive a formula of Perron for expressing the partial sums of a Dirichlet series as an integral of the sum function. We shall require a lemma on contour integrals.

Lemma 4 *If $c > 0$, define $\int_{c-\infty i}^{c+\infty i}$ to mean $\lim_{T\to\infty} \int_{c-iT}^{c+iT}$. Then if a is any positive real number, we have*

$$\frac{1}{2\pi i}\int_{c-\infty i}^{c+\infty i} a^z \frac{dz}{z} = \begin{cases} 1 & \text{if } a > 1, \\ \frac{1}{2} & \text{if } a = 1, \\ 0 & \text{if } 0 < a < 1. \end{cases}$$

Moreover, we have

$$\left|\frac{1}{2\pi i}\int_{c-iT}^{c+iT} a^z \frac{dz}{z}\right| \le \frac{a^c}{\pi T \log\left(\frac{1}{a}\right)} \quad \text{if } 0 < a < 1, \tag{20}$$

$$\left|\frac{1}{2\pi i}\int_{c-iT}^{c+iT} a^z \frac{dz}{z} - 1\right| \le \frac{a^c}{\pi T \log a} \quad \text{if } a > 1, \tag{21}$$

and

$$\left|\frac{1}{2\pi i}\int_{c-iT}^{c+iT} \frac{dz}{z} - \frac{1}{2}\right| \le \frac{c}{\pi T} \quad \text{if } a = 1. \tag{22}$$

PROOF. Suppose first that $0 < a < 1$ and consider the rectangular contour R shown in Figure 11.4. Since a^z/z is analytic inside R we have $\int_R a^z/z\, dz = 0$. Hence

$$\int_{c-iT}^{c+iT} = \int_{b+iT}^{c+iT} + \int_{b-iT}^{b+iT} + \int_{c-iT}^{b-iT},$$

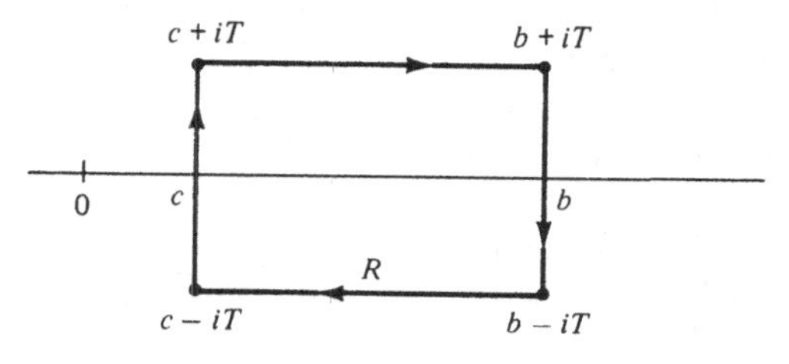

Figure 11.4

so

$$\left| \int_{c-iT}^{c+iT} a^z \frac{dz}{z} \right| \le \int_c^b \frac{a^x}{T} dx + \frac{2Ta^b}{b} + \int_c^b \frac{a^x}{T} dx$$
$$\le \frac{2}{T} \int_c^\infty a^x \, dx + \frac{2Ta^b}{b} = \frac{2}{T} \left(\frac{-a^c}{\log a} \right) + \frac{2Ta^b}{b}.$$

Let $b \to \infty$. Then $a^b \to 0$, hence

$$\left| \int_{c-iT}^{c+iT} a^z \frac{dz}{z} \right| \le \frac{2a^c}{T \log\left(\frac{1}{a}\right)}.$$

This proves (20).

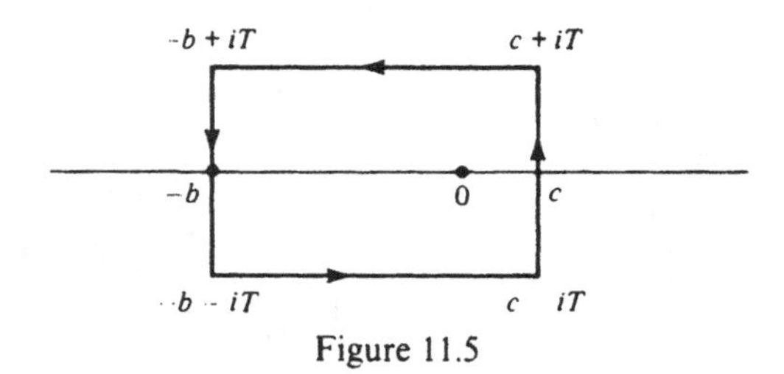

Figure 11.5

If $a > 1$ we use instead the contour R shown in Figure 11.5. Here $b > c > 0$ and $T > c$. Now a^z/z has a first order pole at $z = 0$ with residue 1 since

$$a^z = e^{z \log a} = 1 + z \log a + O(|z|^2) \quad \text{as } z \to 0.$$

Therefore

$$2\pi i = \left(\int_{c-iT}^{c+iT} + \int_{c+iT}^{-b+iT} + \int_{-b+iT}^{-b-iT} + \int_{-b-iT}^{c-iT} \right) a^z \frac{dz}{z},$$

hence

$$\frac{1}{2\pi i} \int_{c-iT}^{c+iT} a^z \frac{dz}{z} - 1 = \frac{1}{2\pi i} \left(\int_{-b+iT}^{c+iT} + \int_{-b-iT}^{-b+iT} + \int_{c-iT}^{-b-iT} \right) a^z \frac{dz}{z}.$$

We now estimate the integrals on the right. We have

$$\left| \int_{-b+iT}^{c+iT} a^z \frac{dz}{z} \right| \le \int_{-b}^{c} \frac{a^x \, dx}{T} \le \frac{1}{T} \int_{-\infty}^{c} a^x \, dx = \frac{1}{T} \frac{a^c}{\log a},$$

$$\left| \int_{-b-iT}^{-b+iT} a^z \frac{dz}{z} \right| \le 2T \frac{a^{-b}}{b},$$

$$\left| \int_{c-iT}^{-b-iT} a^z \frac{dz}{z} \right| \le \int_{-b}^{c} \frac{a^x \, dx}{T} \le \frac{1}{T} \frac{a^c}{\log a}.$$

As $b \to \infty$ the second integral tends to 0 and we obtain (21).

When $a = 1$, we can treat the integral directly. We have

$$\int_{c-iT}^{c+iT} \frac{dz}{z} = \int_{-T}^{T} \frac{i\,dy}{c+iy} = \int_{-T}^{T} \frac{y}{c^2+y^2}\,dy + ic\int_{-T}^{T} \frac{dy}{c^2+y^2}$$

$$= 2ic\int_{0}^{T} \frac{dy}{c^2+y^2},$$

the other integral vanishing because the integrand is an odd function. Hence

$$\frac{1}{2\pi i}\int_{c-iT}^{c+iT} \frac{dz}{z} = \frac{c}{\pi}\int_{0}^{T} \frac{dy}{c^2+y^2} = \frac{1}{\pi}\arctan\frac{T}{c} = \frac{1}{2} - \frac{1}{\pi}\arctan\frac{c}{T}.$$

Since $\arctan c/T < c/T$ this proves (22), and the proof of Lemma 4 is complete. □

Theorem 11.18 Perron's formula. *Let $F(s) = \sum_{n=1}^{\infty} f(n)/n^s$ be absolutely convergent for $\sigma > \sigma_a$; let $c > 0$, $x > 0$ be arbitrary. Then if $\sigma > \sigma_a - c$ we have:*

$$\frac{1}{2\pi i}\int_{c-\infty i}^{c+\infty i} F(s+z)\frac{x^z}{z}\,dz = \sum_{n \le x}{}^{*} \frac{f(n)}{n^s}$$

where $\sum^*$ means that the last term in the sum must be multiplied by 1/2 when x is an integer.

Proof. In the integral, c is the real part of z, so the series for $F(s+z)$ is absolutely and uniformly convergent on compact subsets of the half-plane $\sigma + c > \sigma_a$. Therefore

$$\int_{c-iT}^{c+iT} F(s+z)\frac{x^z}{z}\,dz = \int_{c-iT}^{c+iT} \sum_{n=1}^{\infty} \frac{f(n)}{n^{s+z}}\frac{x^z}{z}\,dz$$

$$= \sum_{n=1}^{\infty} \frac{f(n)}{n^s}\int_{c-iT}^{c+iT} \left(\frac{x}{n}\right)^z \frac{dz}{z}$$

$$= \sum_{n<x} \frac{f(n)}{n^s}\int_{c-iT}^{c+iT} \left(\frac{x}{n}\right)^z \frac{dz}{z} + \sum_{n>x} \frac{f(n)}{n^s}\int_{c-iT}^{c+iT} \left(\frac{x}{n}\right)^z \frac{dz}{z}$$

$$+' \frac{f(x)}{x^s}\int_{c-iT}^{c+iT} \frac{dz}{z},$$

the symbol $+'$ indicating that the last term appears only if x is an integer. In the finite sum $\sum_{n<x}$ we can pass to the limit $T \to \infty$ term by term, and the integral is $2\pi i$ by Lemma 4. (Here $a = x/n$, $a > 1$.) The last term (if it appears) yields $\pi i f(x)x^{-s}$ and the theorem will be proved if we show that

$$\lim_{T\to\infty} \sum_{n>x} \frac{f(n)}{n^s}\int_{c-iT}^{c+iT} \left(\frac{x}{n}\right)^z \frac{dz}{z} = 0. \tag{23}$$

We know that $\int_{c-i\infty}^{c+i\infty} (x/n)^z(dz/z) = 0$ if $n > x$ but to prove (23) we must estimate the rate at which $\int_{c-iT}^{c+iT}$ tends to zero.

From Lemma 4 we have the estimate

$$\left|\int_{c-iT}^{c+iT} a^z \frac{dz}{z}\right| \le \frac{2}{T}\frac{a^c}{\left(\log \dfrac{1}{a}\right)} \quad \text{if } 0 < a < 1.$$

Here $a = x/n$, with $n > x$. In fact, $n \ge [x] + 1$, so that $1/a = n/x \ge ([x] + 1)/x$. Hence

$$\left|\sum_{n>x} \frac{f(n)}{n^s}\int_{c-iT}^{c+iT}\left(\frac{x}{n}\right)^z \frac{dz}{z}\right| \le \sum_{n>x} \frac{|f(n)|}{n^\sigma}\frac{2}{T}\left(\frac{x}{n}\right)^c \frac{1}{\log\left(\dfrac{[x]+1}{x}\right)}$$

$$= \frac{2}{T}\frac{x^c}{\log\left(\dfrac{[x]+1}{x}\right)}\sum_{n>x}\frac{|f(n)|}{n^{\sigma+c}} \to 0 \quad \text{as } T \to \infty.$$

This proves Perron's formula. □

Note. If $c > \sigma_a$ Perron's formula is valid for $s = 0$ and we obtain the following integral representation for the partial sums of the coefficients:

$$\frac{1}{2\pi i}\int_{c-\infty i}^{c+\infty i} F(z)\frac{x^z}{z}\,dz = \sum_{n\le x}{}^{*} f(n).$$

Exercises for Chapter 11

1. Derive the following identities, valid for $\sigma > 1$.

(a) $\zeta(s) = s\displaystyle\int_1^\infty \frac{[x]}{x^{s+1}}\,dx.$

(b) $\displaystyle\sum_p \frac{1}{p^s} = s\int_1^\infty \frac{\pi(x)}{x^{s+1}}\,dx,$ where the sum is extended over all primes.

(c) $\displaystyle\frac{1}{\zeta(s)} = s\int_1^\infty \frac{M(x)}{x^{s+1}}\,dx,$ where $M(x) = \sum_{n\le x}\mu(n).$

(d) $\displaystyle -\frac{\zeta'(s)}{\zeta(s)} = s\int_1^\infty \frac{\psi(x)}{x^{s+1}}\,dx,$ where $\psi(x) = \sum_{n\le x}\Lambda(n).$

(e) $\displaystyle L(s,\chi) = s\int_1^\infty \frac{A(x)}{x^{s+1}}\,dx,$ where $A(x) = \sum_{n\le x}\chi(n).$

Show that (e) is also valid for $\sigma > 0$ if χ is a nonprincipal character. [*Hint:* Theorem 4.2.]

2. Assume that the series $\sum_{n=1}^{\infty} f(n)$ converges with sum A, and let $A(x) = \sum_{n \le x} f(n)$.

   (a) Prove that the Dirichlet series $F(s) = \sum_{n=1}^{\infty} f(n)n^{-s}$ converges for each s with $\sigma > 0$ and that
$$\sum_{n=1}^{\infty} \frac{f(n)}{n^s} = A - s\int_1^{\infty} \frac{R(x)}{x^{s+1}}\,dx,$$
   where $R(x) = A - A(x)$. [*Hint*: Theorem 4.2.]

   (b) Deduce that $F(\sigma) \to A$ as $\sigma \to 0+$.

   (c) If $\sigma > 0$ and $N \ge 1$ is an integer, prove that
$$F(s) = \sum_{n=1}^{N} \frac{f(n)}{n^s} - \frac{A(N)}{N^s} + s\int_N^{\infty} \frac{A(y)}{y^{s+1}}\,dy.$$

   (d) Write $s = \sigma + it$, take $N = 1 + [|t|]$ in part (c) and show that
$$|F(\sigma + it)| = O(|t|^{1-\sigma}) \quad \text{if } 0 < \sigma < 1.$$

3. (a) Prove that the series $\sum n^{-1-it}$ has bounded partial sums if $t \ne 0$. When $t = 0$ the partial sums are unbounded.

   (b) Prove that the series $\sum n^{-1-it}$ diverges for all real t. In other words, the Dirichlet series for $\zeta(s)$ diverges everywhere on the line $\sigma = 1$.

4. Let $F(s) = \sum_{n=1}^{\infty} f(n)n^{-s}$ where $f(n)$ is completely multiplicative and the series converges absolutely for $\sigma > \sigma_a$. Prove that if $\sigma > \sigma_a$ we have
$$\frac{F'(s)}{F(s)} = -\sum_{n=1}^{\infty} \frac{f(n)\Lambda(n)}{n^s}.$$

In the following exercises, $\lambda(n)$ is Liouville's function, $d(n)$ is the number of divisors of n, $\nu(n)$ and $\kappa(n)$ are defined as follows: $\nu(1) = 0$, $\kappa(1) = 1$; if $n = p_1^{a_1} \cdots p_k^{a_k}$ then $\nu(n) = k$ and $\kappa(n) = a_1 a_2 \cdots a_k$.

Prove that the identities in Exercises 5 through 10 are valid for $\sigma > 1$.

5. $\displaystyle\sum_{n=1}^{\infty} \frac{d(n^2)}{n^s} = \frac{\zeta^3(s)}{\zeta(2s)}.$

6. $\displaystyle\sum_{n=1}^{\infty} \frac{\nu(n)}{n^s} = \zeta(s)\sum_p \frac{1}{p^s}.$

7. $\displaystyle\sum_{n=1}^{\infty} \frac{2^{\nu(n)}}{n^s} = \frac{\zeta^2(s)}{\zeta(2s)}.$

8. $\displaystyle\sum_{n=1}^{\infty} \frac{2^{\nu(n)}\lambda(n)}{n^s} = \frac{\zeta(2s)}{\zeta^2(s)}.$

9. $\displaystyle\sum_{n=1}^{\infty} \frac{\kappa(n)}{n^s} = \frac{\zeta(s)\zeta(2s)\zeta(3s)}{\zeta(6s)}.$

10. $\displaystyle\sum_{n=1}^{\infty} \frac{3^{\nu(n)}\kappa(n)}{n^s} = \frac{\zeta^3(s)}{\zeta(3s)}.$

11. Express the sum of the series $\sum_{n=1}^{\infty} 3^{\nu(n)}\kappa(n)\lambda(n)n^{-s}$ in terms of the Riemann zeta function.

12. Let f be a completely multiplicative function such that $f(p) = f(p)^2$ for each prime p. If the series $\sum f(n)n^{-s}$ converges absolutely for $\sigma > \sigma_a$ and has sum $F(s)$, prove that $F(s) \ne 0$ and that
$$\sum_{n=1}^{\infty} \frac{f(n)\lambda(n)}{n^s} = \frac{F(2s)}{F(s)} \quad \text{if } \sigma > \sigma_a.$$

13. Let f be a multiplicative function such that $f(p) = f(p)^2$ for each prime p. If the series $\sum \mu(n)f(n)n^{-s}$ converges absolutely for $\sigma > \sigma_a$ and has sum $F(s)$, prove that whenever $F(s) \neq 0$ we have

$$\sum_{n=1}^{\infty} \frac{f(n)|\mu(n)|}{n^s} = \frac{F(2s)}{F(s)} \quad \text{if } \sigma > \sigma_a.$$

14. Let f be a multiplicative function such that $\sum f(n)n^{-s}$ converges absolutely for $\sigma > \sigma_a$. If p is prime and $\sigma > \sigma_a$ prove that

$$(1 + f(p)p^{-s})\sum_{n=1}^{\infty} \frac{f(n)\mu(n)}{n^s} = (1 - f(p)p^{-s})\sum_{n=1}^{\infty} \frac{f(n)\mu(n)\mu(p, n)}{n^s},$$

where $\mu(p, n)$ is the Möbius function evaluated at the gcd of p and n.
[*Hint:* Euler products.]

15. Prove that

$$\sum_{m=1}^{\infty} \sum_{\substack{n=1 \\ (m,n)=1}}^{\infty} \frac{1}{m^2n^2} = \frac{\zeta^2(2)}{\zeta(4)}.$$

More generally, if each s_i has real part $\sigma_i > 1$, express the multiple sum

$$\sum_{m_1=1}^{\infty} \cdots \sum_{\substack{m_r=1 \\ (m_1, \dots, m_r)=1}}^{\infty} m_1^{-s_1} \cdots m_r^{-s_r}$$

in terms of the Riemann zeta function.

16. Integrals of the form

$$f(s) = \int_1^{\infty} \frac{A(x)}{x^s}\, dx, \tag{24}$$

where $A(x)$ is Riemann-integrable on every compact interval $[1, a]$, have some properties analogous to those of Dirichlet series. For example, they possess a half-plane of absolute convergence $\sigma > \sigma_a$ and a half-plane of convergence $\sigma > \sigma_c$ in which $f(s)$ is analytic. This exercise describes an analogue of Theorem 11.13 (*Landau's theorem*).

Let $f(s)$ be represented in the half-plane $\sigma > \sigma_c$ by (24), where σ_c is finite, and assume that $A(x)$ is real-valued and does not change sign for $x \geq x_0$. Prove that $f(s)$ has a singularity on the real axis at the point $s = \sigma_c$.

17. Let $\lambda_a(n) = \sum_{d|n} d^a\lambda(d)$ where $\lambda(n)$ is Liouville's function. Prove that if $\sigma > \max\{1, \operatorname{Re}(a) + 1\}$, we have

$$\sum_{n=1}^{\infty} \frac{\lambda_a(n)}{n^s} = \frac{\zeta(s)\zeta(2s - 2a)}{\zeta(s - a)}$$

and

$$\sum_{n=1}^{\infty} \frac{\lambda(n)\lambda_a(n)}{n^s} = \frac{\zeta(2s)\zeta(s - a)}{\zeta(s)}.$$

The Functions $\zeta(s)$ and $L(s, \chi)$ 12

12.1 Introduction

This chapter develops further properties of the Riemann zeta function $\zeta(s)$ and the Dirichlet L-functions $L(s, \chi)$ defined for $\sigma > 1$ by the series

$$\zeta(s) = \sum_{n=1}^{\infty} \frac{1}{n^s} \qquad \text{and } L(s, \chi) = \sum_{n=1}^{\infty} \frac{\chi(n)}{n^s}.$$

As in the last chapter we write $s = \sigma + it$. The treatment of both $\zeta(s)$ and $L(s, \chi)$ can be unified by introducing the Hurwitz zeta function $\zeta(s, a)$, defined for $\sigma > 1$ by the series

$$\zeta(s, a) = \sum_{n=0}^{\infty} \frac{1}{(n + a)^s}.$$

Here a is a fixed real number, $0 < a \le 1$. When $a = 1$ this reduces to the Riemann zeta function, $\zeta(s) = \zeta(s, 1)$. We can also express $L(s, \chi)$ in terms of Hurwitz zeta functions. If χ is a character mod k we rearrange the terms in the series for $L(s, \chi)$ according to the residue classes mod k. That is, we write

$$n = qk + r, \quad \text{where } 1 \le r \le k \text{ and } q = 0, 1, 2, \ldots,$$

and obtain

$$L(s, \chi) = \sum_{n=1}^{\infty} \frac{\chi(n)}{n^s} = \sum_{r=1}^{k} \sum_{q=0}^{\infty} \frac{\chi(qk + r)}{(qk + r)^s} = \frac{1}{k^s} \sum_{r=1}^{k} \chi(r) \sum_{q=0}^{\infty} \frac{1}{\left(q + \frac{r}{k}\right)^s}$$

$$= k^{-s} \sum_{r=1}^{k} \chi(r) \zeta\left(s, \frac{r}{k}\right).$$

This representation of $L(s, \chi)$ as a linear combination of Hurwitz zeta functions shows that the properties of L-functions depend ultimately on those of $\zeta(s, a)$.

Our first goal is to obtain the analytic continuation of $\zeta(s, a)$ beyond the line $\sigma = 1$. This is done through an integral representation for $\zeta(s, a)$ obtained from the integral formula for the gamma function $\Gamma(s)$.

12.2 Properties of the gamma function

Throughout the chapter we shall require some basic properties of the gamma function $\Gamma(s)$. They are listed here for easy reference, although not all of them will be needed. Proofs can be found in most textbooks on complex function theory.

For $\sigma > 0$ we have the integral representation

$$\Gamma(s) = \int_0^\infty x^{s-1} e^{-x}\, dx. \tag{1}$$

The function so defined for $\sigma > 0$ can be continued beyond the line $\sigma = 0$, and $\Gamma(s)$ exists as a function which is analytic everywhere in the s-plane except for simple poles at the points

$$s = 0, -1, -2, -3, \ldots,$$

with residue $(-1)^n/n!$ at $s = -n$. We also have the representation

$$\Gamma(s) = \lim_{n \to \infty} \frac{n^s n!}{s(s+1)\cdots(s+n)} \qquad \text{for } s \neq 0, -1, -2, \ldots,$$

and the product formula

$$\frac{1}{\Gamma(s)} = s e^{Cs} \prod_{n=1}^{\infty} \left(1 + \frac{s}{n}\right) e^{-s/n} \quad \text{for all } s,$$

where C is Euler's constant. Since the product converges for all s, $\Gamma(s)$ is never zero. The gamma function satisfies two functional equations,

$$\Gamma(s+1) = s\Gamma(s) \tag{2}$$

and

$$\Gamma(s)\Gamma(1-s) = \frac{\pi}{\sin \pi s}, \tag{3}$$

valid for all s, and a multiplication formula

$$\Gamma(s)\Gamma\left(s + \frac{1}{m}\right) \cdots \Gamma\left(s + \frac{m-1}{m}\right) = (2\pi)^{(m-1)/2} m^{(1/2)-ms} \Gamma(ms), \tag{4}$$

valid for all s and all integers $m \geq 1$.

We will use the integral representation (1), the functional equations (2) and (3), and the fact that $\Gamma(s)$ exists in the whole plane, with simple poles at the integers $s = 0, -1, -2, \ldots$ We also note that $\Gamma(n + 1) = n!$ if n is a nonnegative integer.

12.3 Integral representation for the Hurwitz zeta function

The Hurwitz zeta function $\zeta(s, a)$ is initially defined for $\sigma > 1$ by the series

$$\zeta(s, a) = \sum_{n=0}^{\infty} \frac{1}{(n + a)^s}.$$

Theorem 12.1 *The series for $\zeta(s, a)$ converges absolutely for $\sigma > 1$. The convergence is uniform in every half-plane $\sigma \geq 1 + \delta$, $\delta > 0$, so $\zeta(s, a)$ is an analytic function of s in the half-plane $\sigma > 1$.*

PROOF. All these statements follow from the inequalities

$$\sum_{n=1}^{\infty} |(n + a)^{-s}| = \sum_{n=1}^{\infty} (n + a)^{-\sigma} \leq \sum_{n=1}^{\infty} (n + a)^{-(1+\delta)}. \qquad \square$$

Theorem 12.2 *For $\sigma > 1$ we have the integral representation*

$$\Gamma(s)\zeta(s, a) = \int_0^{\infty} \frac{x^{s-1}e^{-ax}}{1 - e^{-x}}\, dx. \tag{5}$$

In particular, when $a = 1$ we have

$$\Gamma(s)\zeta(s) = \int_0^{\infty} \frac{x^{s-1}e^{-x}}{1 - e^{-x}}\, dx.$$

PROOF. First we keep s real, $s > 1$, and then extend the result to complex s by analytic continuation.

In the integral for $\Gamma(s)$ we make the change of variable $x = (n + a)t$, where $n \geq 0$, to obtain

$$\Gamma(s) = \int_0^{\infty} e^{-x}x^{s-1}\, dx = (n + a)^s \int_0^{\infty} e^{-(n+a)t}t^{s-1}\, dt,$$

or

$$(n + a)^{-s}\Gamma(s) = \int_0^{\infty} e^{-nt}e^{-at}t^{s-1}\, dt.$$

Summing over all $n \geq 0$ we find

$$\zeta(s, a)\Gamma(s) = \sum_{n=0}^{\infty} \int_0^{\infty} e^{-nt}e^{-at}t^{s-1}\, dt,$$

the series on the right being convergent if $\sigma > 1$. Now we wish to interchange the sum and integral. The simplest way to justify this is to regard the integral as a Lebesgue integral. Since the integrand is nonnegative, Levi's convergence theorem (Theorem 10.25 in Reference [2]) tells us that the series

$$\sum_{n=0}^{\infty} e^{-nt}e^{-at}t^{s-1}$$

converges almost everywhere to a sum function which is Lebesgue-integrable on $[0, +\infty)$ and that

$$\zeta(s, a)\Gamma(s) = \sum_{n=0}^{\infty} \int_0^{\infty} e^{-nt}e^{-at}t^{s-1}\, dt = \int_0^{\infty} \sum_{n=0}^{\infty} e^{-nt}e^{-at}t^{s-1}\, dt.$$

But if $t > 0$ we have $0 < e^{-t} < 1$ and hence

$$\sum_{n=0}^{\infty} e^{-nt} = \frac{1}{1 - e^{-t}},$$

the series being a geometric series. Therefore we have

$$\sum_{n=0}^{\infty} e^{-nt}e^{-at}t^{s-1} = \frac{e^{-at}t^{s-1}}{1 - e^{-t}}$$

almost everywhere on $[0, +\infty)$, in fact everywhere except at 0, so

$$\zeta(s, a)\Gamma(s) = \int_0^{\infty} \sum_{n=0}^{\infty} e^{-nt}e^{-at}t^{s-1}\, dt = \int_0^{\infty} \frac{e^{-at}t^{s-1}}{1 - e^{-t}}\, dt.$$

This proves (5) for real $s > 1$. To extend it to all complex s with $\sigma > 1$ we note that both members are analytic for $\sigma > 1$. To show that the right member is analytic we assume $1 + \delta \le \sigma \le c$, where $c > 1$ and $\delta > 0$ and write

$$\int_0^{\infty} \left|\frac{e^{-at}t^{s-1}}{1 - e^{-t}}\right| dt \le \int_0^{\infty} \frac{e^{-at}t^{\sigma-1}}{1 - e^{-t}}\, dt = \left(\int_0^1 + \int_1^{\infty}\right) \frac{e^{-at}t^{\sigma-1}}{1 - e^{-t}}\, dt.$$

If $0 \le t \le 1$ we have $t^{\sigma-1} \le t^{\delta}$, and if $t \ge 1$ we have $t^{\sigma-1} \le t^{c-1}$. Also, since $e^t - 1 \ge t$ for $t \ge 0$ we have

$$\int_0^1 \frac{e^{-at}t^{\sigma-1}}{1 - e^{-t}}\, dt \le \int_0^1 \frac{e^{(1-a)t}t^{\delta}}{e^t - 1}\, dt \le e^{(1-a)} \int_0^1 t^{\delta-1}\, dt = \frac{e^{1-a}}{\delta},$$

and

$$\int_1^{\infty} \frac{e^{-at}t^{\sigma-1}}{1 - e^{-t}}\, dt \le \int_1^{\infty} \frac{e^{-at}t^{c-1}}{1 - e^{-t}}\, dt \le \int_0^{\infty} \frac{e^{-at}t^{c-1}}{1 - e^{-t}}\, dt = \Gamma(c)\zeta(c, a).$$

This shows that the integral in (5) converges uniformly in every strip $1 + \delta \le \sigma \le c$, where $\delta > 0$, and therefore represents an analytic function in every such strip, hence also in the half-plane $\sigma > 1$. Therefore, by analytic continuation, (5) holds for all s with $\sigma > 1$. □

12.4 A contour integral representation for the Hurwitz zeta function

To extend $\zeta(s, a)$ beyond the line $\sigma = 1$ we derive another representation in terms of a contour integral. The contour C is a loop around the negative real axis, as shown in Figure 12.1. The loop is composed of three parts C_1, C_2, C_3. C_2 is a positively oriented circle of radius $c < 2\pi$ about the origin, and C_1, C_3 are the lower and upper edges of a "cut" in the z-plane along the negative real axis, traversed as shown in Figure 12.1.

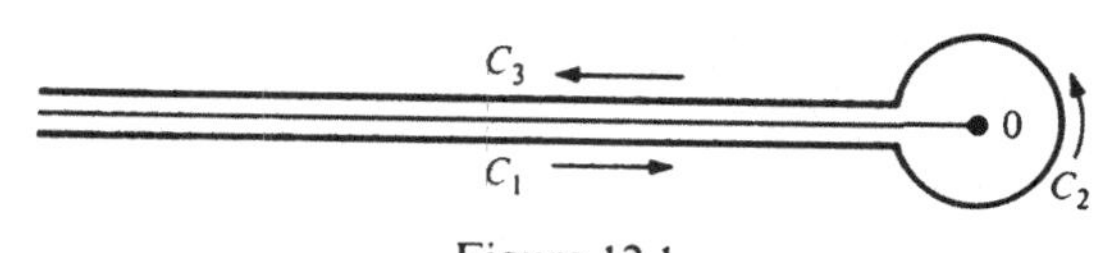

Figure 12.1

This means that we use the parametrizations $z = re^{-\pi i}$ on C_1 and $z = re^{\pi i}$ on C_3 where r varies from c to $+\infty$.

Theorem 12.3 *If* $0 < a \leq 1$ *the function defined by the contour integral*

$$I(s, a) = \frac{1}{2\pi i} \int_C \frac{z^{s-1}e^{az}}{1 - e^z}\, dz$$

is an entire function of s. *Moreover, we have*

$$\zeta(s, a) = \Gamma(1 - s)I(s, a) \quad \textit{if } \sigma > 1. \tag{6}$$

Proof. Here z^s means $r^s e^{-\pi i s}$ on C_1 and $r^s e^{\pi i s}$ on C_3. We consider an arbitrary compact disk $|s| \leq M$ and prove that the integrals along C_1 and C_3 converge uniformly on every such disk. Since the integrand is an entire function of s this will prove that $I(s, a)$ is entire.

Along C_1 we have, for $r \geq 1$,

$$|z^{s-1}| = r^{\sigma-1}|e^{-\pi i(\sigma - 1 + it)}| = r^{\sigma-1}e^{\pi t} \leq r^{M-1}e^{\pi M}$$

since $|s| \leq M$. Similarly, along C_3 we have, for $r \geq 1$,

$$|z^{s-1}| = r^{\sigma-1}|e^{\pi i(\sigma - 1 + it)}| = r^{\sigma-1}e^{-\pi t} \leq r^{M-1}e^{\pi M}.$$

Hence on either C_1 or C_3 we have, for $r \geq 1$,

$$\left|\frac{z^{s-1}e^{az}}{1 - e^z}\right| \leq \frac{r^{M-1}e^{\pi M}e^{-ar}}{1 - e^{-r}} = \frac{r^{M-1}e^{\pi M}e^{(1-a)r}}{e^r - 1}.$$

But $e^r - 1 > e^r/2$ when $r > \log 2$ so the integrand is bounded by $Ar^{M-1}e^{-ar}$ where A is a constant depending on M but not on r. Since $\int_c^\infty r^{M-1}e^{-ar}\, dr$ converges if $c > 0$ this shows that the integrals along C_1 and C_3 converge uniformly on every compact disk $|s| \leq M$, and hence $I(s, a)$ is an entire function of s.

To prove (6) we write

$$2\pi i I(s, a) = \left(\int_{C_1} + \int_{C_2} + \int_{C_3}\right) z^{s-1} g(z)\, dz$$

where $g(z) = e^{az}/(1 - e^z)$. On C_1 and C_3 we have $g(z) = g(-r)$, and on C_2 we write $z = ce^{i\theta}$, where $-\pi \leq \theta \leq \pi$. This gives us

$$\begin{aligned} 2\pi i I(s, a) &= \int_{\infty}^{c} r^{s-1} e^{-\pi i s} g(-r)\, dr + i \int_{-\pi}^{\pi} c^{s-1} e^{(s-1)i\theta} ce^{i\theta} g(ce^{i\theta})\, d\theta \\ &\quad + \int_{c}^{\infty} r^{s-1} e^{\pi i s} g(-r)\, dr \\ &= 2i \sin(\pi s) \int_{c}^{\infty} r^{s-1} g(-r)\, dr + ic^s \int_{-\pi}^{\pi} e^{is\theta} g(ce^{i\theta})\, d\theta. \end{aligned}$$

Dividing by $2i$, we get

$$\pi I(s, a) = \sin(\pi s) I_1(s, c) + I_2(s, c)$$

say. Now let $c \to 0$. We find

$$\lim_{c \to 0} I_1(s, c) = \int_0^{\infty} \frac{r^{s-1} e^{-ar}}{1 - e^{-r}}\, dr = \Gamma(s)\zeta(s, a),$$

if $\sigma > 1$. We show next that $\lim_{c \to 0} I_2(s, c) = 0$. To do this note that $g(z)$ is analytic in $|z| < 2\pi$ except for a first order pole at $z = 0$. Therefore $zg(z)$ is analytic everywhere inside $|z| < 2\pi$ and hence is bounded there, say $|g(z)| \leq A/|z|$, where $|z| = c < 2\pi$ and A is a constant. Therefore we have

$$|I_2(s, c)| \leq \frac{c^\sigma}{2} \int_{-\pi}^{\pi} e^{-t\theta} \frac{A}{c}\, d\theta \leq Ae^{\pi|t|} c^{\sigma - 1}.$$

If $\sigma > 1$ and $c \to 0$ we find $I_2(s, c) \to 0$ hence $\pi I(s, a) = \sin(\pi s)\Gamma(s)\zeta(s, a)$. Since $\Gamma(s)\Gamma(1 - s) = \pi/\sin \pi s$ this proves (6). □

12.5 The analytic continuation of the Hurwitz zeta function

In the equation $\zeta(s, a) = \Gamma(1 - s)I(s, a)$, valid for $\sigma > 1$, the functions $I(s, a)$ and $\Gamma(1 - s)$ are meaningful for every complex s. Therefore we can use this equation to define $\zeta(s, a)$ for $\sigma \leq 1$.

Definition If $\sigma \leq 1$ we define $\zeta(s, a)$ by the equation

(7) $$\zeta(s, a) = \Gamma(1 - s)I(s, a).$$

This equation provides the analytic continuation of $\zeta(s, a)$ in the entire s-plane.

Theorem 12.4 *The function $\zeta(s, a)$ so defined is analytic for all s except for a simple pole at $s = 1$ with residue 1.*

PROOF. Since $I(s, a)$ is entire the only possible singularities of $\zeta(s, a)$ are the poles of $\Gamma(1 - s)$, that is, the points $s = 1, 2, 3, \ldots$ But Theorem 12.1 shows that $\zeta(s, a)$ is analytic at $s = 2, 3, \ldots$, so $s = 1$ is the only possible pole of $\zeta(s, a)$.

Now we show that there is a pole at $s = 1$ with residue 1. If s is any integer, say $s = n$, the integrand in the contour integral for $I(s, a)$ takes the same values on C_1 as on C_3 and hence the integrals along C_1 and C_3 cancel, leaving

$$I(n, a) = \frac{1}{2\pi i}\int_{C_2} \frac{z^{n-1}e^{az}}{1 - e^z}\,dz = \operatorname*{Res}_{z=0} \frac{z^{n-1}e^{az}}{1 - e^z}.$$

In particular when $s = 1$ we have

$$I(1, a) = \operatorname*{Res}_{z=0} \frac{e^{az}}{1 - e^z} = \lim_{z\to 0} \frac{ze^{az}}{1 - e^z} = \lim_{z\to 0} \frac{z}{1 - e^z} = \lim_{z\to 0} \frac{-1}{e^z} = -1.$$

To find the residue of $\zeta(s, a)$ at $s = 1$ we compute the limit

$$\begin{aligned}\lim_{s\to 1}(s - 1)\zeta(s, a) &= -\lim_{s\to 1}(1 - s)\Gamma(1 - s)I(s, a) = -I(1, a)\lim_{s\to 1}\Gamma(2 - s)\\ &= \Gamma(1) = 1.\end{aligned}$$

This proves that $\zeta(s, a)$ has a simple pole at $s = 1$ with residue 1. □

Note. Since $\zeta(s, a)$ is analytic at $s = 2, 3, \ldots$ and $\Gamma(1 - s)$ has poles at these points, Equation (7) implies that $I(s, a)$ vanishes at these points.

12.6 Analytic continuation of $\zeta(s)$ and $L(s, \chi)$

In the introduction we proved that for $\sigma > 1$ we have

$$\zeta(s) = \zeta(s, 1)$$

and

$$L(s, \chi) = k^{-s}\sum_{r=1}^{k} \chi(r)\zeta\left(s, \frac{r}{k}\right), \tag{8}$$

where χ is any Dirichlet character mod k. Now we use these formulas as *definitions* of the functions $\zeta(s)$ and $L(s, \chi)$ for $\sigma \le 1$. In this way we obtain the analytic continuation of $\zeta(s)$ and $L(s, \chi)$ beyond the line $\sigma = 1$.

Theorem 12.5 (a) *The Riemann zeta function $\zeta(s)$ is analytic everywhere except for a simple pole at $s = 1$ with residue 1.*

(b) *For the principal character χ_1* mod k, *the L-function $L(s, \chi_1)$ is analytic everywhere except for a simple pole at $s = 1$ with residue $\varphi(k)/k$.*

(c) *If $\chi \ne \chi_1$, $L(s, \chi)$ is an entire function of s.*

Proof. Part (a) follows at once from Theorem 12.4. To prove (b) and (c) we use the relation

$$\sum_{r \bmod k} \chi(r) = \begin{cases} 0 & \text{if } \chi \neq \chi_1, \\ \varphi(k) & \text{if } \chi = \chi_1. \end{cases}$$

Since $\zeta(s, r/k)$ has a simple pole at $s = 1$ with residue 1, the function $\chi(r)\zeta(s, r/k)$ has a simple pole at $s = 1$ with residue $\chi(r)$. Therefore

$$\operatorname*{Res}_{s=1} L(s, \chi) = \lim_{s \to 1}(s - 1)L(s, \chi) = \lim_{s \to 1}(s - 1)k^{-s}\sum_{r=1}^{k} \chi(r)\zeta\left(s, \frac{r}{k}\right)$$

$$= \frac{1}{k}\sum_{r=1}^{k} \chi(r) = \begin{cases} 0 & \text{if } \chi \neq \chi_1, \\ \dfrac{\varphi(k)}{k} & \text{if } \chi = \chi_1. \end{cases}$$

□

12.7 Hurwitz's formula for $\zeta(s, a)$

The function $\zeta(s, a)$ was originally defined for $\sigma > 1$ by an infinite series. Hurwitz obtained another series representation for $\zeta(s, a)$ valid in the half-plane $\sigma < 0$. Before we state this formula we discuss a lemma that will be used in its proof.

Lemma 1 *Let $S(r)$ denote the region that remains when we remove from the z-plane all open circular disks of radius r, $0 < r < \pi$, with centers at $z = 2n\pi i$, $n = 0, \pm 1, \pm 2, \ldots$ Then if $0 < a \leq 1$ the function*

$$g(z) = \frac{e^{az}}{1 - e^{z}}$$

is bounded in $S(r)$. (The bound depends on r.)

Proof. Write $z = x + iy$ and consider the punctured rectangle

$$Q(r) = \{z : |x| \leq 1, |y| \leq \pi, |z| \geq r\},$$

shown in Figure 12.2.

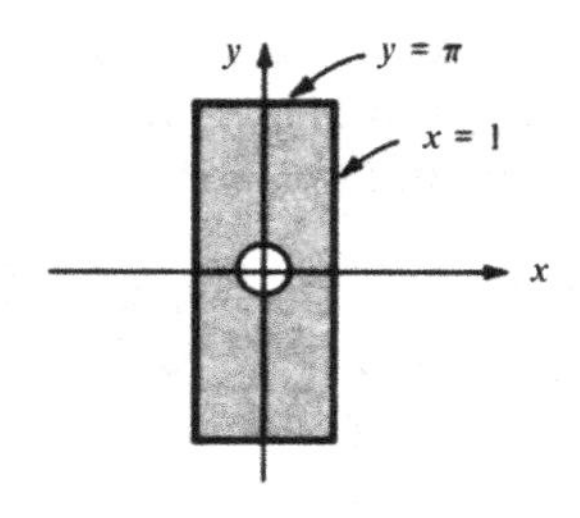

Figure 12.2

This is a compact set so g is bounded on $Q(r)$. Also, since $|g(z + 2\pi i)| = |g(z)|$, g is bounded in the punctured infinite strip

$$\{z : |x| \le 1, |z - 2n\pi i| \ge r, n = 0, \pm 1, \pm 2, \ldots\}.$$

Now we show that g is bounded outside this strip. Suppose $|x| \ge 1$ and consider

$$|g(z)| = \left|\frac{e^{az}}{1 - e^z}\right| = \frac{e^{ax}}{|1 - e^z|} \le \frac{e^{ax}}{|1 - e^x|}.$$

For $x \ge 1$ we have $|1 - e^x| = e^x - 1$ and $e^{ax} \le e^x$, so

$$|g(z)| \le \frac{e^x}{e^x - 1} = \frac{1}{1 - e^{-x}} \le \frac{1}{1 - e^{-1}} = \frac{e}{e - 1}.$$

Also, when $x \le -1$ we have $|1 - e^x| = 1 - e^x$ so

$$|g(z)| \le \frac{e^{ax}}{1 - e^x} \le \frac{1}{1 - e^x} \le \frac{1}{1 - e^{-1}} = \frac{e}{e - 1}.$$

Therefore $|g(z)| \le e/(e - 1)$ for $|x| \ge 1$ and the proof of the lemma is complete. □

We turn now to Hurwitz's formula. This involves another Dirichlet series $F(x, s)$ given by

$$F(x, s) = \sum_{n=1}^{\infty} \frac{e^{2\pi i n x}}{n^s}, \tag{9}$$

where x is real and $\sigma > 1$. Note that $F(x, s)$ is a periodic function of x with period 1 and that $F(1, s) = \zeta(s)$. The series converges absolutely if $\sigma > 1$. If x is not an integer the series also converges (conditionally) for $\sigma > 0$ because for each fixed nonintegral x the coefficients have bounded partial sums.

Note. We shall refer to $F(x, s)$ as the *periodic zeta function.*

Theorem 12.6 Hurwitz's formula. *If* $0 < a \le 1$ *and* $\sigma > 1$ *we have*

$$\zeta(1 - s, a) = \frac{\Gamma(s)}{(2\pi)^s} \{e^{-\pi i s/2} F(a, s) + e^{\pi i s/2} F(-a, s)\}. \tag{10}$$

If $a \ne 1$ *this representation is also valid for* $\sigma > 0$.

PROOF. Consider the function

$$I_N(s, a) = \frac{1}{2\pi i} \int_{C(N)} \frac{z^{s-1} e^{az}}{1 - e^z}\, dz,$$

where $C(N)$ is the contour shown in Figure 12.3, N being an integer.

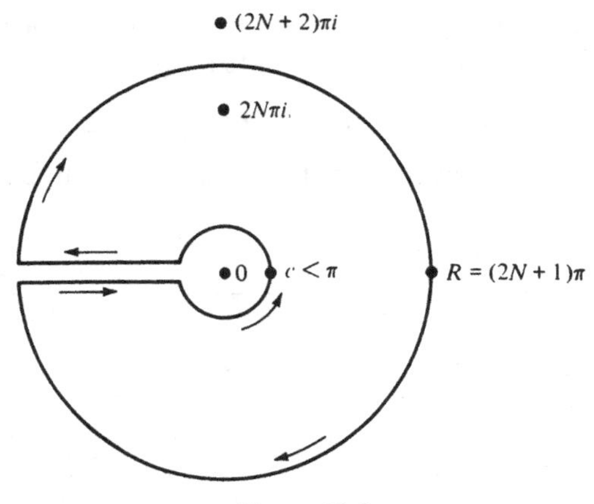

Figure 12.3

First we prove that $\lim_{N\to\infty} I_N(s, a) = I(s, a)$ if $\sigma < 0$. For this it suffices to show that the integral along the outer circle tends to 0 as $N \to \infty$.

On the outer circle we have $z = Re^{i\theta}$, $-\pi \le \theta \le \pi$, hence

$$|z^{s-1}| = |R^{s-1}e^{i\theta(s-1)}| = R^{\sigma-1}e^{-t\theta} \le R^{\sigma-1}e^{\pi|t|}.$$

Since the outer circle lies in the set $S(r)$ of Lemma 1, the integrand is bounded by $Ae^{\pi|t|}R^{\sigma-1}$, where A is the bound for $|g(z)|$ implied by Lemma 1; hence the integral is bounded by

$$2\pi Ae^{\pi|t|}R^{\sigma},$$

and this $\to 0$ as $R \to \infty$ if $\sigma < 0$. Therefore, replacing s by $1 - s$ we see that

$$\lim_{N\to\infty} I_N(1-s, a) = I(1-s, a) \quad \text{if } \sigma > 1. \tag{11}$$

Now we compute $I_N(1-s, a)$ explicitly by Cauchy's residue theorem. We have

$$I_N(1-s, a) = -\sum_{\substack{n=-N \\ n\neq 0}}^{N} R(n) = -\sum_{n=1}^{N} \{R(n) + R(-n)\}$$

where

$$R(n) = \operatorname*{Res}_{z=2n\pi i}\left(\frac{z^{-s}e^{az}}{1-e^z}\right).$$

Now

$$R(n) = \lim_{z\to 2n\pi i}(z - 2n\pi i)\frac{z^{-s}e^{az}}{1-e^z} = \frac{e^{2n\pi ia}}{(2n\pi i)^s}\lim_{z\to 2n\pi i}\frac{z-2n\pi i}{1-e^z} = -\frac{e^{2n\pi ia}}{(2n\pi i)^s},$$

hence

$$I_N(1-s, a) = \sum_{n=1}^{N} \frac{e^{2n\pi i a}}{(2n\pi i)^s} + \sum_{n=1}^{N} \frac{e^{-2n\pi i a}}{(-2n\pi i)^s}.$$

But $i^{-s} = e^{-\pi i s/2}$ and $(-i)^{-s} = e^{\pi i s/2}$ so

$$I_N(1-s, a) = \frac{e^{-\pi i s/2}}{(2\pi)^s} \sum_{n=1}^{N} \frac{e^{2n\pi i a}}{n^s} + \frac{e^{\pi i s/2}}{(2\pi)^s} \sum_{n=1}^{N} \frac{e^{-2n\pi i a}}{n^s}.$$

Letting $N \to \infty$ and using (11) we obtain

$$I(1-s, a) = \frac{e^{-\pi i s/2}}{(2\pi)^s} F(a, s) + \frac{e^{\pi i s/2}}{(2\pi)^s} F(-a, s).$$

Hence

$$\zeta(1-s, a) = \Gamma(s)I(1-s, a) = \frac{\Gamma(s)}{(2\pi)^s} \{e^{-\pi i s/2}F(a, s) + e^{\pi i s/2}F(-a, s)\}. \quad \square$$

12.8 The functional equation for the Riemann zeta function

The first application of Hurwitz's formula is Riemann's functional equation for $\zeta(s)$.

Theorem 12.7 *For all s we have*

$$\zeta(1-s) = 2(2\pi)^{-s}\Gamma(s)\cos\left(\frac{\pi s}{2}\right)\zeta(s) \tag{12}$$

or, equivalently,

$$\zeta(s) = 2(2\pi)^{s-1}\Gamma(1-s)\sin\left(\frac{\pi s}{2}\right)\zeta(1-s). \tag{13}$$

PROOF. Taking $a = 1$ in the Hurwitz formula we obtain, for $\sigma > 1$,

$$\zeta(1-s) = \frac{\Gamma(s)}{(2\pi)^s} \{e^{-\pi i s/2}\zeta(s) + e^{\pi i s/2}\zeta(s)\} = \frac{\Gamma(s)}{(2\pi)^s} 2 \cos\left(\frac{\pi s}{2}\right)\zeta(s).$$

This proves (12) for $\sigma > 1$ and the result holds for all s by analytic continuation. To deduce (13) from (12) replace s by $1 - s$. $\square$

Note. Taking $s = 2n + 1$ in (12) where $n = 1, 2, 3, \ldots$, the factor $\cos(\pi s/2)$ vanishes and we find the so-called *trivial zeros* of $\zeta(s)$,

$$\zeta(-2n) = 0 \quad \text{for } n = 1, 2, 3, \ldots$$

The functional equation can be put in a simpler form if we use Legendre's duplication formula for the gamma function,

$$2\pi^{1/2}2^{-2s}\Gamma(2s) = \Gamma(s)\Gamma\left(s + \frac{1}{2}\right),$$

which is the special case $m = 2$ of Equation (4). When s is replaced by $(1 - s)/2$ this becomes

$$2^s\pi^{1/2}\Gamma(1 - s) = \Gamma\left(\frac{1 - s}{2}\right)\Gamma\left(1 - \frac{s}{2}\right).$$

Since

$$\Gamma\left(\frac{s}{2}\right)\Gamma\left(1 - \frac{s}{2}\right) = \frac{\pi}{\sin \frac{\pi s}{2}}$$

this gives us

$$\Gamma(1 - s) \sin \frac{\pi s}{2} = \frac{2^{-s}\pi^{1/2}\Gamma\left(\frac{1 - s}{2}\right)}{\Gamma\left(\frac{s}{2}\right)}.$$

Using this to replace the product $\Gamma(1 - s)\sin(\pi s/2)$ in (13) we obtain

$$\pi^{-s/2}\Gamma\left(\frac{s}{2}\right)\zeta(s) = \pi^{-(1-s)/2}\Gamma\left(\frac{1 - s}{2}\right)\zeta(1 - s).$$

In other words, the functional equation takes the form

$$\Phi(s) = \Phi(1 - s),$$

where

$$\Phi(s) = \pi^{-s/2}\Gamma\left(\frac{s}{2}\right)\zeta(s).$$

The function $\Phi(s)$ has simple poles at $s = 0$ and $s = 1$. Following Riemann, we multiply $\Phi(s)$ by $s(1 - s)/2$ to remove the poles and define

$$\xi(s) = \frac{1}{2} s(1 - s)\Phi(s).$$

Then $\xi(s)$ is an entire function of s and satisfies the functional equation

$$\xi(s) = \xi(1 - s).$$

12.9 A functional equation for the Hurwitz zeta function

The functional equation for $\zeta(s)$ is a special case of a functional equation for $\zeta(s, a)$ when a is rational.

Theorem 12.8 *If h and k are integers, $1 \le h \le k$, then for all s we have*

$$\zeta\left(1 - s, \frac{h}{k}\right) = \frac{2\Gamma(s)}{(2\pi k)^s} \sum_{r=1}^{k} \cos\left(\frac{\pi s}{2} - \frac{2\pi rh}{k}\right)\zeta\left(s, \frac{r}{k}\right). \tag{14}$$

PROOF. This comes from the fact that the function $F(x, s)$ is a linear combination of Hurwitz zeta functions when x is rational. In fact, if $x = h/k$ we can rearrange the terms in (9) according to the residue classes mod k by writing

$$n = qk + r, \quad \text{where } 1 \le r \le k \text{ and } q = 0, 1, 2, \ldots$$

This gives us, for $\sigma > 1$,

$$F\left(\frac{h}{k}, s\right) = \sum_{n=1}^{\infty} \frac{e^{2\pi inh/k}}{n^s} = \sum_{r=1}^{k} \sum_{q=0}^{\infty} \frac{e^{2\pi irh/k}}{(qk + r)^s} = \frac{1}{k^s} \sum_{r=1}^{k} e^{2\pi irh/k} \sum_{q=0}^{\infty} \frac{1}{\left(q + \frac{r}{k}\right)^s}$$

$$= k^{-s} \sum_{r=1}^{k} e^{2\pi irh/k} \zeta\left(s, \frac{r}{k}\right).$$

Therefore if we take $a = h/k$ in Hurwitz's formula we obtain

$$\zeta\left(1 - s, \frac{h}{k}\right) = \frac{\Gamma(s)}{(2\pi k)^s} \sum_{r=1}^{k} (e^{-\pi is/2} e^{2\pi irh/k} + e^{\pi is/2} e^{-2\pi irh/k}) \zeta\left(s, \frac{r}{k}\right)$$

$$= \frac{2\Gamma(s)}{(2\pi k)^s} \sum_{r=1}^{k} \cos\left(\frac{\pi s}{2} - \frac{2\pi rh}{k}\right)\zeta\left(s, \frac{r}{k}\right),$$

which proves (14) for $\sigma > 1$. The result holds for all s by analytic continuation. □

It should be noted that when $h = k = 1$ there is only one term in the sum in (14) and we obtain Riemann's functional equation.

12.10 The functional equation for L-functions

Hurwitz's formula can also be used to deduce a functional equation for the Dirichlet L-functions. First we show that it suffices to consider only the primitive characters mod k.

Theorem 12.9 *Let χ be any Dirichlet character* mod k, *let d be any induced modulus, and write*

$$\chi(n) = \psi(n)\chi_1(n),$$

where ψ is a character mod d *and χ_1 is the principal character* mod k. *Then for all s we have*

$$L(s, \chi) = L(s, \psi)\prod_{p|k}\left(1 - \frac{\psi(p)}{p^s}\right).$$

Proof. First keep $\sigma > 1$ and use the Euler product

$$L(s, \chi) = \prod_p \frac{1}{1 - \dfrac{\chi(p)}{p^s}}.$$

Since $\chi(p) = \psi(p)\chi_1(p)$ and since $\chi_1(p) = 0$ if $p|k$ and $\chi_1(p) = 1$ if $p \nmid k$ we find

$$\begin{aligned} L(s, \chi) &= \prod_{p\nmid k} \frac{1}{1 - \dfrac{\psi(p)}{p^s}} = \prod_p \frac{1}{1 - \dfrac{\psi(p)}{p^s}} \cdot \prod_{p|k}\left(1 - \frac{\psi(p)}{p^s}\right) \\ &= L(s, \psi)\prod_{p|k}\left(1 - \frac{\psi(p)}{p^s}\right). \end{aligned}$$

This proves the theorem for $\sigma > 1$ and we extend it to all s by analytic continuation. □

Note. If we choose d in the foregoing theorem to be the conductor of χ, then ψ is a primitive character modulo d. This shows that every L-series $L(s, \chi)$ is equal to the L-series $L(s, \psi)$ of a primitive character, multiplied by a finite number of factors.

To deduce the functional equation for L-functions from Hurwitz's formula we first express $L(s, \chi)$ in terms of the periodic zeta function $F(x, s)$.

Theorem 12.10 *Let χ be a primitive character* mod k. *Then for $\sigma > 1$ we have*

$$G(1, \bar{\chi})L(s, \chi) = \sum_{h=1}^{k} \bar{\chi}(h)F\left(\frac{h}{k}, s\right), \tag{15}$$

where $G(m, \chi)$ is the Gauss sum associated with χ,

$$G(m, \chi) = \sum_{r=1}^{k} \chi(r)e^{2\pi irm/k}.$$

Proof. Take $x = h/k$ in (9), multiply by $\bar{\chi}(h)$ and sum on h to obtain

$$\sum_{h=1}^{k} \bar{\chi}(h)F\left(\frac{h}{k}, s\right) = \sum_{h=1}^{k} \sum_{n=1}^{\infty} \bar{\chi}(h)e^{2\pi inh/k}n^{-s} = \sum_{n=1}^{\infty} n^{-s} \sum_{h=1}^{k} \bar{\chi}(h)e^{2\pi inh/k}$$
$$= \sum_{n=1}^{\infty} n^{-s}G(n, \bar{\chi}).$$

But $G(n, \bar{\chi})$ is separable because $\bar{\chi}$ is primitive, so $G(n, \bar{\chi}) = \chi(n)G(1, \bar{\chi})$, hence

$$\sum_{h=1}^{k} \bar{\chi}(h)F\left(\frac{h}{k}, s\right) = G(1, \bar{\chi}) \sum_{n=1}^{\infty} \chi(n)n^{-s} = G(1, \bar{\chi})L(s, \chi). \qquad \square$$

Theorem 12.11 Functional equation for Dirichlet L-functions. *If χ is any primitive character* mod k *then for all s we have*

$$(16) \qquad L(1 - s, \chi) = \frac{k^{s-1}\Gamma(s)}{(2\pi)^s} \{e^{-\pi is/2} + \chi(-1)e^{\pi is/2}\}G(1, \chi)L(s, \bar{\chi}).$$

Proof. We take $x = h/k$ in Hurwitz's formula then multiply each member by $\chi(h)$ and sum on h. This gives us

$$\sum_{h=1}^{k} \chi(h)\zeta\left(1 - s, \frac{h}{k}\right) = \frac{\Gamma(s)}{(2\pi)^s} \left\{e^{-\pi is/2} \sum_{h=1}^{k} \chi(h)F\left(\frac{h}{k}, s\right)\right.$$
$$\left. + e^{\pi is/2} \sum_{h=1}^{k} \chi(h)F\left(\frac{-h}{k}, s\right)\right\}.$$

Since $F(x, s)$ is periodic in x with period 1 and $\chi(h) = \chi(-1)\chi(-h)$ we can write

$$\sum_{h \bmod k} \chi(h)F\left(\frac{-h}{k}, s\right) = \chi(-1) \sum_{h \bmod k} \chi(-h)F\left(\frac{-h}{k}, s\right)$$
$$= \chi(-1) \sum_{h \bmod k} \chi(k - h)F\left(\frac{k - h}{k}, s\right)$$
$$= \chi(-1) \sum_{h \bmod k} \chi(h)F\left(\frac{h}{k}, s\right),$$

and the previous formula becomes

$$\sum_{h=1}^{k} \chi(h)\zeta\left(1 - s, \frac{h}{k}\right) = \frac{\Gamma(s)}{(2\pi)^s} \{e^{-\pi is/2} + \chi(-1)e^{\pi is/2}\} \sum_{h=1}^{k} \chi(h)F\left(\frac{h}{k}, s\right).$$

Now we multiply both members by k^{s-1} and use (15) to obtain (16). $\square$

12.11 Evaluation of $\zeta(-n, a)$

The value of $\zeta(-n, a)$ can be calculated explicitly if n is a nonnegative integer. Taking $s = -n$ in the relation $\zeta(s, a) = \Gamma(1 - s)I(s, a)$ we find

$$\zeta(-n, a) = \Gamma(1 + n)I(-n, a) = n!\,I(-n, a).$$

We also have

$$I(-n, a) = \operatorname*{Res}_{z=0}\left(\frac{z^{-n-1}e^{az}}{1 - e^z}\right).$$

The calculation of this residue leads to an interesting class of functions known as *Bernoulli polynomials.*

Definition For any complex x we define the functions $B_n(x)$ by the equation

$$\frac{ze^{xz}}{e^z - 1} = \sum_{n=0}^{\infty} \frac{B_n(x)}{n!} z^n, \quad \text{where } |z| < 2\pi.$$

The numbers $B_n(0)$ are called Bernoulli numbers and are denoted by B_n. Thus,

$$\frac{z}{e^z - 1} = \sum_{n=0}^{\infty} \frac{B_n}{n!} z^n, \quad \text{where } |z| < 2\pi.$$

Theorem 12.12 *The functions $B_n(x)$ are polynomials in x given by*

$$B_n(x) = \sum_{k=0}^{n} \binom{n}{k} B_k x^{n-k}.$$

Proof. We have

$$\sum_{n=0}^{\infty} \frac{B_n(x)}{n!} z^n = \frac{z}{e^z - 1} \cdot e^{xz} = \left(\sum_{n=0}^{\infty} \frac{B_n}{n!} z^n\right)\left(\sum_{n=0}^{\infty} \frac{x^n}{n!} z^n\right).$$

Equating coefficients of z^n we find

$$\frac{B_n(x)}{n!} = \sum_{k=0}^{n} \frac{B_k}{k!} \frac{x^{n-k}}{(n - k)!}$$

from which the theorem follows. □

Theorem 12.13 *For every integer $n \geq 0$ we have*

$$\zeta(-n, a) = -\frac{B_{n+1}(a)}{n + 1}. \tag{17}$$

PROOF. As noted earlier, we have $\zeta(-n, a) = n!I(-n, a)$. Now

$$I(-n, a) = \operatorname*{Res}_{z=0}\left(\frac{z^{-n-1}e^{az}}{1 - e^z}\right) = -\operatorname*{Res}_{z=0}\left(z^{-n-2}\frac{ze^{az}}{e^z - 1}\right)$$

$$= -\operatorname*{Res}_{z=0}\left(z^{-n-2}\sum_{m=0}^{\infty}\frac{B_m(a)}{m!}z^m\right) = -\frac{B_{n+1}(a)}{(n+1)!},$$

from which we obtain (17). □

12.12 Properties of Bernoulli numbers and Bernoulli polynomials

Theorem 12.14 *The Bernoulli polynomials $B_n(x)$ satisfy the difference equation*

$$B_n(x + 1) - B_n(x) = nx^{n-1} \quad \text{if } n \geq 1. \tag{18}$$

Therefore we have

$$B_n(0) = B_n(1) \quad \text{if } n \geq 2. \tag{19}$$

PROOF. We have the identity

$$z\frac{e^{(x+1)z}}{e^z - 1} - z\frac{e^{xz}}{e^z - 1} = ze^{xz}$$

from which we find

$$\sum_{n=0}^{\infty}\frac{B_n(x + 1) - B_n(x)}{n!}z^n = \sum_{n=0}^{\infty}\frac{x^n}{n!}z^{n+1}.$$

Equating coefficients of z^n we obtain (18). Taking $x = 0$ in (18) we obtain (19).

Theorem 12.15 *If $n \geq 2$ we have*

$$B_n = \sum_{k=0}^{n}\binom{n}{k}B_k.$$

PROOF. This follows by taking $x = 1$ in Theorem 12.12 and using (19). □

Theorem 12.15 gives a recursion formula for computing Bernoulli numbers. The definition gives $B_0 = 1$, and Theorem 12.15 yields in succession the values

$$B_0 = 1, \quad B_1 = -\frac{1}{2}, \quad B_2 = \frac{1}{6}, \quad B_3 = 0, \quad B_4 = -\frac{1}{30},$$

$$B_5 = 0, \quad B_6 = \frac{1}{42}, \quad B_7 = 0, \quad B_8 = -\frac{1}{30}, \quad B_9 = 0,$$

$$B_{10} = \frac{5}{66}, \quad B_{11} = 0.$$

From a knowledge of the B_k we can compute the polynomials $B_n(x)$ by using Theorem 12.12. The first few are:

$$B_0(x) = 1, \qquad B_1(x) = x - \frac{1}{2}, \qquad B_2(x) = x^2 - x + \frac{1}{6},$$

$$B_3(x) = x^3 - \frac{3}{2}x^2 + \frac{1}{2}x, \qquad B_4(x) = x^4 - 2x^3 + x^2 - \frac{1}{30}.$$

We observe that Theorems 12.12 and 12.15 can be written symbolically as follows:

$$B_n(x) = (B + x)^n, \qquad B_n = (B + 1)^n.$$

In these symbolic formulas the right members are to be expanded by the binomial theorem, then each power B^k is to be replaced by B_k.

Theorem 12.16 *If $n \geq 0$ we have*

$$\zeta(-n) = -\frac{B_{n+1}(1)}{n+1} = \begin{cases} -\dfrac{1}{2} & \text{if } n = 0, \\ -\dfrac{B_{n+1}}{n+1} & \text{if } n \geq 1. \end{cases} \tag{20}$$

Also, if $n \geq 1$ we have $\zeta(-2n) = 0$, hence $B_{2n+1} = 0$.

PROOF. To evaluate $\zeta(-n)$ we simply take $a = 1$ in Theorem 12.13. We have already noted that the functional equation

$$\zeta(1 - s) = 2(2\pi)^{-s}\Gamma(s)\cos\left(\frac{\pi s}{2}\right)\zeta(s) \tag{21}$$

implies $\zeta(-2n) = 0$ for $n \geq 1$, hence $B_{2n+1} = 0$ by (20). □

Note. The result $B_{2n+1} = 0$ also follows by noting that the left member of

$$\frac{z}{e^z - 1} + \frac{1}{2}z = 1 + \sum_{n=2}^{\infty} \frac{B_n}{n!} z^n$$

is an even function of z.

Theorem 12.17 *If k is a positive integer we have*

$$\zeta(2k) = (-1)^{k+1} \frac{(2\pi)^{2k} B_{2k}}{2(2k)!}. \tag{22}$$

PROOF. We take $s = 2k$ in the functional equation for $\zeta(s)$ to obtain

$$\zeta(1 - 2k) = 2(2\pi)^{-2k}\Gamma(2k)\cos(\pi k)\zeta(2k),$$

or

$$-\frac{B_{2k}}{2k} = 2(2\pi)^{-2k}(2k - 1)!(-1)^k\zeta(2k).$$

This implies (22). □

Note. If we put $s = 2k + 1$ in (21) both members vanish and we get no information about $\zeta(2k + 1)$. As yet no simple formula analogous to (22) is known for $\zeta(2k + 1)$ or even for any special case such as $\zeta(3)$. It is not even known whether $\zeta(2k + 1)$ is rational or irrational for any k except for $\zeta(3)$, which was recently proved irrational by R. Apéry (Astérisque 61 (1979), 11–13).

Theorem 12.18 *The Bernoulli numbers B_{2k} alternate in sign. That is,*

$$(-1)^{k+1}B_{2k} > 0.$$

Moreover, $|B_{2k}| \to \infty$ as $k \to \infty$. In fact

$$(-1)^{k+1}B_{2k} \sim \frac{2(2k)!}{(2\pi)^{2k}} \quad \text{as } k \to \infty. \tag{23}$$

PROOF. Since $\zeta(2k) > 0$, (22) shows that the numbers B_{2k} alternate in sign. The asymptotic relation (23) follows from the fact that $\zeta(2k) \to 1$ as $k \to \infty$.

Note. From (23) it follows that $|B_{2k+2}/B_{2k}| \sim k^2/\pi^2$ as $k \to \infty$. Also, by invoking Stirling's formula, $n! \sim (n/e)^n\sqrt{2\pi n}$ we find

$$(-1)^{k+1}B_{2k} \sim 4\pi\sqrt{e}\left(\frac{k}{\pi e}\right)^{2k+1/2} \quad \text{as } k \to \infty.$$

The next theorem gives the Fourier expansion of the polynomial $B_n(x)$ in the interval $0 < x \leq 1$.

Theorem 12.19 *If $n = 1$ and $0 < x < 1$, or if $n \geq 2$ and $0 \leq x \leq 1$, we have*

$$B_n(x) = -\frac{n!}{(2\pi i)^n}\sum_{\substack{k=-\infty \\ k \neq 0}}^{+\infty}\frac{e^{2\pi ikx}}{k^n}, \tag{24}$$

and hence

$$B_{2n}(x) = (-1)^{n+1}\frac{2(2n)!}{(2\pi)^{2n}}\sum_{k=1}^{\infty}\frac{\cos 2\pi kx}{k^{2n}},$$

$$B_{2n+1}(x) = (-1)^{n+1}\frac{2(2n+1)!}{(2\pi)^{2n+1}}\sum_{k=1}^{\infty}\frac{\sin 2\pi kx}{k^{2n+1}}.$$

PROOF. Equation (24) follows at once by taking $s = n$ in Hurwitz's formula and applying Theorem 12.13. The other two formulas are special cases of (24).

Note. The function $\bar{B}_n(x)$ defined for all real x by the right member of (24) is called the nth *Bernoulli periodic function.* It is periodic with period 1 and agrees with the Bernoulli polynomial $B_n(x)$ in the interval $0 < x \leq 1$. Thus we have

$$\bar{B}_n(x) = B_n(x - [x]).$$

12.13 Formulas for $L(0, \chi)$

Theorem 12.13 implies

$$\zeta(0, a) = -B_1(a) = \frac{1}{2} - a.$$

In particular $\zeta(0) = \zeta(0, 1) = -1/2$. We can also calculate $L(0, \chi)$ for every Dirichlet character χ.

Theorem 12.20 *Let χ be any Dirichlet character* mod k.

(a) *If $\chi = \chi_1$ (the principal character), then $L(0, \chi_1) = 0$.*

(b) *If $\chi \neq \chi_1$ we have*

$$L(0, \chi) = -\frac{1}{k}\sum_{r=1}^{k} r\chi(r).$$

Moreover, $L(0, \chi) = 0$ if $\chi(-1) = 1$.

PROOF. If $\chi = \chi_1$ we use the formula

$$L(s, \chi_1) = \zeta(s)\prod_{p|k}(1 - p^{-s})$$

proved for $\sigma > 1$ in Chapter 11. This also holds for all s by analytic continuation. When $s = 0$ the product vanishes so $L(0, \chi_1) = 0$.

If $\chi \neq \chi_1$ we have

$$L(0, \chi) = \sum_{r=1}^{k} \chi(r)\zeta\left(0, \frac{r}{k}\right) = \sum_{r=1}^{k} \chi(r)\left(\frac{1}{2} - \frac{r}{k}\right) = -\frac{1}{k}\sum_{r=1}^{k} r\chi(r).$$

Now

$$\sum_{r=1}^{k} r\chi(r) = \sum_{r=1}^{k} (k - r)\chi(k - r) = k\sum_{r=1}^{k} \chi(k - r) - \sum_{r=1}^{k} r\chi(-r)$$

$$= -\chi(-1)\sum_{r=1}^{k} r\chi(r).$$

Therefore if $\chi(-1) = 1$ we have $\sum_{r=1}^{k} r\chi(r) = 0$. □

12.14 Approximation of $\zeta(s, a)$ by finite sums

Some applications require estimates on the rate of growth of $\zeta(\sigma + it, a)$ as a function of t. These will be deduced from another representation of $\zeta(s, a)$ obtained from Euler's summation formula. This relates $\zeta(s, a)$ to the partial sums of its series in the half-plane $\sigma > 0$ and also gives an alternate way to extend $\zeta(s, a)$ analytically beyond the line $\sigma = 1$.

Theorem 12.21 *For any integer $N \geq 0$ and $\sigma > 0$ we have*

$$\zeta(s, a) = \sum_{n=0}^{N} \frac{1}{(n+a)^s} + \frac{(N+a)^{1-s}}{s-1} - s\int_N^\infty \frac{x-[x]}{(x+a)^{s+1}}\,dx. \tag{25}$$

PROOF. We apply Euler's summation formula (Theorem 3.1) with $f(t) = (t+a)^{-s}$ and with integers x and y to obtain

$$\sum_{y<n\leq x} \frac{1}{(n+a)^s} = \int_y^x \frac{dt}{(t+a)^s} - s\int_y^x \frac{t-[t]}{(t+a)^{s+1}}\,dt.$$

Take $y = N$ and let $x \to \infty$, keeping $\sigma > 1$. This gives us

$$\sum_{n=N+1}^{\infty} \frac{1}{(n+a)^s} = \int_N^\infty \frac{dt}{(t+a)^s} - s\int_N^\infty \frac{t-[t]}{(t+a)^{s+1}}\,dt,$$

or

$$\zeta(s, a) - \sum_{n=0}^{N} \frac{1}{(n+a)^s} = \frac{(N+a)^{1-s}}{s-1} - s\int_N^\infty \frac{t-[t]}{(t+a)^{s+1}}\,dt.$$

This proves (25) for $\sigma > 1$. If $\sigma \geq \delta > 0$ the integral is dominated by $\int_N^\infty (t+a)^{-\delta-1}\,dt$ so it converges uniformly for $\sigma \geq \delta$ and hence represents an analytic function in the half-plane $\sigma > 0$. Therefore (25) holds for $\sigma > 0$ by analytic continuation. □

The integral on the right of (25) can also be written as a series. We split the integral into a sum of integrals in which $[x]$ is constant, say $[x] = n$, and we obtain

$$\int_N^\infty \frac{x-[x]}{(x+a)^{s+1}}\,dx = \sum_{n=N}^{\infty} \int_n^{n+1} \frac{x-n}{(x+a)^{s+1}}\,dx = \sum_{n=N}^{\infty} \int_0^1 \frac{u}{(u+n+a)^{s+1}}\,du.$$

Therefore (25) can also be written in the form

$$\zeta(s, a) - \sum_{n=0}^{N} \frac{1}{(n+a)^s} = \frac{(N+a)^{1-s}}{s-1} - s\sum_{n=N}^{\infty} \int_0^1 \frac{u}{(u+n+a)^{s+1}}\,du \tag{26}$$

if $\sigma > 0$. Integration by parts leads to similar representations in successively larger half-planes, as indicated in the next theorem.

Theorem 12.22 *If $\sigma > -1$ we have*

$$\begin{aligned}\zeta(s, a) - \sum_{n=0}^{N} \frac{1}{(n+a)^s} = {} & \frac{(N+a)^{1-s}}{s-1} \\ & - \frac{s}{2!}\left\{\zeta(s+1, a) - \sum_{n=0}^{N} \frac{1}{(n+a)^{s+1}}\right\} \\ & - \frac{s(s+1)}{2!}\sum_{n=N}^{\infty} \int_0^1 \frac{u^2}{(n+a+u)^{s+2}}\,du.\end{aligned} \tag{27}$$

More generally, if $\sigma > -m$, where $m = 1, 2, 3, \ldots$, we have

$$(28)\quad \zeta(s, a) - \sum_{n=0}^{N} \frac{1}{(n+a)^s} = \frac{(N+a)^{1-s}}{s-1} - \sum_{r=1}^{m} \frac{s(s+1)\cdots(s+r-1)}{(r+1)!} \times \left\{\zeta(s+r, a) - \sum_{n=0}^{N} \frac{1}{(n+a)^{s+r}}\right\} - \frac{s(s+1)\cdots(s+m)}{(m+1)!} \times \sum_{n=N}^{\infty} \int_0^1 \frac{u^{m+1}}{(n+a+u)^{s+m+1}}\,du.$$

PROOF. Integration by parts implies

$$\int \frac{u\,du}{(n+a+u)^{s+1}} = \frac{u^2}{2(n+a+u)^{s+1}} + \frac{s+1}{2}\int \frac{u^2\,du}{(n+a+u)^{s+2}},$$

so if $\sigma > 0$ we have

$$\sum_{n=N}^{\infty} \int_0^1 \frac{u\,du}{(n+a+u)^{s+1}} = \frac{1}{2}\sum_{n=N}^{\infty} \frac{1}{(n+a+1)^{s+1}} + \frac{s+1}{2}\sum_{n=N}^{\infty} \int_0^1 \frac{u^2\,du}{(n+a+u)^{s+2}}.$$

But if $\sigma > 0$ the first sum on the right is $\zeta(s+1, a) - \sum_{n=0}^{N} (n+a)^{-s-1}$ and (26) implies (27). The result is also valid for $\sigma > -1$ by analytic continuation. By repeated integration by parts we obtain the more general representation in (28). □

12.15 Inequalities for $|\zeta(s, a)|$

The formulas in the foregoing section yield upper bounds for $|\zeta(\sigma + it, a)|$ as a function of t.

Theorem 12.23 (a) *If $\delta > 0$ we have*

$$(29)\qquad |\zeta(s, a) - a^{-s}| \le \zeta(1+\delta) \quad \text{if } \sigma \ge 1+\delta.$$

(b) *If $0 < \delta < 1$ there is a positive constant $A(\delta)$, depending on δ but not on s or a, such that*

$$(30)\quad |\zeta(s, a) - a^{-s}| \le A(\delta)|t|^{\delta} \quad \text{if } 1-\delta \le \sigma \le 2 \text{ and } |t| \ge 1,$$

$$(31)\quad |\zeta(s, a) - a^{-s}| \le A(\delta)|t|^{1+\delta} \quad \text{if } -\delta \le \sigma \le \delta \text{ and } |t| \ge 1,$$

$$(32)\quad |\zeta(s, a)| \le A(\delta)|t|^{m+1+\delta} \quad \text{if } -m-\delta \le \sigma \le -m+\delta \text{ and } |t| \ge 1,$$

where $m = 1, 2, 3, \ldots$

Proof. For part (a) we use the defining series for $\zeta(s, a)$ to obtain

$$|\zeta(s, a) - a^{-s}| \leq \sum_{n=1}^{\infty} \frac{1}{(n + a)^{\sigma}} \leq \sum_{n=1}^{\infty} \frac{1}{n^{1+\delta}} = \zeta(1 + \delta),$$

which implies (29).

For part (b) we use the representation in (25) when $1 - \delta \leq \sigma \leq 2$ to obtain

$$\begin{aligned} |\zeta(s, a) - a^{-s}| &\leq \sum_{n=1}^{N} \frac{1}{(n + a)^{\sigma}} + \frac{(N + a)^{1-\sigma}}{|s - 1|} + |s| \int_N^{\infty} \frac{dx}{(x + a)^{\sigma+1}} \\ &< 1 + \int_1^N \frac{dx}{(x + a)^{\sigma}} + \frac{(N + a)^{1-\sigma}}{|s - 1|} + \frac{|s|}{\sigma}(N + a)^{-\sigma}. \end{aligned}$$

Since $\sigma \geq 1 - \delta > 0$ we have $(x + a)^{\sigma} \geq (x + a)^{1-\delta} > x^{1-\delta}$ so

$$\int_1^N \frac{dx}{(x + a)^{\sigma}} \leq \int_1^N \frac{dx}{x^{1-\delta}} < \frac{N^{\delta}}{\delta}.$$

Also, since $|s - 1| = |\sigma - 1 + it| \geq |t| \geq 1$ we have

$$\frac{(N + a)^{1-\sigma}}{|s - 1|} \leq (N + a)^{\delta} \leq (N + 1)^{\delta}.$$

Finally, since $|s| \leq |\sigma| + |t| \leq 2 + |t|$ we find

$$\frac{|s|}{\sigma}(N + a)^{-\sigma} < \frac{2 + |t|}{1 - \delta}(N + a)^{\delta-1} < \frac{2 + |t|}{1 - \delta}\frac{1}{N^{1-\delta}}.$$

These give us

$$|\zeta(s, a) - a^{-s}| < 1 + \frac{N^{\delta}}{\delta} + (N + 1)^{\delta} + \frac{2 + |t|}{1 - \delta}\frac{N^{\delta}}{N}.$$

Now take $N = 1 + [|t|]$. Then the last three terms are $O(|t|^{\delta})$, where the constant implied by the O-symbol depends only on δ. This proves (30).

To prove (31) we use the representation in (27). This gives us

$$\begin{aligned} |\zeta(s, a) - a^{-s}| \leq \sum_{n=1}^{N} \frac{1}{(n + a)^{\sigma}} &+ \frac{(N + a)^{1-\sigma}}{|s - 1|} + \frac{1}{2}|s|\{|\zeta(s + 1), a) - a^{-s-1}|\} \\ &+ \frac{1}{2}|s|\sum_{n=1}^{N} \frac{1}{(n + a)^{\sigma+1}} + \frac{1}{2}|s||s + 1|\sum_{n=N}^{\infty} \frac{1}{(n + a)^{\sigma+2}}. \end{aligned}$$

As in the proof of (30) we take $N = 1 + [|t|]$ so that $N = O(|t|)$ and we show that each term on the right is $O(|t|^{1+\delta})$, where the constant implied by the O-symbol depends only on δ. The inequalities $-\delta \le \sigma \le \delta$ imply $1 - \delta \le 1 - \sigma \le 1 + \delta$, hence

$$\sum_{n=1}^{N} \frac{1}{(n+a)^\sigma} < 1 + \int_1^N \frac{dx}{(x+a)^\sigma} < 1 + \frac{(N+a)^{1-\sigma}}{1-\sigma}$$
$$\le 1 + \frac{(N+1)^{1+\delta}}{1-\delta} = O(|t|^{1+\delta}).$$

Since $|s - 1| \ge |t| \ge 1$ the second term is also $O(|t|^{1+\delta})$. For the third term we use (30), noting that $1 - \delta \le \sigma + 1 \le 1 + \delta$ and $|s| = O(|t|)$, and we find that this term is also $O(|t|^{1+\delta})$. Next, we have

$$|s| \sum_{n=1}^{N} \frac{1}{(n+a)^{\sigma+1}} = O\left(|t| \int_1^N \frac{dx}{(x+a)^{1-\delta}}\right)$$
$$= O(|t| N^\delta) = O(|t|^{1+\delta}).$$

Finally,

$$|s||s+1| \sum_{n=N}^{\infty} \frac{1}{(n+a)^{\sigma+2}} = O\left(|t|^2 \int_N^\infty \frac{dx}{(x+a)^{\sigma+2}}\right) = O(|t|^2 N^{-\sigma-1})$$
$$= O(|t|^2 N^{\delta-1}) = O(|t|^{1+\delta}).$$

This completes the proof of (31).

The proof of (32) is similar, except that we use (28) and note that $a^{-\sigma} = O(1)$ when $\sigma < 0$. □

12.16 Inequalities for $|\zeta(s)|$ and $|L(s, \chi)|$

When $a = 1$ the estimates in Theorem 12.23 give corresponding estimates for $|\zeta(s)|$. They also lead to bounds for Dirichlet L-series. If $\sigma \ge 1 + \delta$, where $\delta > 0$, both $|\zeta(s)|$ and $|L(s, \chi)|$ are dominated by $\zeta(1 + \delta)$ so we consider only $\sigma \le 1 + \delta$.

Theorem 12.24 *Let χ be any Dirichlet character* mod k *and assume* $0 < \delta < 1$. *Then there is a positive constant $A(\delta)$, depending on δ but not on s or k, such that for $s = \sigma + it$ with $|t| \ge 1$ we have*

(33) $$|L(s, \chi)| \le A(\delta)|kt|^{m+1+\delta} \quad \text{if } -m-\delta \le \sigma \le -m+\delta,$$

where $m = -1, 0, 1, 2, \ldots$

PROOF. We recall the relation

$$L(s, \chi) = k^{-s} \sum_{r=1}^{k-1} \chi(r) \zeta\left(s, \frac{r}{k}\right).$$

If $m = 1, 2, 3, \ldots$ we use (32) to obtain

$$|L(s, \chi)| \leq k^{-\sigma} \sum_{r=1}^{k-1} \left|\zeta\left(s, \frac{r}{k}\right)\right| < k^{m+\delta} k A(\delta) |t|^{m+1+\delta}$$

which proves (33) for $m \geq 1$. If $m = 0$ or -1 we write

$$L(s, \chi) = \sum_{r=1}^{k-1} \frac{\chi(r)}{r^s} + k^{-s} \sum_{r=1}^{k-1} \chi(r) \left\{\zeta\left(s, \frac{r}{k}\right) - \left(\frac{r}{k}\right)^{-s}\right\}. \tag{34}$$

Since $-m - \delta \leq \sigma \leq -m + \delta$ we can use (30) and (31) to obtain

$$k^{-\sigma} \left|\zeta\left(s, \frac{r}{k}\right) - \left(\frac{r}{k}\right)^{-s}\right| \leq k^{m+\delta} A(\delta) |t|^{m+1+\delta},$$

so the second sum in (34) is dominated by $A(\delta)|kt|^{m+1+\delta}$. The first sum is dominated by

$$\sum_{r=1}^{k-1} \frac{1}{r^\sigma} \leq \sum_{r=1}^{k-1} r^{m+\delta} < 1 + \int_1^k x^{m+\delta}\, dx = \frac{k^{m+1+\delta}}{m+1+\delta} \leq \frac{k^{m+1+\delta}}{\delta},$$

and this sum can also be absorbed in the estimate $A(\delta)|kt|^{m+1+\delta}$. □

Exercises for Chapter 12

1. Let $f(n)$ be an arithmetical function which is periodic modulo k.

   (a) Prove that the Dirichlet series $\sum f(n) n^{-s}$ converges absolutely for $\sigma > 1$ and that
   $$\sum_{n=1}^{\infty} \frac{f(n)}{n^s} = k^{-s} \sum_{r=1}^{k} f(r) \zeta\left(s, \frac{r}{k}\right) \quad \text{if } \sigma > 1.$$

   (b) If $\sum_{r=1}^{k} f(r) = 0$ prove that the Dirichlet series $\sum f(n) n^{-s}$ converges for $\sigma > 0$ and that there is an entire function $F(s)$ such that $F(s) = \sum f(n) n^{-s}$ for $\sigma > 0$.

2. If x is real and $\sigma > 1$, let $F(x, s)$ denote the periodic zeta function,
   $$F(x, s) = \sum_{n=1}^{\infty} \frac{e^{2\pi i n x}}{n^s}.$$
   If $0 < a < 1$ and $\sigma > 1$ prove that Hurwitz's formula implies
   $$F(a, s) = \frac{\Gamma(1-s)}{(2\pi)^{1-s}} \{e^{\pi i(1-s)/2} \zeta(1-s, a) + e^{\pi i(s-1)/2} \zeta(1-s, 1-a)\}.$$

3. The formula in Exercise 2 can be used to extend the definition of $F(a, s)$ over the entire s-plane if $0 < a < 1$. Prove that $F(a, s)$, so extended, is an entire function of s.

4. If $0 < a < 1$ and $0 < b < 1$ let
   $$\Phi(a, b, s) = \frac{\Gamma(s)}{(2\pi)^s} \{\zeta(s, a) F(b, 1+s) + \zeta(s, 1-a) F(1-b, 1+s)\},$$

where F is the function in Exercise 2. Prove that

$$\frac{\Phi(a, b, s)}{\Gamma(s)\Gamma(-s)} = e^{\pi is/2}\{\zeta(s, a)\zeta(-s, 1-b) + \zeta(s, 1-a)\zeta(-s, b)\}$$

$$+ e^{-\pi is/2}\{\zeta(-s, 1-b)\zeta(s, 1-a) + \zeta(-s, b)\zeta(s, a)\},$$

and deduce that $\Phi(a, b, s) = \Phi(1-b, a, -s)$. This functional equation is useful in the theory of elliptic modular functions.

In Exercises 5, 6 and 7, $\xi(s)$ denotes the entire function introduced in Section 12.8,

$$\xi(s) = \frac{1}{2}s(s-1)\pi^{-s/2}\Gamma\left(\frac{s}{2}\right)\zeta(s).$$

5. Prove that $\xi(s)$ is real on the lines $t = 0$ and $\sigma = 1/2$, and that $\xi(0) = \xi(1) = 1/2$.
6. Prove that the zeros of $\xi(s)$ (if any exist) are all situated in the strip $0 \le \sigma \le 1$ and lie symmetrically about the lines $t = 0$ and $\sigma = 1/2$.
7. Show that the zeros of $\zeta(s)$ in the critical strip $0 < \sigma < 1$ (if any exist) are identical in position and order of multiplicity with those of $\xi(s)$.
8. Let χ be a primitive character mod k. Define

$$a = a(\chi) = \begin{cases} 0 & \text{if } \chi(-1) = 1, \\ 1 & \text{if } \chi(-1) = -1. \end{cases}$$

(a) Show that the functional equation for $L(s, \chi)$ has the form

$$L(1-s, \bar{\chi}) = \varepsilon(\chi)2(2\pi)^{-s}k^{s-\frac{1}{2}}\cos\left(\frac{\pi(s-a)}{2}\right)\Gamma(s)L(s, \chi), \text{ where } |\varepsilon(\chi)| = 1.$$

(b) Let

$$\xi(s, \chi) = \left(\frac{k}{\pi}\right)^{(s+a)/2}\Gamma\left(\frac{s+a}{2}\right)L(s, \chi).$$

Show that $\xi(1-s, \bar{\chi}) = \varepsilon(\chi)\xi(s, \chi)$.

9. Refer to Exercise 8.

(a) Prove that $\xi(s, \chi) \ne 0$ if $\sigma > 1$ or $\sigma < 0$.
(b) Describe the location of the zeros of $L(s, \chi)$ in the half-plane $\sigma < 0$.

10. Let χ be a nonprimitive character modulo k. Describe the location of the zeros of $L(s, \chi)$ in the half-plane $\sigma < 0$.
11. Prove that the Bernoulli polynomials satisfy the relations

$$B_n(1-x) = (-1)^n B_n(x) \text{ and } B_{2n+1}(\tfrac{1}{2}) = 0 \text{ for every } n \ge 0.$$

12. Let B_n denote the nth Bernoulli number. Note that

$$B_2 = \tfrac{1}{6} = 1 - \tfrac{1}{2} - \tfrac{1}{3}, \qquad B_4 = \tfrac{-1}{30} = 1 - \tfrac{1}{2} - \tfrac{1}{3} - \tfrac{1}{5},$$
$$B_6 = \tfrac{1}{42} = 1 - \tfrac{1}{2} - \tfrac{1}{3} - \tfrac{1}{7}.$$

These formulas illustrate a theorem discovered in 1840 by von Staudt and Clausen (independently). If $n \geq 1$ we have

$$B_{2n} = I_n - \sum_{p-1|2n} \frac{1}{p},$$

where I_n is an integer and the sum is over all primes p such that $p - 1$ divides $2n$. This exercise outlines a proof due to Lucas.

(a) Prove that

$$B_n = \sum_{k=0}^{n} \frac{1}{k+1} \sum_{r=0}^{k} (-1)^r \binom{k}{r} r^n.$$

[*Hint:* Write $x = \log\{1 + (e^x - 1)\}$ and use the power series for $x/(e^x - 1)$.]

(b) Prove that

$$B_n = \sum_{k=0}^{n} \frac{k!}{k+1} c(n, k),$$

where $c(n, k)$ is an integer.

(c) If a, b are integers with $a \geq 2$, $b \geq 2$ and $ab > 4$, prove that $ab\,|\,(ab - 1)!$. This shows that in the sum of part (b), every term with $k + 1$ composite, $k > 3$, is an integer.

(d) If p is prime, prove that

$$\sum_{r=0}^{p-1} (-1)^r \binom{p-1}{r} r^n \equiv \begin{cases} -1 \pmod p & \text{if } p - 1\,|\,n,\ n > 0, \\ \ \ 0 \pmod p & \text{if } p - 1 \nmid n. \end{cases}$$

(e) Use the above results or some other method to prove the von Staudt–Clausen theorem.

13. Prove that the derivative of the Bernoulli polynomial $B_n(x)$ is $nB_{n-1}(x)$ if $n \geq 2$.

14. Prove that the Bernoulli polynomials satisfy the addition formula

$$B_n(x + y) = \sum_{k=0}^{n} \binom{n}{k} B_k(x) y^{n-k}.$$

15. Prove that the Bernoulli polynomials satisfy the multiplication formula

$$B_p(mx) = m^{p-1} \sum_{k=0}^{m-1} B_p\left(x + \frac{k}{m}\right).$$

16. Prove that if $r \geq 1$ the Bernoulli numbers satisfy the relation

$$\sum_{k=0}^{r} \frac{2^{2k} B_{2k}}{(2k)!(2r + 1 - 2k)!} = \frac{1}{(2r)!}.$$

17. Calculate the integral $\int_0^1 xB_p(x)\,dx$ in two ways and deduce the formula

$$\sum_{r=0}^{p} \binom{p}{r} \frac{B_r}{p + 2 - r} = \frac{B_{p+1}}{p + 1}.$$

18. (a) Verify the identity

$$\frac{uv}{(e^u - 1)(e^v - 1)} \frac{e^{u+v} - 1}{u + v} = \frac{uv}{u + v}\left(1 + \frac{1}{e^u - 1} + \frac{1}{e^v - 1}\right)$$

$$= 1 + \sum_{n=2}^{\infty} \frac{uv}{n!}\left(\frac{u^{n-1} + v^{n-1}}{u + v}\right)B_n.$$

(b) Let $J = \int_0^1 B_p(x)B_q(x)\,dx$. Show that J is the coefficient of $p!q!u^pv^q$ in the expansion of part (a). Use this to deduce that

$$\int_0^1 B_p(x)B_q(x)\,dx = \begin{cases} (-1)^{p+1}\dfrac{p!q!}{(p+q)!}B_{p+q} & \text{if } p \geq 1, q \geq 1, \\ 1 & \text{if } p = q = 0, \\ 0 & \text{if } p \geq 1, q = 0; \text{ or } p = 0, q \geq 1. \end{cases}$$

19. (a) Use a method similar to that in Exercise 18 to derive the identity

$$(u + v)\sum_{m=0}^{\infty}\sum_{n=0}^{\infty} B_m(x)B_n(x)\frac{u^mv^n}{m!n!} = \sum_{m=0}^{\infty}\sum_{n=0}^{\infty} B_{m+n}(x)\frac{u^mv^n}{m!n!}\sum_{r=0}^{\infty}\frac{B_{2r}}{(2r)!}(u^{2r}v + uv^{2r}).$$

(b) Compare coefficients in (a) and integrate the result to obtain the formula

$$B_m(x)B_n(x) = \sum_r \left\{\binom{m}{2r}n + \binom{n}{2r}m\right\}\frac{B_{2r}B_{m+n-2r}(x)}{m + n - 2r} + (-1)^{m+1}\frac{m!n!}{(m+n)!}B_{m+n}$$

for $m \geq 1, n \geq 1$. Indicate the range of the index r.

20. Show that if $m \geq 1, n \geq 1$ and $p \geq 1$, we have

$$\int_0^1 B_m(x)B_n(x)B_p(x)\,dx$$

$$= (-1)^{p+1}p!\sum_r \left\{\binom{m}{2r}n + \binom{n}{2r}m\right\}\frac{(m + n - 2r - 1)!}{(m + n + p - 2r)!}B_{2r}B_{m+n+p-2r}.$$

In particular, compute $\int_0^1 B_2{}^3(x)\,dx$ from this formula.

21. Let $f(n)$ be an arithmetical function which is periodic mod k, and let

$$g(n) = \frac{1}{k}\sum_{m \bmod k} f(m)e^{-2\pi imn/k}$$

denote the finite Fourier coefficients of f. If

$$F(s) = k^{-s}\sum_{r=1}^{k} f(r)\zeta\left(s, \frac{r}{k}\right),$$

prove that

$$F(1 - s) = \frac{\Gamma(s)}{(2\pi)^s}\left\{e^{\pi is/2}\sum_{r=1}^{k} g(r)\zeta\left(s, \frac{r}{k}\right) + e^{-\pi is/2}\sum_{r=1}^{k} g(-r)\zeta\left(s, \frac{r}{k}\right)\right\}.$$

22. Let χ be any nonprincipal character mod k and let $S(x) = \sum_{n \le x} \chi(n)$.

(a) If $N \ge 1$ and $\sigma > 0$ prove that

$$L(s, \chi) = \sum_{n=1}^{N} \frac{\chi(n)}{n^s} + s \int_N^{\infty} \frac{S(x) - S(N)}{x^{s+1}}\, dx.$$

(b) If $s = \sigma + it$ with $\sigma \ge \delta > 0$ and $|t| \ge 0$, use part (a) to show that there is a constant $A(\delta)$ such that, if $\delta < 1$,

$$|L(s, \chi)| \le A(\delta)B(k)(|t| + 1)^{1-\delta}$$

where $B(k)$ is an upper bound for $|S(x)|$. In Theorem 13.15 it is shown that $B(k) = O(\sqrt{k} \log k)$.

(c) Prove that for some constant $A > 0$ we have

$$|L(s, \chi)| \le A \log k \quad \text{if } \sigma \ge 1 - \frac{1}{\log k} \text{ and } 0 \le |t| \le 2.$$

[*Hint*: Take $N = k$ in part (a).]

13 Analytic Proof of the Prime Number Theorem

13.1 The plan of the proof

The prime number theorem is equivalent to the statement

$$\psi(x) \sim x \quad \text{as } x \to \infty, \tag{1}$$

where $\psi(x)$ is Chebyshev's function,

$$\psi(x) = \sum_{n \leq x} \Lambda(n).$$

This chapter gives an analytic proof of (1) based on properties of the Riemann zeta function. The analytic proof is shorter than the elementary proof sketched in Chapter 4 and its principal ideas are easier to comprehend. This section outlines the main features of the proof.

The function ψ is a step function and it is more convenient to deal with its integral, which we denote by ψ_1. Thus, we consider

$$\psi_1(x) = \int_1^x \psi(t)\, dt.$$

The integral ψ_1 is a continuous piecewise linear function. We show first that the asymptotic relation

$$\psi_1(x) \sim \frac{1}{2} x^2 \quad \text{as } x \to \infty \tag{2}$$

implies (1) and then prove (2). For this purpose we express $\psi_1(x)/x^2$ in terms of the Riemann zeta function by means of a contour integral,

$$\frac{\psi_1(x)}{x^2} = \frac{1}{2\pi i} \int_{c-\infty i}^{c+\infty i} \frac{x^{s-1}}{s(s+1)} \left(-\frac{\zeta'(s)}{\zeta(s)} \right) ds, \quad \text{where } c > 1.$$

The quotient $-\zeta'(s)/\zeta(s)$ has a first order pole at $s = 1$ with residue 1. If we subtract this pole we get the formula

$$\frac{\psi_1(x)}{x^2} - \frac{1}{2}\left(1 - \frac{1}{x}\right)^2 = \frac{1}{2\pi i}\int_{c-\infty i}^{c+\infty i} \frac{x^{s-1}}{s(s+1)}\left(-\frac{\zeta'(s)}{\zeta(s)} - \frac{1}{s-1}\right) ds, \quad \text{for } c > 1.$$

We let

$$h(s) = \frac{1}{s(s+1)}\left(-\frac{\zeta'(s)}{\zeta(s)} - \frac{1}{s-1}\right)$$

and rewrite the last equation in the form

$$(3) \qquad \frac{\psi_1(x)}{x^2} - \frac{1}{2}\left(1 - \frac{1}{x}\right)^2 = \frac{1}{2\pi i}\int_{c-\infty i}^{c+\infty i} x^{s-1}h(s)\, ds$$
$$= \frac{x^{c-1}}{2\pi}\int_{-\infty}^{+\infty} h(c + it)e^{it \log x}\, dt.$$

To complete the proof we are required to show that

$$(4) \qquad \lim_{x\to\infty} \frac{x^{c-1}}{2\pi}\int_{-\infty}^{+\infty} h(c + it)e^{it \log x}\, dt = 0.$$

Now the Riemann–Lebesgue lemma in the theory of Fourier series states that

$$\lim_{x\to\infty} \int_{-\infty}^{+\infty} f(t)e^{itx}\, dt = 0$$

if the integral $\int_{-\infty}^{+\infty} |f(t)|\, dt$ converges. The integral in (4) is of this type, with x replaced by $\log x$, and we can easily show that the integral $\int_{-\infty}^{+\infty} |h(c + it)|\, dt$ converges if $c > 1$, so the integral in (4) tends to 0 as $x \to \infty$. However, the factor x^{c-1} outside the integral tends to ∞ when $c > 1$, so we are faced with an indeterminate form, $\infty \cdot 0$. Equation (3) holds for every $c > 1$. If we could put $c = 1$ in (3) the troublesome factor x^{c-1} would disappear. But then $h(c + it)$ becomes $h(1 + it)$ and the integrand involves $\zeta'(s)/\zeta(s)$ on the line $\sigma = 1$. In this case it is more difficult to prove that the integral $\int_{-\infty}^{+\infty} |h(1 + it)|\, dt$ converges, a fact which needs to be verified before we can apply the Riemann–Lebesgue lemma. The last and most difficult part of the proof is to show that it *is* possible to replace c by 1 in (3) and that the integral $\int_{-\infty}^{+\infty} |h(1 + it)|\, dt$ converges. This requires a more detailed study of the Riemann zeta function in the vicinity of the line $\sigma = 1$.

Now we proceed to carry out the plan outlined above. We begin with some lemmas.

13.2 Lemmas

Lemma 1 *For any arithmetical function $a(n)$ let*

$$A(x) = \sum_{n\le x} a(n),$$

where $A(x) = 0$ *if* $x < 1$. *Then*

$$\sum_{n \le x} (x - n)a(n) = \int_1^x A(t)\,dt. \tag{5}$$

PROOF. We apply Abel's identity (Theorem 4.2) which states that

$$\sum_{n \le x} a(n)f(n) = A(x)f(x) - \int_1^x A(t)f'(t)\,dt \tag{6}$$

if f has a continuous derivative on $[1, x]$. Taking $f(t) = t$ we have

$$\sum_{n \le x} a(n)f(n) = \sum_{n \le x} na(n) \quad \text{and } A(x)f(x) = x \sum_{n \le x} a(n)$$

so (6) reduces to (5). □

The next lemma is a form of L'Hôspital's rule for increasing piecewise linear functions.

Lemma 2 *Let* $A(x) = \sum_{n \le x} a(n)$ *and let* $A_1(x) = \int_1^x A(t)\,dt$. *Assume also that* $a(n) \ge 0$ *for all* n. *If we have the asymptotic formula*

$$A_1(x) \sim Lx^c \quad \text{as } x \to \infty \tag{7}$$

for some $c > 0$ *and* $L > 0$, *then we also have*

$$A(x) \sim cLx^{c-1} \quad \text{as } x \to \infty. \tag{8}$$

In other words, formal differentiation of (7) *gives a correct result.*

PROOF. The function $A(x)$ is increasing since the $a(n)$ are nonnegative. Choose any $\beta > 1$ and consider the difference $A_1(\beta x) - A_1(x)$. We have

$$A_1(\beta x) - A_1(x) = \int_x^{\beta x} A(u)\,du \ge \int_x^{\beta x} A(x)\,du = A(x)(\beta x - x)$$
$$= x(\beta - 1)A(x).$$

This gives us

$$xA(x) \le \frac{1}{\beta - 1}\{A_1(\beta x) - A_1(x)\}$$

or

$$\frac{A(x)}{x^{c-1}} \le \frac{1}{\beta - 1}\left\{\frac{A_1(\beta x)}{(\beta x)^c}\beta^c - \frac{A_1(x)}{x^c}\right\}.$$

Keep β fixed and let $x \to \infty$ in this inequality. We find

$$\limsup_{x \to \infty} \frac{A(x)}{x^{c-1}} \le \frac{1}{\beta - 1}(L\beta^c - L) = L\frac{\beta^c - 1}{\beta - 1}.$$

Now let $\beta \to 1+$. The quotient on the right is the difference quotient for the derivative of x^c at $x = 1$ and has the limit c. Therefore

$$\limsup_{x \to \infty} \frac{A(x)}{x^{c-1}} \leq cL. \tag{9}$$

Now consider any α with $0 < \alpha < 1$ and consider the difference $A_1(x) - A_1(\alpha x)$. An argument similar to the above shows that

$$\liminf_{x \to \infty} \frac{A(x)}{x^{c-1}} \geq L \frac{1 - \alpha^c}{1 - \alpha}.$$

As $\alpha \to 1-$ the right member tends to cL. This, together with (9) shows that $A(x)/x^{c-1}$ tends to the limit cL as $x \to \infty$. □

When $a(n) = \Lambda(n)$ we have $A(x) = \psi(x)$, $A_1(x) = \psi_1(x)$, and $a(n) \geq 0$. Therefore we can apply Lemmas 1 and 2 and immediately obtain:

Theorem 13.1 *We have*

$$\psi_1(x) = \sum_{n \leq x} (x - n)\Lambda(n). \tag{10}$$

Also, the asymptotic relation $\psi_1(x) \sim x^2/2$ *implies* $\psi(x) \sim x$ *as* $x \to \infty$.

Our next task is to express $\psi_1(x)/x^2$ as a contour integral involving the zeta function. For this we will require the special cases $k = 1$ and $k = 2$ of the following lemma on contour integrals. (Compare with Lemma 4 in Chapter 11.)

Lemma 3 *If* $c > 0$ *and* $u > 0$, *then for every integer* $k \geq 1$ *we have*

$$\frac{1}{2\pi i} \int_{c-\infty i}^{c+\infty i} \frac{u^{-z}}{z(z+1)\cdots(z+k)}\, dz = \begin{cases} \dfrac{1}{k!}(1-u)^k & \text{if } 0 < u \leq 1, \\ 0 & \text{if } u > 1, \end{cases}$$

the integral being absolutely convergent.

PROOF. First we note that the integrand is equal to $u^{-z}\Gamma(z)/\Gamma(z + k + 1)$. This follows by repeated use of the functional equation $\Gamma(z + 1) = z\Gamma(z)$. To prove the lemma we apply Cauchy's residue theorem to the integral

$$\frac{1}{2\pi i} \int_{C(R)} \frac{u^{-z}\Gamma(z)}{\Gamma(z + k + 1)}\, dz,$$

where $C(R)$ is the contour shown in Figure 13.1(a) if $0 < u \leq 1$, and that in Figure 13.1(b) if $u > 1$. The radius R of the circle is greater than $2k + c$ so all the poles at $z = 0, -1, \ldots, -k$ lie inside the circle.

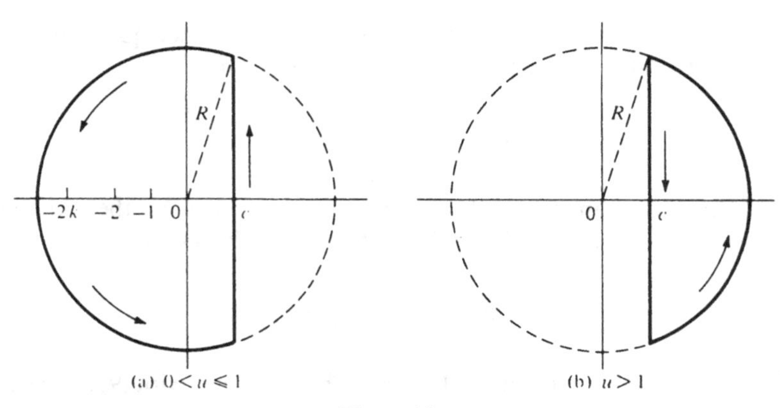

Figure 13.1

Now we show that the integral along each of the circular arcs tends to 0 as $R \to \infty$. If $z = x + iy$ and $|z| = R$ the integrand is dominated by

$$\left|\frac{u^{-z}}{z(z+1)\cdots(z+k)}\right| = \frac{u^{-x}}{|z|\,|z+1|\cdots|z+k|} \leq \frac{u^{-c}}{R|z+1|\cdots|z+k|}.$$

The inequality $u^{-x} \leq u^{-c}$ follows from the fact that u^{-x} is an increasing function of x if $0 < u \leq 1$ and a decreasing function if $u > 1$. Now if $1 \leq n \leq k$ we have

$$|z+n| \geq |z| - n = R - n \geq R - k \geq R/2$$

since $R > 2k$. Therefore the integral along each circular arc is dominated by

$$\frac{2\pi R u^{-c}}{R(\frac{1}{2}R)^k} = O(R^{-k})$$

and this $\to 0$ as $R \to \infty$ since $k \geq 1$.

If $u > 1$ the integrand is analytic inside $C(R)$ hence $\int_{C(R)} = 0$. Letting $R \to \infty$ we find that the lemma is proved in this case.

If $0 < u \leq 1$ we evaluate the integral around $C(R)$ by Cauchy's residue theorem. The integrand has poles at the integers $n = 0, -1, \ldots, -k$, hence

$$\begin{aligned}\frac{1}{2\pi i}\int_{C(R)} \frac{u^{-z}\Gamma(z)}{\Gamma(z+k+1)}\,dz &= \sum_{n=0}^{k} \operatorname*{Res}_{z=-n} \frac{u^{-z}\Gamma(z)}{\Gamma(z+k+1)} \\ &= \sum_{n=0}^{k} \frac{u^n}{\Gamma(k+1-n)} \operatorname*{Res}_{z=-n} \Gamma(z) = \sum_{n=0}^{k} \frac{u^n(-1)^n}{(k-n)!\,n!} \\ &= \frac{1}{k!}\sum_{n=0}^{k}\binom{k}{n}(-u)^n = \frac{(1-u)^k}{k!}.\end{aligned}$$

Letting $R \to \infty$ we obtain the lemma. □

13.3 A contour integral representation for $\psi_1(x)/x^2$

Theorem 13.2 *If $c > 1$ and $x \geq 1$ we have*

$$\frac{\psi_1(x)}{x^2} = \frac{1}{2\pi i}\int_{c-\infty i}^{c+\infty i} \frac{x^{s-1}}{s(s+1)}\left(-\frac{\zeta'(s)}{\zeta(s)}\right) ds. \tag{11}$$

PROOF. From Equation (10) we have $\psi_1(x)/x = \sum_{n \leq x} (1 - n/x)\Lambda(n)$. Now use Lemma 3 with $k = 1$ and $u = n/x$. If $n \leq x$ we obtain

$$1 - \frac{n}{x} = \frac{1}{2\pi i}\int_{c-\infty i}^{c+\infty i} \frac{(x/n)^s}{s(s+1)}\, ds.$$

Multiplying this relation by $\Lambda(n)$ and summing over all $n \leq x$ we find

$$\frac{\psi_1(x)}{x} = \sum_{n \leq x} \frac{1}{2\pi i}\int_{c-\infty i}^{c+\infty i} \frac{\Lambda(n)(x/n)^s}{s(s+1)}\, ds = \sum_{n=1}^{\infty} \frac{1}{2\pi i}\int_{c-\infty i}^{c+\infty i} \frac{\Lambda(n)(x/n)^s}{s(s+1)}\, ds$$

since the integral vanishes if $n > x$. This can be written as

$$\frac{\psi_1(x)}{x} = \sum_{n=1}^{\infty} \int_{c-\infty i}^{c+\infty i} f_n(s)\, ds, \tag{12}$$

where $2\pi i f_n(s) = \Lambda(n)(x/n)^s/(s^2 + s)$. Next we wish to interchange the sum and integral in (12). For this it suffices to prove that the series

$$\sum_{n=1}^{\infty} \int_{c-\infty i}^{c+\infty i} |f_n(s)|\, ds \tag{13}$$

is convergent. (See Theorem 10.26 in [2].) The partial sums of this series satisfy the inequality

$$\sum_{n=1}^{N} \int_{c-\infty i}^{c+\infty i} \frac{\Lambda(n)(x/n)^c}{|s||s+1|}\, ds = \sum_{n=1}^{N} \frac{\Lambda(n)}{n^c}\int_{c-\infty i}^{c+\infty i} \frac{x^c}{|s||s+1|}\, ds \leq A \sum_{n=1}^{\infty} \frac{\Lambda(n)}{n^c},$$

where A is a constant, so (13) converges. Hence we can interchange the sum and integral in (12) to obtain

$$\frac{\psi_1(x)}{x} = \int_{c-\infty i}^{c+\infty i} \sum_{n=1}^{\infty} f_n(s)\, ds = \frac{1}{2\pi i}\int_{c-\infty i}^{c+\infty i} \frac{x^s}{s(s+1)} \sum_{n=1}^{\infty} \frac{\Lambda(n)}{n^s}\, ds$$

$$= \frac{1}{2\pi i}\int_{c-\infty i}^{c+\infty i} \frac{x^s}{s(s+1)}\left(-\frac{\zeta'(s)}{\zeta(s)}\right) ds.$$

Now divide by x to obtain (11). □

Theorem 13.3 *If $c > 1$ and $x \geq 1$ we have*

$$\frac{\psi_1(x)}{x^2} - \frac{1}{2}\left(1 - \frac{1}{x}\right)^2 = \frac{1}{2\pi i}\int_{c-\infty i}^{c+\infty i} x^{s-1}h(s)\,ds, \tag{14}$$

where

$$h(s) = \frac{1}{s(s+1)}\left(-\frac{\zeta'(s)}{\zeta(s)} - \frac{1}{s-1}\right). \tag{15}$$

PROOF. This time we use Lemma 3 with $k = 2$ to get

$$\frac{1}{2}\left(1 - \frac{1}{x}\right)^2 = \frac{1}{2\pi i}\int_{c-\infty i}^{c+\infty i} \frac{x^s}{s(s+1)(s+2)}\,ds,$$

where $c > 0$. Replace s by $s - 1$ in the integral (keeping $c > 1$) and subtract the result from (11) to obtain Theorem 13.3. □

If we parameterize the path of integration by writing $s = c + it$, we find $x^{s-1} = x^{c-1}x^{it} = x^{c-1}e^{it\log x}$ and Equation (14) becomes

$$\frac{\psi_1(x)}{x^2} - \frac{1}{2}\left(1 - \frac{1}{x}\right)^2 = \frac{x^{c-1}}{2\pi}\int_{-\infty}^{+\infty} h(c+it)e^{it\log x}\,dt. \tag{16}$$

Our next task is to show that the right member of (16) tends to 0 as $x \to \infty$. As mentioned earlier, we first show that we can put $c = 1$ in (16). For this purpose we need to study $\zeta(s)$ in the neighborhood of the line $\sigma = 1$.

13.4 Upper bounds for $|\zeta(s)|$ and $|\zeta'(s)|$ near the line $\sigma = 1$

To study $\zeta(s)$ near the line $\sigma = 1$ we use the representation obtained from Theorem 12.21 which is valid for $\sigma > 0$,

$$\zeta(s) = \sum_{n=1}^{N} \frac{1}{n^s} - s\int_N^{\infty} \frac{x - [x]}{x^{s+1}}\,dx + \frac{N^{1-s}}{s-1}. \tag{17}$$

We also use the formula for $\zeta'(s)$ obtained by differentiating each member of (17),

$$\zeta'(s) = -\sum_{n=1}^{N} \frac{\log n}{n^s} + s\int_N^{\infty} \frac{(x - [x])\log x}{x^{s+1}}\,dx - \int_N^{\infty} \frac{x - [x]}{x^{s+1}}\,dx - \frac{N^{1-s}\log N}{s-1} - \frac{N^{1-s}}{(s-1)^2}. \tag{18}$$

The next theorem uses these relations to obtain upper bounds for $|\zeta(s)|$ and $|\zeta'(s)|$.

Theorem 13.4 *For every $A > 0$ there exists a constant M (depending on A) such that*

$$(19) \qquad |\zeta(s)| \leq M \log t \qquad \textit{and } |\zeta'(s)| \leq M \log^2 t$$

for all s with $\sigma \geq 1/2$ satisfying

$$(20) \qquad \sigma > 1 - \frac{A}{\log t} \qquad \textit{and } t \geq e.$$

Note. The inequalities (20) describe a region of the type shown in Figure 13.2.

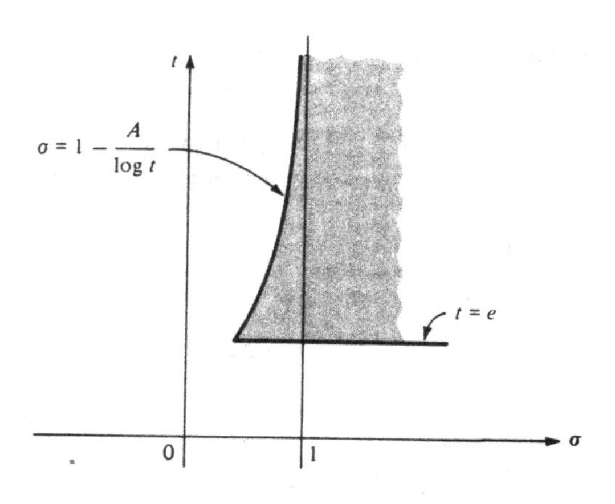

Figure 13.2

Proof. If $\sigma \geq 2$ we have $|\zeta(s)| \leq \zeta(2)$ and $|\zeta'(s)| \leq |\zeta'(2)|$ and the inequalities in (19) are trivially satisfied. Therefore we can assume $\sigma < 2$ and $t \geq e$. We then have

$$|s| \leq \sigma + t \leq 2 + t < 2t \qquad \text{and } |s - 1| \geq t$$

so $1/|s - 1| \leq 1/t$. Estimating $|\zeta(s)|$ by using (17) we find

$$|\zeta(s)| \leq \sum_{n=1}^{N} \frac{1}{n^\sigma} + 2t \int_N^\infty \frac{1}{x^{\sigma+1}}\, dx + \frac{N^{1-\sigma}}{t} = \sum_{n=1}^{N} \frac{1}{n^\sigma} + \frac{2t}{\sigma N^\sigma} + \frac{N^{1-\sigma}}{t}.$$

Now we make N depend on t by taking $N = [t]$. Then $N \leq t < N + 1$ and $\log n \leq \log t$ if $n \leq N$. The inequality (20) implies $1 - \sigma < A/\log t$ so

$$\frac{1}{n^\sigma} = \frac{n^{1-\sigma}}{n} = \frac{1}{n} e^{(1-\sigma)\log n} < \frac{1}{n} e^{A \log n/\log t} \leq \frac{1}{n} e^A = O\left(\frac{1}{n}\right).$$

Therefore

$$\frac{2t}{\sigma N^{\sigma}} = O\left(\frac{N+1}{N}\right) = O(1) \qquad \text{and} \quad \frac{N^{1-\sigma}}{t} = \frac{N}{t}\frac{1}{N^{\sigma}} = O\left(\frac{1}{N}\right) = O(1),$$

so

$$|\zeta(s)| = O\left(\sum_{n=1}^{N} \frac{1}{n}\right) + O(1) = O(\log N) + O(1) = O(\log t).$$

This proves the inequality for $|\zeta(s)|$ in (19). To obtain the inequality for $|\zeta'(s)|$ we apply the same type of argument to (18). The only essential difference is that an extra factor $\log N$ appears on the right. But $\log N = O(\log t)$ so we get $|\zeta'(s)| = O(\log^2 t)$ in the specified region. □

13.5 The nonvanishing of $\zeta(s)$ on the line $\sigma = 1$

In this section we prove that $\zeta(1 + it) \neq 0$ for every real t. The proof is based on an inequality which will also be needed in the next section.

Theorem 13.5 *If* $\sigma > 1$ *we have*

$$\zeta^3(\sigma)|\zeta(\sigma + it)|^4|\zeta(\sigma + 2it)| \geq 1. \tag{21}$$

PROOF. We recall the identity $\zeta(s) = e^{G(s)}$ proved in Section 11.9, Example 1, where

$$G(s) = \sum_{n=2}^{\infty} \frac{\Lambda(n)}{\log n} n^{-s} = \sum_{p}\sum_{m=1}^{\infty} \frac{1}{mp^{ms}} \qquad (\sigma > 1).$$

This can be written as

$$\zeta(s) = \exp\left\{\sum_{p}\sum_{m=1}^{\infty} \frac{1}{mp^{ms}}\right\} = \exp\left\{\sum_{p}\sum_{m=1}^{\infty} \frac{e^{-imt\log p}}{mp^{m\sigma}}\right\},$$

from which we find

$$|\zeta(s)| = \exp\left\{\sum_{p}\sum_{m=1}^{\infty} \frac{\cos(mt\log p)}{mp^{m\sigma}}\right\}.$$

We apply this formula repeatedly with $s = \sigma$, $s = \sigma + it$ and $s = \sigma + 2it$, and obtain

$$\zeta^3(\sigma)|\zeta(\sigma + it)|^4|\zeta(\sigma + 2it)|$$
$$= \exp\left\{\sum_{p}\sum_{m=1}^{\infty} \frac{3 + 4\cos(mt\log p) + \cos(2mt\log p)}{mp^{m\sigma}}\right\}.$$

But we have the trigonometric inequality

$$3 + 4\cos\theta + \cos 2\theta \geq 0$$

which follows from the identity

$$3 + 4\cos\theta + \cos 2\theta = 3 + 4\cos\theta + 2\cos^2\theta - 1 = 2(1 + \cos\theta)^2.$$

Therefore each term in the last infinite series is nonnegative so we obtain (21). □

Theorem 13.6 *We have $\zeta(1 + it) \neq 0$ for every real t.*

PROOF. We need only consider $t \neq 0$. Rewrite (21) in the form

$$\{(\sigma - 1)\zeta(\sigma)\}^3 \left|\frac{\zeta(\sigma + it)}{\sigma - 1}\right|^4 |\zeta(\sigma + 2it)| \geq \frac{1}{\sigma - 1}. \tag{22}$$

This is valid if $\sigma > 1$. Now let $\sigma \to 1+$ in (22). The first factor approaches 1 since $\zeta(s)$ has residue 1 at the pole $s = 1$. The third factor tends to $|\zeta(1 + 2it)|$. If $\zeta(1 + it)$ were equal to 0 the middle factor could be written as

$$\left|\frac{\zeta(\sigma + it) - \zeta(1 + it)}{\sigma - 1}\right|^4 \to |\zeta'(1 + it)|^4 \quad \text{as } \sigma \to 1+.$$

Therefore, if for some $t \neq 0$ we had $\zeta(1 + it) = 0$ the left member of (22) would approach the limit $|\zeta'(1 + it)|^4 |\zeta(1 + 2it)|$ as $\sigma \to 1+$. But the right member tends to ∞ as $\sigma \to 1+$ and this gives a contradiction. □

13.6 Inequalities for $|1/\zeta(s)|$ and $|\zeta'(s)/\zeta(s)|$

Now we apply Theorem 13.5 once more to obtain the following inequalities for $|1/\zeta(s)|$ and $|\zeta'(s)/\zeta(s)|$.

Theorem 13.7 *There is a constant $M > 0$ such that*

$$\left|\frac{1}{\zeta(s)}\right| < M \log^7 t \qquad \text{and} \qquad \left|\frac{\zeta'(s)}{\zeta(s)}\right| < M \log^9 t$$

whenever $\sigma \geq 1$ and $t \geq e$.

PROOF. For $\sigma \geq 2$ we have

$$\left|\frac{1}{\zeta(s)}\right| = \left|\sum_{n=1}^{\infty} \frac{\mu(n)}{n^s}\right| \leq \sum_{n=1}^{\infty} \frac{1}{n^2} \leq \zeta(2)$$

and

$$\left|\frac{\zeta'(s)}{\zeta(s)}\right| \leq \sum_{n=1}^{\infty} \frac{\Lambda(n)}{n^2},$$

so the inequalities hold trivially if $\sigma \geq 2$. Suppose, then, that $1 \leq \sigma \leq 2$ and $t \geq e$. Rewrite inequality (21) as follows:

$$\frac{1}{|\zeta(\sigma + it)|} \leq \zeta(\sigma)^{3/4} |\zeta(\sigma + 2it)|^{1/4}.$$

Now $(\sigma - 1)\zeta(\sigma)$ is bounded in the interval $1 \leq \sigma \leq 2$, say $(\sigma - 1)\zeta(\sigma) \leq M$, where M is an absolute constant. Then

$$\zeta(\sigma) \leq \frac{M}{\sigma - 1} \quad \text{if } 1 < \sigma \leq 2.$$

Also, $\zeta(\sigma + 2it) = O(\log t)$ if $1 \leq \sigma \leq 2$ (by Theorem 13.4), so for $1 < \sigma \leq 2$ we have

$$\frac{1}{|\zeta(\sigma + it)|} \leq \frac{M^{3/4}(\log t)^{1/4}}{(\sigma - 1)^{3/4}} = \frac{A(\log t)^{1/4}}{(\sigma - 1)^{3/4}},$$

where A is an absolute constant. Therefore for some constant $B > 0$ we have

$$|\zeta(\sigma + it)| > \frac{B(\sigma - 1)^{3/4}}{(\log t)^{1/4}}, \quad \text{if } 1 < \sigma \leq 2, \text{ and } t \geq e. \tag{23}$$

This also holds trivially for $\sigma = 1$. Let α be any number satisfying $1 < \alpha < 2$. Then if $1 \leq \sigma \leq \alpha$, $t \geq e$, we may use Theorem 13.4 to write

$$\begin{aligned} |\zeta(\sigma + it) - \zeta(\alpha + it)| &\leq \int_\sigma^\alpha |\zeta'(u + it)|\, du \leq (\alpha - \sigma)M \log^2 t \\ &\leq (\alpha - 1)M \log^2 t. \end{aligned}$$

Hence, by the triangle inequality,

$$\begin{aligned} |\zeta(\sigma + it)| &\geq |\zeta(\alpha + it)| - |\zeta(\sigma + it) - \zeta(\alpha + it)| \\ &\geq |\zeta(\alpha + it)| - (\alpha - 1)M \log^2 t \geq \frac{B(\alpha - 1)^{3/4}}{(\log t)^{1/4}} - (\alpha - 1)M \log^2 t. \end{aligned}$$

This holds if $1 \leq \sigma \leq \alpha$, and by (23) it also holds for $\alpha \leq \sigma \leq 2$ since $(\sigma - 1)^{3/4} \geq (\alpha - 1)^{3/4}$. In other words, if $1 \leq \sigma \leq 2$ and $t \geq e$ we have the inequality

$$|\zeta(\sigma + it)| \geq \frac{B(\alpha - 1)^{3/4}}{(\log t)^{1/4}} - (\alpha - 1)M \log^2 t$$

for any α satisfying $1 < \alpha < 2$. Now we make α depend on t and choose α so the first term on the right is twice the second. This requires

$$\alpha = 1 + \left(\frac{B}{2M}\right)^4 \frac{1}{(\log t)^9}.$$

Clearly $\alpha > 1$ and also $\alpha < 2$ if $t \geq t_0$ for some t_0. Thus, if $t \geq t_0$ and $1 \leq \sigma \leq 2$ we have

$$|\zeta(\sigma + it)| \geq (\alpha - 1)M \log^2 t = \frac{C}{(\log t)^7}.$$

The inequality also holds with (perhaps) a different C if $e \leq t \leq t_0$.

This proves that $|\zeta(s)| \geq C \log^{-7} t$ for all $\sigma \geq 1$, $t \geq e$, giving us a corresponding upper bound for $|1/\zeta(s)|$. To get the inequality for $|\zeta'(s)/\zeta(s)|$ we apply Theorem 13.4 and obtain an extra factor $\log^2 t$. □

13.7 Completion of the proof of the prime number theorem

Now we are almost ready to complete the proof of the prime number theorem. We need one more fact from complex function theory which we state as a lemma.

Lemma 4 *If $f(s)$ has a pole of order k at $s = \alpha$ then the quotient $f'(s)/f(s)$ has a first order pole at $s = \alpha$ with residue $-k$.*

PROOF. We have $f(s) = g(s)/(s - \alpha)^k$, where g is analytic at α and $g(\alpha) \neq 0$. Hence for all s in a neighborhood of α we have

$$f'(s) = \frac{g'(s)}{(s-\alpha)^k} - \frac{kg(s)}{(s-\alpha)^{k+1}} = \frac{g(s)}{(s-\alpha)^k}\left\{\frac{-k}{s-\alpha} + \frac{g'(s)}{g(s)}\right\}.$$

Thus

$$\frac{f'(s)}{f(s)} = \frac{-k}{s-\alpha} + \frac{g'(s)}{g(s)}.$$

This proves the lemma since $g'(s)/g(s)$ is analytic at α. □

Theorem 13.8 *The function*

$$F(s) = -\frac{\zeta'(s)}{\zeta(s)} - \frac{1}{s-1}$$

is analytic at $s = 1$.

PROOF. By Lemma 4, $-\zeta'(s)/\zeta(s)$ has a first order pole at 1 with residue 1, as does $1/(s - 1)$. Hence their difference is analytic at $s = 1$. □

Theorem 13.9 *For $x \geq 1$ we have*

$$\frac{\psi_1(x)}{x^2} - \frac{1}{2}\left(1 - \frac{1}{x}\right)^2 = \frac{1}{2\pi}\int_{-\infty}^{\infty} h(1 + it)e^{it \log x}\, dt,$$

where the integral $\int_{-\infty}^{\infty} |h(1 + it)|\, dt$ converges. Therefore, by the Riemann–Lebesgue lemma we have

$$\psi_1(x) \sim x^2/2 \tag{24}$$

and hence

$$\psi(x) \sim x \quad \textit{as } x \to \infty.$$

PROOF. In Theorem 13.3 we proved that if $c > 1$ and $x \geq 1$ we have

$$\frac{\psi_1(x)}{x^2} - \frac{1}{2}\left(1 - \frac{1}{x}\right)^2 = \frac{1}{2\pi i}\int_{c-\infty i}^{c+\infty i} x^{s-1}h(s)\,ds,$$

where

$$h(s) = \frac{1}{s(s+1)}\left(-\frac{\zeta'(s)}{\zeta(s)} - \frac{1}{s-1}\right).$$

Our first task is to show that we can move the path of integration to the line $\sigma = 1$. To do this we apply Cauchy's theorem to the rectangle R shown in Figure 13.3. The integral of $x^{s-1}h(s)$ around R is 0 since the integrand

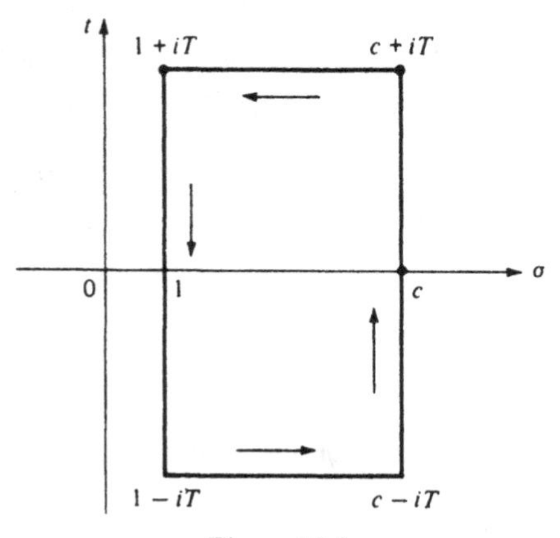

Figure 13.3

is analytic inside and on R. Now we show that the integrals along the horizontal segments tend to 0 as $T \to \infty$. Since the integrand has the same absolute value at conjugate points, it suffices to consider only the upper segment, $t = T$. On this segment we have the estimates

$$\left|\frac{1}{s(s+1)}\right| \leq \frac{1}{T^2} \quad \text{and} \quad \left|\frac{1}{s(s+1)(s-1)}\right| \leq \frac{1}{T^3} \leq \frac{1}{T^2}.$$

Also, there is a constant M such that $|\zeta'(s)/\zeta(s)| \leq M \log^9 t$ if $\sigma \geq 1$ and $t \geq e$. Hence if $T \geq e$ we have

$$|h(s)| \leq \frac{M \log^9 T}{T^2}$$

so that

$$\left|\int_{1+iT}^{c+iT} x^{s-1}h(s)\,ds\right| \leq \int_1^c x^{c-1}\frac{M\log^9 T}{T^2}\,d\sigma = Mx^{c-1}\frac{\log^9 T}{T^2}(c-1).$$

Therefore the integrals along the horizontal segments tend to 0 as $T \to \infty$, and hence we have

$$\int_{c-\infty i}^{c+\infty i} x^{s-1}h(s)\,ds = \int_{1-\infty i}^{1+\infty i} x^{s-1}h(s)\,ds.$$

On the line $\sigma = 1$ we write $s = 1 + it$ to obtain

$$\frac{1}{2\pi i}\int_{1-\infty i}^{1+\infty i} x^{s-1}h(s)\,ds = \frac{1}{2\pi}\int_{-\infty}^{\infty} h(i + it)e^{it\log x}\,dt.$$

Now we note that

$$\int_{-\infty}^{\infty} |h(1 + it)|\,dt = \int_{-e}^{e} + \int_{e}^{\infty} + \int_{-\infty}^{-e}.$$

In the integral from e to ∞ we have

$$|h(1 + it)| \le \frac{M\log^9 t}{t^2}$$

so $\int_e^\infty |h(1 + it)|\,dt$ converges. Similarly, $\int_{-\infty}^{-e}$ converges, so $\int_{-\infty}^{\infty} |h(1 + it)|\,dt$ converges. Thus we may apply the Riemann–Lebesgue lemma to obtain $\psi_1(x) \sim x^2/2$. By Theorem 13.1 this implies $\psi(x) \sim x$ as $x \to \infty$, and this completes the proof of the prime number theorem. □

13.8 Zero-free regions for $\zeta(s)$

The inequality $|1/\zeta(s)| < M\log^7 t$ which we proved in Theorem 13.7 for $\sigma \ge 1$ and $t \ge e$ can be extended to the left of the line $\sigma = 1$. The estimate is not obtained in a vertical strip but rather in a region somewhat like that shown in Figure 13.2 where the left boundary curve approaches the line $\sigma = 1$ asymptotically as $t \to \infty$. The inequality implies the nonvanishing of $\zeta(s)$ in this region. More precisely, we have:

Theorem 13.10 *Assume $\sigma \ge 1/2$. Then there exist constants $A > 0$ and $C > 0$ such that*

$$|\zeta(\sigma + it)| > \frac{C}{\log^7 t}$$

whenever

$$1 - \frac{A}{\log^9 t} < \sigma \le 1 \qquad \text{and } t \ge e. \tag{25}$$

This implies that $\zeta(\sigma + it) \ne 0$ if σ and t satisfy (25).

Proof. The triangle inequality, used in conjunction with Theorem 13.7, gives us

$$(26)\qquad |\zeta(\sigma + it)| \geq |\zeta(1 + it)| - |\zeta(1 + it) - \zeta(\sigma + it)|$$
$$> \frac{B}{\log^7 t} - |\zeta(1 + it) - \zeta(\sigma + it)|,$$

for some $B > 0$. To estimate the last term we write

$$|\zeta(1 + it) - \zeta(\sigma + it)| = \left|\int_\sigma^1 \zeta'(u + it)\, du\right| \leq \int_\sigma^1 |\zeta'(u + it)|\, du.$$

Since $t \geq e$ we have $\log^9 t \geq \log t$ so $1 - (A/\log^9 t) \geq 1 - (A/\log t)$. Thus, if σ satisfies (25) for any $A > 0$ we can apply Theorem 13.4 to estimate $|\zeta'(u + it)|$, giving us

$$|\zeta(1 + it) - \zeta(\sigma + it)| \leq M(1 - \sigma)\log^2 t < M \log^2 t \frac{A}{\log^9 t} = \frac{MA}{\log^7 t}.$$

Using this in (26) we find

$$|\zeta(\sigma + it)| > \frac{B - MA}{\log^7 t}.$$

This holds for some $B > 0$, any $A > 0$ and some $M > 0$ depending on A. A value of M that works for some A also works for every smaller A. Therefore we can choose A small enough so that $B - MA > 0$. If we let $C = B - MA$ the last inequality becomes $|\zeta(\sigma + it)| > C \log^{-7} t$ which proves the theorem for all σ and t satisfying

$$1 - \frac{A}{\log^9 t} < \sigma < 1, \qquad \text{and } t \geq e.$$

But the result also holds for $\sigma = 1$ by Theorem 13.7 so the proof is complete. □

We know that $\zeta(s) \neq 0$ if $\sigma \geq 1$, and the functional equation

$$\zeta(s) = 2(2\pi)^{1-s}\Gamma(1 - s)\sin\left(\frac{\pi s}{2}\right)\zeta(1 - s)$$

shows that $\zeta(s) \neq 0$ if $\sigma \leq 0$ except for the zeros at $s = -2, -4, -6, \ldots$ which arise from the vanishing of $\sin(\pi s/2)$. These are called the "trivial" zeros of $\zeta(s)$. The next theorem shows that, aside from the trivial zeros, $\zeta(s)$ has no further zeros on the real axis.

Theorem 13.11 *If $\sigma > 0$ we have*

$$(27)\qquad (1 - 2^{1-s})\zeta(s) = \sum_{n=1}^{\infty} \frac{(-1)^{n-1}}{n^s}.$$

This implies that $\zeta(s) < 0$ if s is real and $0 < s < 1$.

PROOF. First assume that $\sigma > 1$. Then we have

$$\begin{aligned}(1 - 2^{1-s})\zeta(s) &= \sum_{n=1}^{\infty} \frac{1}{n^s} - 2\sum_{n=1}^{\infty} \frac{1}{(2n)^s} \\ &= (1 + 2^{-s} + 3^{-s} + \cdots) - 2(2^{-s} + 4^{-s} + 6^{-s} + \cdots) \\ &= 1 - 2^{-s} + 3^{-s} - 4^{-s} + 5^{-s} - 6^{-s} + \cdots,\end{aligned}$$

which proves (27) for $\sigma > 1$. However, if $\sigma > 0$ the series on the right converges, so (27) also holds for $\sigma > 0$ by analytic continuation.

When s is real the series in (27) is an alternating series with a positive sum. If $0 < s < 1$ the factor $(1 - 2^{1-s})$ is negative hence $\zeta(s)$ is also negative. □

13.9 The Riemann hypothesis

In his famous 8-page memoir on $\pi(x)$ published in 1859, Riemann [58] stated that it seems likely that the nontrivial zeros of $\zeta(s)$ all lie on the line $\sigma = 1/2$, although he could not prove this. The assertion that all the nontrivial zeros have real part 1/2 is now called the *Riemann hypothesis*. In 1900 Hilbert listed the problem of proving or disproving the Riemann hypothesis as one of the most important problems confronting twentieth century mathematicians. To this day it remains unsolved.

The Riemann hypothesis has attracted the attention of many eminent mathematicians and a great deal has been discovered about the distribution of the zeros of $\zeta(s)$. The functional equation shows that all the nontrivial zeros (if any exist) must lie in the strip $0 < \sigma < 1$, the so-called "critical strip." It is easy to show that the zeros are symmetrically located about the real axis and about the "critical line" $\sigma = 1/2$.

In 1915 Hardy proved that an infinite number of zeros are located on the critical line. In 1921 Hardy and Littlewood showed that the number of zeros on the line segment joining 1/2 to $(1/2) + iT$ is at least AT for some positive constant A, if T is sufficiently large. In 1942 Selberg improved this by showing that the number is at least $AT \log T$ for some $A > 0$. It is also known that the number in the critical strip with $0 < t < T$ is asymptotic to $T \log T/2\pi$ as $T \to \infty$, so Selberg's result shows that a positive fraction of the zeros lie on the critical line. Recently (1974) Levinson showed that this fraction is at least 1/3. That is, the constant in Selberg's theorem satisfies $A \geq 1/(6\pi)$.

Extensive calculations by Gram, Backlund, Lehmer, Haselgrove, Rosser, Yohe, Schoenfeld, and others have shown that the first three-and-a-half million zeros above the real axis are on the critical line. In spite of all this evidence in favor of the Riemann hypothesis, the calculations also reveal certain phenomena which suggest that counterexamples to the Riemann hypothesis might very well exist. For a fascinating account of the story of large-scale calculations concerning $\zeta(s)$ the reader should consult [17].

13.10 Application to the divisor function

The prime number theorem can sometimes be used to estimate the order of magnitude of multiplicative arithmetical functions. In this section we use it to derive inequalities for $d(n)$, the number of divisors of n.

In Chapter 3 we proved that the average order of $d(n)$ is $\log n$. When n is prime we have $d(n) = 2$ so the growth of $d(n)$ is most pronounced when n has many divisors. Suppose n is the product of all the primes $\le x$, say

$$n = 2 \cdot 3 \cdot 5 \cdot \cdots \cdot p_{\pi(x)}. \tag{28}$$

Since $d(n)$ is multiplicative we have

$$d(n) = d(2)d(3) \cdots d(p_{\pi(x)}) = 2^{\pi(x)}.$$

For large x, $\pi(x)$ is approximately $x/\log x$ and (28) implies that

$$\log n = \sum_{p \le x} \log p = \vartheta(x) \sim x$$

so $2^{\pi(x)}$ is approximately $2^{\log n/\log\log n}$. Now

$$2^{a \log n} = e^{a \log n \log 2} = n^{a \log 2}$$

hence $2^{\log n/\log\log n} = n^{\log 2/\log\log n}$. In other words, when n is of the form (28) then $d(n)$ is approximately $2^{\log n/\log\log n} = n^{\log 2/\log\log n}$.

By pursuing this idea with a little more care we obtain the following inequalities for $d(n)$.

Theorem 13.12 *Let $\varepsilon > 0$ be given. Then we have:*

(a) *There exists an integer $N(\varepsilon)$ such that $n \ge N(\varepsilon)$ implies*

$$d(n) < 2^{(1+\varepsilon)\log n/\log\log n} = n^{(1+\varepsilon)\log 2/\log\log n}.$$

(b) *For infinitely many n we have*

$$d(n) > 2^{(1-\varepsilon)\log n/\log\log n} = n^{(1-\varepsilon)\log 2/\log\log n}.$$

Note. These inequalities are equivalent to the relation

$$\limsup_{n\to\infty} \frac{\log d(n) \log\log n}{\log n} = \log 2.$$

PROOF. Write $n = p_1^{a_1} \cdots p_k^{a_k}$, so that $d(n) = \prod_{i=1}^k (a_i + 1)$. We split the product into two parts, separating those prime divisors $< f(n)$ from those $\ge f(n)$, where $f(n)$ will be specified later. Then $d(n) = P_1(n)P_2(n)$ where

$$P_1(n) = \prod_{p_i < f(n)} (a_i + 1) \qquad \text{and } P_2(n) = \prod_{p_i \ge f(n)} (a_i + 1).$$

In the product $P_2(n)$ we use the inequality $(a + 1) \leq 2^a$ to obtain $P_2(n) \leq 2^{S(n)}$, where

$$S(n) = \sum_{\substack{i=1 \\ p_i \geq f(n)}}^{k} a_i.$$

Now

$$n = \prod_{i=1}^{k} p_i^{a_i} \geq \prod_{p_i \geq f(n)} p_i^{a_i} \geq \prod_{p_i \geq f(n)} f(n)^{a_i} = f(n)^{S(n)},$$

hence

$$\log n \geq S(n)\log f(n), \qquad \text{or } S(n) \leq \frac{\log n}{\log f(n)}.$$

This gives us

$$P_2(n) \leq 2^{\log n/\log f(n)}. \tag{29}$$

To estimate $P_1(n)$ we write

$$P_1(n) = \exp\left\{\sum_{p_i < f(n)} \log(a_i + 1)\right\}$$

and show that $\log(a_i + 1) < 2 \log\log n$ if n is sufficiently large. In fact, we have

$$n \geq p_i^{a_i} \geq 2^{a_i}$$

hence

$$\log n \geq a_i \log 2, \qquad \text{or } a_i \leq \log n/\log 2.$$

Therefore

$$1 + a_i \leq 1 + \frac{\log n}{\log 2} < (\log n)^2 \quad \text{if } n \geq n_1$$

for some n_1. Thus $n \geq n_1$ implies $\log(1 + a_i) < \log(\log n)^2 = 2 \log\log n$. This gives us

$$P_1(n) < \exp\left\{2 \log\log n \sum_{p_i < f(n)} 1\right\} \leq \exp\{2 \log\log n\, \pi(f(n))\}.$$

Using the inequality $\pi(x) < 6x/\log x$ (see Theorem 4.6) we obtain

$$P_1(n) < \exp\left\{\frac{12f(n)\log\log n}{\log f(n)}\right\} = 2^{cf(n)\log\log n/\log f(n)}, \tag{30}$$

where $c = 12/\log 2$. Combining (29) and (30) we obtain $d(n) = P_1(n)P_2(n) < 2^{g(n)}$ where

$$g(n) = \frac{\log n + cf(n)\log\log n}{\log f(n)} = \frac{\log n}{\log\log n} \, \frac{1 + c\dfrac{f(n)\log\log n}{\log n}}{\dfrac{\log f(n)}{\log\log n}}.$$

Now we choose $f(n)$ to make $f(n)\log\log n/\log n \to 0$ and also to make $\log f(n)/\log\log n \to 1$ as $n \to \infty$. For this it suffices to take

$$f(n) = \frac{\log n}{(\log\log n)^2}.$$

Then

$$g(n) = \frac{\log n}{\log\log n}\frac{1 + o(1)}{1 + o(1)} = \frac{\log n}{\log\log n}(1 + o(1)) < (1 + \varepsilon)\frac{\log n}{\log\log n}$$

if $n \geq N(\varepsilon)$ for some $N(\varepsilon)$. This proves part (a).

To prove part (b) we pick a set of integers n with a large number of prime factors. In fact, we take n to be the product of all the primes $\leq x$. Then $n \to \infty$ if and only if $x \to \infty$. For such n we have, by the prime number theorem,

$$d(n) = 2^{\pi(x)} = 2^{(1 + o(1))x/\log x}$$

Also for such n we have

$$\log n = \sum_{p \leq x} \log p = \vartheta(x) = x(1 + o(1))$$

so

$$x = \frac{\log n}{1 + o(1)} = (1 + o(1))\log n$$

hence

$$\log x = \log\log n + \log(1 + o(1)) = \log\log n\left(1 + \frac{\log(1 + o(1))}{\log\log n}\right)$$
$$= (1 + o(1))\log\log n.$$

Therefore $x/\log x = (1 + o(1))\log n/\log\log n$ and

$$d(n) = 2^{(1 + o(1))\log n/\log\log n}$$

for such n. But $1 + o(1) > 1 - \varepsilon$ if $n \geq N(\varepsilon)$ for some $N(\varepsilon)$, and this proves (b). □

Note. As a corollary of Theorem 13.12 we obtain the relation

$$d(n) = o(n^\delta) \tag{31}$$

for every $\delta > 0$. This result can also be derived without the use of the prime number theorem. (See Exercise 13.13.)

13.11 Application to Euler's totient

The type of argument used in the foregoing section can also be used to obtain inequalities for $\varphi(n)$. When n is prime we have $\varphi(n) = n - 1$. When n has a large number of prime factors $\varphi(n)$ will be much smaller. In fact, if n is the product of all primes $\le x$ we have

$$\varphi(n) = n \prod_{p \le x} \left(1 - \frac{1}{p}\right).$$

The next theorem gives the asymptotic behavior of this product for large x.

Theorem 13.13 *There is a positive constant c such that, for $x \ge 2$,*

$$\prod_{p \le x} \left(1 - \frac{1}{p}\right) = \frac{c}{\log x} + O\left(\frac{1}{\log^2 x}\right). \tag{32}$$

Note. It can be shown that $c = e^{-C}$, where C is Euler's constant. (See [31].)

PROOF. Let $P(x)$ denote the product in (32). Then $\log P(x) = \sum_{p \le x} \log(1 - 1/p)$. To estimate this sum we use the power series expansion

$$-\log(1 - t) = t + \frac{t^2}{2} + \frac{t^3}{3} + \cdots + \frac{t^n}{n} + \cdots \quad (|t| < 1)$$

with $t = 1/p$. Transposing one term we find, with $a_p = -\log(1 - 1/p) - 1/p$,

$$0 < a_p = \frac{1}{2p^2} + \frac{1}{3p^3} + \cdots < \frac{1}{2}\left(\frac{1}{p^2} + \frac{1}{p^3} + \cdots\right) = \frac{1}{2p(p-1)}.$$

This inequality shows that the infinite series

$$\sum_p a_p = \sum_p \left\{-\log\left(1 - \frac{1}{p}\right) - \frac{1}{p}\right\} \tag{33}$$

converges, since it is dominated by $\sum_{n=2}^{\infty} 1/n(n-1)$. If B denotes the sum of the series in (33) we have

$$0 < B - \sum_{p \le x} a_p = \sum_{p > x} a_p \le \sum_{n \ge x} \frac{1}{n(n-1)} = -\sum_{n \ge x} \left(\frac{1}{n} - \frac{1}{n-1}\right) = O\left(\frac{1}{x}\right).$$

Hence

$$\sum_{p \le x} a_p = B + O\left(\frac{1}{x}\right),$$

or

$$-\log P(x) = \sum_{p \le x} \frac{1}{p} + B + O\left(\frac{1}{x}\right).$$

But by Theorem 4.12 the sum on the right is $\log \log x + A + O(1/\log x)$ so

$$\log P(x) = -\log \log x - B - A + O\left(\frac{1}{\log x}\right).$$

Therefore

$$P(x) = \exp\{\log P(x)\} = e^{-B-A}e^{-\log \log x}e^{O(1/\log x)}.$$

Now let $c = e^{-B-A}$ and use the inequality $e^u = 1 + O(u)$ for $0 < u < 1$ to obtain

$$P(x) = \frac{c}{\log x}\left\{1 + O\left(\frac{1}{\log x}\right)\right\} = \frac{c}{\log x} + O\left(\frac{1}{\log^2 x}\right).$$

This completes the proof. □

Theorem 13.14 *Let c be the constant of Theorem* 13.13, *and let* $\varepsilon > 0$ *be given.*

(a) *There exists an* $N(\varepsilon)$ *such that*

$$\varphi(n) \geq (1 - \varepsilon)\frac{cn}{\log \log n} \quad \textit{for all } n \geq N(\varepsilon).$$

(b) *For infinitely many n we have*

$$\varphi(n) \leq (1 + \varepsilon)\frac{cn}{\log \log n}.$$

In other words,

$$\liminf_{n\to\infty} \frac{\varphi(n)\log \log n}{n} = c.$$

Proof. We prove part (b) first. Take $n = \prod_{p \leq x} p$. Then

$$\frac{\varphi(n)}{n} = \prod_{p\leq x}\left(1 - \frac{1}{p}\right) = \frac{c}{\log x} + O\left(\frac{1}{\log^2 x}\right).$$

But $\log n = \vartheta(x) = (1 + o(1))x$, so $\log \log n = (1 + o(1))\log x$, hence

$$\frac{\varphi(n)}{n} = \frac{c(1 + o(1))}{\log \log n} + O\left(\frac{1}{(\log \log n)^2}\right) = \frac{c(1 + o(1))}{\log \log n} \leq (1 + \varepsilon)\frac{c}{\log \log n}$$

if $n \geq N(\varepsilon)$ for some $N(\varepsilon)$. This proves (b).

To prove (a) take any $n > 1$ and write

$$\frac{\varphi(n)}{n} = \prod_{p|n}\left(1 - \frac{1}{p}\right) = P_1(n)P_2(n)$$

where

$$P_1(n) = \prod_{\substack{p|n \\ p \leq \log n}}\left(1 - \frac{1}{p}\right), \quad \text{and} \quad P_2(n) = \prod_{\substack{p|n \\ p > \log n}}\left(1 - \frac{1}{p}\right).$$

Then

$$P_2(n) > \prod_{\substack{p|n \\ p > \log n}} \left(1 - \frac{1}{\log n}\right) = \left(1 - \frac{1}{\log n}\right)^{f(n)} \tag{34}$$

where $f(n)$ is the number of primes which divide n and exceed $\log n$. Since

$$n \geq \prod_{p|n} p > \prod_{\substack{p|n \\ p > \log n}} p \geq (\log n)^{f(n)}$$

we find $\log n > f(n)\log\log n$, so $f(n) < \log n/\log\log n$. Since $1 - (1/\log n) < 1$, inequality (34) gives us

$$P_2(n) > \left(1 - \frac{1}{\log n}\right)^{\log n/\log\log n} = \left\{\left(1 - \frac{1}{\log n}\right)^{\log n}\right\}^{1/\log\log n} \tag{35}$$

Now $(1 - (1/u))^u \to e^{-1}$ as $u \to \infty$ so the last member in (35) tends to 1 as $n \to \infty$. Hence (35) gives us

$$P_2(n) > 1 + o(1) \quad \text{as } n \to \infty.$$

Therefore

$$\frac{\varphi(n)}{n} = P_1(n)P_2(n) > (1 + o(1)) \prod_{\substack{p|n \\ p \leq \log n}} \left(1 - \frac{1}{p}\right) \geq (1 + o(1)) \prod_{p \leq \log n} \left(1 - \frac{1}{p}\right)$$

$$= (1 + o(1)) \frac{c}{\log\log n}(1 + o(1)) \geq (1 - \varepsilon)\frac{c}{\log\log n}$$

if $n \geq N(\varepsilon)$. This proves part (a). □

13.12 Extension of Pólya's inequality for character sums

We conclude this chapter by extending Pólya's inequality (Theorem 8.21) to arbitrary nonprincipal characters. The proof makes use of the estimate for the divisor function,

$$d(n) = O(n^\delta)$$

obtained in (31).

Theorem 13.15 *If χ is any nonprincipal character* mod k, *then for all $x \geq 2$ we have*

$$\sum_{m \leq x} \chi(m) = O(\sqrt{k}\log k).$$

Proof. If χ is primitive, Theorem 8.21 shows that

$$\sum_{m \le x} \chi(m) < \sqrt{k} \log k.$$

Now consider any nonprincipal character χ mod k and let c denote the conductor of χ. Then $c \mid k$, $c < k$, and we can write

$$\chi(m) = \psi(m)\chi_1(m)$$

where χ_1 is the principal character mod k and ψ is a primitive character mod c. Then

$$\sum_{m \le x} \chi(m) = \sum_{\substack{m \le x \\ (m,k)=1}} \psi(m) = \sum_{m \le x} \psi(m) \sum_{d \mid (m,k)} \mu(d) = \sum_{m \le x} \sum_{\substack{d \mid k \\ d \mid m}} \mu(d)\psi(m)$$

$$= \sum_{d \mid k} \mu(d) \sum_{q \le x/d} \psi(qd) = \sum_{d \mid k} \mu(d)\psi(d) \sum_{q \le x/d} \psi(q).$$

Hence

$$\left| \sum_{m \le x} \chi(m) \right| \le \sum_{d \mid k} |\mu(d)\psi(d)| \left| \sum_{q \le x/d} \psi(q) \right| < \sqrt{c} \log c \sum_{d \mid k} |\mu(d)\psi(d)| \tag{36}$$

because ψ is primitive mod c. In the last sum each factor $|\mu(d)\psi(d)|$ is either 0 or 1. If $|\mu(d)\psi(d)| = 1$ then $|\mu(d)| = 1$ so d is a squarefree divisor of k, say

$$d = p_1 p_2 \cdots p_r.$$

Also, $|\psi(d)| = 1$ so $(d, c) = 1$, which means no prime factor p_i divides c. Hence each p_i divides k/c so d divides k/c. In other words,

$$\sum_{d \mid k} |\mu(d)\psi(d)| \le \sum_{d \mid k/c} 1 = d\left(\frac{k}{c}\right) = O\left(\left(\frac{k}{c}\right)^{\delta}\right)$$

for every $\delta > 0$. In particular, $d(k/c) = O(\sqrt{k/c})$ so (36) implies

$$\sum_{m \le x} \chi(m) = O\left(\sqrt{\frac{k}{c}} \sqrt{c} \log c\right) = O(\sqrt{k} \log c) = O(\sqrt{k} \log k). \qquad \square$$

Exercises for Chapter 13

1. Chebyshev proved that if $\psi(x)/x$ tends to a limit as $x \to \infty$ then this limit equals 1. A proof was outlined in Exercise 4.26. This exercise outlines another proof based on the identity

$$-\frac{\zeta'(s)}{\zeta(s)} = s \int_1^{\infty} \frac{\psi(x)}{x^{s+1}}\, dx, \qquad (\sigma > 1) \tag{37}$$

given in Exercise 11.1(d).

(a) Prove that $(1 - s)\zeta'(s)/\zeta(s) \to 1$ as $s \to 1$.

(b) Let $\delta = \limsup_{x\to\infty}(\psi(x)/x)$. Given $\varepsilon > 0$, choose $N = N(\varepsilon)$ so that $x \geq N$ implies $\psi(x) \leq (\delta + \varepsilon)x$. Keep s real, $1 < s \leq 2$, split the integral in (37) into two parts, $\int_1^N + \int_N^\infty$, and estimate each part to obtain the inequality

$$-\frac{\zeta'(s)}{\zeta(s)} \leq C(\varepsilon) + \frac{s(\delta + \varepsilon)}{s - 1},$$

where $C(\varepsilon)$ is a constant independent of s. Use (a) to deduce that $\delta \geq 1$.

(c) Let $\gamma = \liminf_{x\to\infty}(\psi(x)/x)$ and use a similar argument to deduce that $\gamma \leq 1$. Therefore if $\psi(x)/x$ tends to a limit as $x \to \infty$ then $\gamma = \delta = 1$.

2. Let $A(x) = \sum_{n\leq x} a(n)$, where

$$a(n) = \begin{cases} 0 & \text{if } n \neq \text{a prime power}, \\ \dfrac{1}{k} & \text{if } n = p^k. \end{cases}$$

Prove that $A(x) = \pi(x) + O(\sqrt{x}\log\log x)$.

3. (a) If $c > 1$ and $x \neq$ integer, prove that if $x > 1$,

$$\frac{1}{2\pi i}\int_{c-\infty i}^{c+\infty i} \log\zeta(s)\frac{x^s}{s}\,ds = \pi(x) + \frac{1}{2}\pi(x^{1/2}) + \frac{1}{3}\pi(x^{1/3}) + \cdots.$$

(b) Show that the prime number theorem is equivalent to the asymptotic relation

$$\frac{1}{2\pi i}\int_{c-\infty i}^{c+\infty i} \log\zeta(s)\frac{x^s}{s}\,ds \sim \frac{x}{\log x} \quad \text{as } x \to \infty.$$

A proof of the prime number theorem based on this relation was given by Landau in 1903.

4. Let $M(x) = \sum_{n\leq x}\mu(n)$. The exact order of magnitude of $M(x)$ for large x is not known. In Chapter 4 it was shown that the prime number theorem is equivalent to the relation $M(x) = o(x)$ as $x \to \infty$. This exercise relates the order of magnitude of $M(x)$ with the Riemann hypothesis.

Suppose there is a positive constant θ such that

$$M(x) = O(x^\theta) \quad \text{for } x \geq 1.$$

Prove that the formula

$$\frac{1}{\zeta(s)} = s\int_1^\infty \frac{M(x)}{x^{s+1}}\,dx,$$

which holds for $\sigma > 1$ (see Exercise 11.1(c)) would also be valid for $\sigma > \theta$. Deduce that $\zeta(s) \neq 0$ for $\sigma > \theta$. In particular, this shows that the relation $M(x) = O(x^{1/2+\varepsilon})$ for every $\varepsilon > 0$ implies the Riemann hypothesis. It can also be shown that the Riemann hypothesis implies $M(x) = O(x^{1/2+\varepsilon})$ for every $\varepsilon > 0$. (See Titchmarsh [69], p. 315.)

5. Prove the following lemma, which is similar to Lemma 2. Let

$$A_1(x) = \int_1^x \frac{A(u)}{u}\,du,$$

where $A(u)$ is a nonnegative increasing function for $u \geq 1$. If we have the asymptotic formula

$$A_1(x) \sim Lx^c \quad \text{as } x \to \infty,$$

for some $c > 0$ and $L > 0$, then we also have

$$A(x) \sim cLx^c \quad \text{as } x \to \infty.$$

6. Prove that

$$\frac{1}{2\pi i}\int_{2-\infty i}^{2+\infty i} \frac{y^s}{s^2}\,ds = 0 \quad \text{if } 0 < y < 1.$$

What is the value of this integral if $y \geq 1$?

7. Express

$$\frac{1}{2\pi i}\int_{2-\infty i}^{2+\infty i} \frac{x^s}{s^2}\left(-\frac{\zeta'(s)}{\zeta(s)}\right) ds$$

as a finite sum involving $\Lambda(n)$.

8. Let χ be any Dirichlet character mod k with χ_1 the principal character. Define

$$F(\sigma, t) = 3\frac{L'}{L}(\sigma, \chi_1) + 4\frac{L'}{L}(\sigma + it, \chi) + \frac{L'}{L}(\sigma + 2it, \chi^2).$$

If $\sigma > 1$ prove that $F(\sigma, t)$ has real part equal to

$$-\sum_{n=1}^{\infty} \frac{\Lambda(n)}{n^\sigma} \operatorname{Re}\{3\chi_1(n) + 4\chi(n)n^{-it} + \chi^2(n)n^{-2it}\}$$

and deduce that $\operatorname{Re} F(\sigma, t) \leq 0$.

9. Assume that $L(s, \chi)$ has a zero of order $m \geq 1$ at $s = 1 + it$. Prove that for this t we have:

(a) $\displaystyle \frac{L'}{L}(\sigma + it, \chi) = \frac{m}{\sigma - 1} + O(1) \quad \text{as } \sigma \to 1+,$

and

(b) there exists an integer $r \geq 0$ such that

$$\frac{L'}{L}(\sigma + 2it, \chi^2) = \frac{r}{\sigma - 1} + O(1) \quad \text{as } \sigma \to 1+,$$

except when $\chi^2 = \chi_1$ and $t = 0$.

10. Use Exercises 8 and 9 to prove that

$$L(1 + it, \chi) \neq 0 \quad \text{for all real } t \text{ if } \chi^2 \neq \chi_1$$

and that

$$L(1 + it, \chi) \neq 0 \quad \text{for all real } t \neq 0 \text{ if } \chi^2 = \chi_1.$$

[*Hint:* Consider $F(\sigma, t)$ as $\sigma \to 1+$.]

11. For any arithmetical function $f(n)$, prove that the following statements are equivalent:

(a) $f(n) = O(n^\varepsilon)$ for every $\varepsilon > 0$ and all $n \geq n_1$.
(b) $f(n) = o(n^\delta)$ for every $\delta > 0$ as $n \to \infty$.

12. Let $f(n)$ be a multiplicative function such that if p is prime then

$$f(p^m) \to 0 \quad \text{as } p^m \to \infty.$$

That is, for every $\varepsilon > 0$ there is an $N(\varepsilon)$ such that $|f(p^m)| < \varepsilon$ whenever $p^m > N(\varepsilon)$. Prove that $f(n) \to 0$ as $n \to \infty$.
[*Hint:* There is a constant $A > 0$ such that $|f(p^m)| < A$ for all primes p and all $m \geq 0$, and a constant $B > 0$ such that $|f(p^m)| < 1$ whenever $p^m > B$.]

13. If $\alpha \geq 0$ let $\sigma_\alpha(n) = \sum_{d|n} d^\alpha$. Prove that for every $\delta > 0$ we have

$$\sigma_\alpha(n) = o(n^{\alpha+\delta}) \quad \text{as } n \to \infty.$$

[*Hint*: Use Exercise 12.]

14 Partitions

14.1 Introduction

Until now this book has been concerned primarily with *multiplicative number theory*, a study of arithmetical functions related to prime factorization of integers. We turn now to another branch of number theory called *additive number theory*. A basic problem here is that of expressing a given positive integer n as a sum of integers from some given set A, say

$$A = \{a_1, a_2, \ldots\},$$

where the elements a_i are special numbers such as primes, squares, cubes, triangular numbers, etc. Each representation of n as a sum of elements of A is called a *partition* of n and we are interested in the arithmetical function $A(n)$ which counts the number of partitions of n into summands taken from A. We illustrate with some famous examples.

Goldbach conjecture *Every even $n > 4$ is the sum of two odd primes.*

In this example $A(n)$ is the number of solutions of the equation

$$n = p_1 + p_2, \tag{1}$$

where the p_i are odd primes. Goldbach's assertion is that $A(n) \geq 1$ for even $n > 4$. This conjecture dates back to 1742 and is undecided to this date. In 1937 the Russian mathematician Vinogradov proved that every sufficiently large odd number is the sum of three odd primes. In 1966 the Chinese mathematician Chen Jing-run proved that every sufficiently large even number is the sum of a prime plus a number with no more than two prime factors. (See [10].)

Representation by squares *For a given integer $k \geq 2$ consider the partition function $r_k(n)$ which counts the number of solutions of the equation*

$$n = x_1^2 + \cdots + x_k^2, \tag{2}$$

where the x_i may be positive, negative or zero, and the order of summands is taken into account.

For $k = 2, 4, 6$, or 8, Jacobi [34] expressed $r_k(n)$ in terms of divisor functions. For example, he proved that

$$r_2(n) = 4\{d_1(n) - d_3(n)\},$$

where $d_1(n)$ and $d_3(n)$ are the number of divisors of n congruent to 1 and 3 mod 4, respectively. Thus, $r_2(5) = 8$ because both divisors, 1 and 5, are congruent to 1 mod 4. In fact there are four representations given by

$$5 = 2^2 + 1^2 = (-2)^2 + 1^2 = (-2)^2 + (-1)^2 = 2^2 + (-1)^2,$$

and four more with the order of summands reversed.

For $k = 4$ Jacobi proved that

$$r_4(n) = 8 \sum_{d|n} d = 8\sigma(n) \quad \text{if } n \text{ is odd},$$

$$= 24 \sum_{\substack{d|n \\ d \text{ odd}}} d \quad \text{if } n \text{ is even}.$$

The formulas for $r_6(n)$ and $r_8(n)$ are a bit more complicated but of the same general type. (See [14].)

Exact formulas for $r_k(n)$ have also been found for $k = 3, 5$, or 7; they involve Jacobi's extension of Legendre's symbol for quadratic residues. For example, if n is odd it is known that

$$r_3(n) = 24 \sum_{m \leq n/4} (m|n) \quad \text{if } n \equiv 1 \pmod 4$$

$$= 8 \sum_{m \leq n/2} (m|n) \quad \text{if } n \equiv 3 \pmod 4,$$

where now the numbers x_1, x_2, x_3 in (2) are taken to be relatively prime.

For larger values of k the analysis of $r_k(n)$ is considerably more complicated. There is a large literature on the subject with contributions by Mordell, Hardy, Littlewood, Ramanujan, and many others. For $k \geq 5$ it is known that $r_k(n)$ can be expressed by an asymptotic formula of the form

$$r_k(n) = \rho_k(n) + R_k(n), \tag{3}$$

where $\rho_k(n)$ is the principal term, given by the infinite series

$$\rho_k(n) = \frac{\pi^{k/2}n^{k/2-1}}{\Gamma\left(\frac{k}{2}\right)} \sum_{q=1}^{\infty} \sum_{\substack{h=1 \\ (h,q)=1}}^{q} \left(\frac{G(h;q)}{q}\right)^k e^{-2\pi inh/q},$$

and $R_k(n)$ is a remainder term of smaller order. The series for $\rho_k(n)$ is called the *singular series* and the numbers $G(h; q)$ are quadratic Gauss sums,

$$G(h;q) = \sum_{r=1}^{q} e^{2\pi ihr^2/q}.$$

In 1917 Mordell noted that $r_k(n)$ is the coefficient of x^n in the power series expansion of the kth power of the series

$$\vartheta = 1 + 2\sum_{n=1}^{\infty} x^{n^2}.$$

The function ϑ is related to elliptic modular functions which play an important role in the derivation of (3).

Waring's problem *To determine whether, for a given positive integer k, there is an integer s (depending only on k) such that the equation*

(4) $$n = x_1^k + x_2^k + \cdots + x_s^k$$

has solutions for every $n \geq 1$.

The problem is named for the English mathematician E. Waring who stated in 1770 (without proof and with limited numerical evidence) that every n is the sum of 4 squares, of 9 cubes, of 19 fourth powers, etc. In this example the partition function $A(n)$ is the number of solutions of (4), and the problem is to decide if there exists an s such that $A(n) \geq 1$ for all n.

If s exists for a given k then there is a least value of s and this is denoted by $g(k)$. Lagrange proved the existence of $g(2)$ in 1770 and, during the next 139 years, the existence of $g(k)$ was shown for $k = 3, 4, 5, 6, 7, 8$ and 10. In 1909 Hilbert proved the existence of $g(k)$ for every k by an inductive argument but did not determine its numerical value for any k. The exact value of $g(k)$ is now known for every k except $k = 4$. Hardy and Littlewood gave an asymptotic formula for the number of solutions of (4) in terms of a singular series analogous to that in (3). For a historical account of Waring's problem see W. J. Ellison [18].

Unrestricted partitions

One of the most fundamental problems in additive number theory is that of unrestricted partitions. The set of summands consists of all positive integers, and the partition function to be studied is the number of ways n

can be written as a sum of positive integers $\leq n$, that is, the number of solutions of

$$n = a_{i_1} + a_{i_2} + \cdots \tag{5}$$

The number of summands is unrestricted, repetition is allowed, and the order of the summands is not taken into account. The corresponding partition function is denoted by $p(n)$ and is called the *unrestricted partition function*, or simply *the partition function*. The summands are called *parts*. For example, there are exactly five partitions of 4, given by

$$4 = 3 + 1 = 2 + 2 = 2 + 1 + 1 = 1 + 1 + 1 + 1,$$

so $p(4) = 5$. Similarly, $p(5) = 7$, the partitions of 5 being

$$\begin{aligned} 5 &= 4 + 1 = 3 + 2 = 3 + 1 + 1 = 2 + 2 + 1 = 2 + 1 + 1 + 1 \\ &= 1 + 1 + 1 + 1 + 1. \end{aligned}$$

The rest of this chapter is devoted to a study of $p(n)$ and related functions.

14.2 Geometric representation of partitions

There is a simple way of representing partitions geometrically by using a display of lattice points called a *graph*. For example, the partition of 15 given by

$$6 + 3 + 3 + 2 + 1$$

can be represented by 15 lattice points arranged in five rows as follows:

```
. . . . . .
. . .
. . .
. .
.
```

If we read this graph vertically we get another partition of 15,

$$5 + 4 + 3 + 1 + 1 + 1.$$

Two such partitions are said to be *conjugate*. Note that the *largest* part in either of these partitions is equal to the *number* of parts in the other. Thus we have the following theorem.

Theorem 14.1 *The number of partitions of n into m parts is equal to the number of partitions of n into parts, the largest of which is m.*

Several theorems can be proved by simple combinatorial arguments involving graphs, and we will return later to a beautiful illustration of this method. However, the deepest results in the theory of partitions require a more analytical treatment to which we turn now.

14.3 Generating functions for partitions

A function $F(s)$ defined by a Dirichlet series $F(s) = \sum f(n)n^{-s}$ is called a *generating function* of the coefficients $f(n)$. Dirichlet series are useful generating functions in multiplicative number theory because of the relation

$$n^{-s}m^{-s} = (nm)^{-s}.$$

In additive number theory it is more convenient to use generating functions represented by power series,

$$F(x) = \sum f(n)x^n$$

because $x^n x^m = x^{n+m}$. The next theorem exhibits a generating function for the partition function $p(n)$.

Theorem 14.2 Euler. *For* $|x| < 1$ *we have*

$$\prod_{m=1}^{\infty} \frac{1}{1 - x^m} = \sum_{n=0}^{\infty} p(n)x^n,$$

where $p(0) = 1$.

PROOF. First we give a formal derivation of this identity, ignoring questions of convergence, then we give a more rigorous proof.

If each factor in the product is expanded into a power series (a geometric series) we get

$$\prod_{n=1}^{\infty} \frac{1}{1 - x^n} = (1 + x + x^2 + \cdots)(1 + x^2 + x^4 + \cdots)(1 + x^3 + x^6 + \cdots)\cdots$$

Now we multiply the series on the right, treating them as though they were polynomials, and collect like powers of x to obtain a power series of the form

$$1 + \sum_{k=1}^{\infty} a(k)x^k.$$

We wish to show that $a(k) = p(k)$. Suppose we take the term x^{k_1} from the first series, the term x^{2k_2} from the second, the term x^{3k_3} from the third, . . . , and the term x^{mk_m} from the mth, where each $k_i \geq 0$. Their product is

$$x^{k_1}x^{2k_2}x^{3k_3}\cdots x^{mk_m} = x^k,$$

say, where

$$k = k_1 + 2k_2 + 3k_3 + \cdots + mk_m.$$

This can also be written as follows:

$$k = (1 + 1 + \cdots + 1) + (2 + 2 + \cdots + 2) + \cdots + (m + m + \cdots + m),$$

where the first parenthesis contains k_1 ones, the second k_2 twos, and so on. This is a partition of k into positive summands. Thus, each partition of k will produce one such term x^k and, conversely, each term x^k comes from a corresponding partition of k. Therefore $a(k)$, the coefficient of x^k, is equal to $p(k)$, the number of partitions of k.

The foregoing argument is not a rigorous proof because we have ignored questions of convergence and we have also multiplied together infinitely many geometric series, treating them as though they were polynomials. However, it is not difficult to transform the above ideas into a rigorous proof.

For this purpose we restrict x to lie in the interval $0 \le x < 1$ and introduce two functions,

$$F_m(x) = \prod_{k=1}^{m} \frac{1}{1 - x^k}, \qquad \text{and } F(x) = \prod_{k=1}^{\infty} \frac{1}{1 - x^k} = \lim_{m \to \infty} F_m(x).$$

The product defining $F(x)$ converges absolutely if $0 \le x < 1$ because its reciprocal $\prod(1 - x^k)$ converges absolutely (since the series $\sum x^k$ converges absolutely). Note also that for each fixed x the sequence $\{F_m(x)\}$ is increasing because

$$F_{m+1}(x) = \frac{1}{1 - x^{m+1}} F_m(x) \ge F_m(x).$$

Thus $F_m(x) \le F(x)$ for each fixed x, $0 \le x < 1$, and every m. Now $F_m(x)$ is the product of a finite number of absolutely convergent series. Therefore it, too, is an absolutely convergent series which we can write as

$$F_m(x) = 1 + \sum_{k=1}^{\infty} p_m(k)x^k.$$

Here $p_m(k)$ is the number of solutions of the equation

$$k = k_1 + 2k_2 + \cdots + mk_m.$$

In other words, $p_m(k)$ is the number of partitions of k into parts not exceeding m. If $m \ge k$, then $p_m(k) = p(k)$. Therefore we always have

$$p_m(k) \le p(k)$$

with equality when $m \ge k$. In other words, we have

$$\lim_{m \to \infty} p_m(k) = p(k).$$

Now we split the series for $F_m(x)$ into two parts,

$$\begin{aligned} F_m(x) &= \sum_{k=0}^{m} p_m(k)x^k + \sum_{k=m+1}^{\infty} p_m(k)x^k \\ &= \sum_{k=0}^{m} p(k)x^k + \sum_{k=m+1}^{\infty} p_m(k)x^k. \end{aligned}$$

Since $x \geq 0$ we have

$$\sum_{k=0}^{m} p(k)x^k \leq F_m(x) \leq F(x).$$

This shows that the series $\sum_{k=0}^{\infty} p(k)x^k$ converges. Moreover, since $p_m(k) \leq p(k)$ we have

$$\sum_{k=0}^{\infty} p_m(k)x^k \leq \sum_{k=0}^{\infty} p(k)x^k \leq F(x)$$

so, for each fixed x, the series $\sum p_m(k)x^k$ converges uniformly in m. Letting $m \to \infty$ we get

$$F(x) = \lim_{m\to\infty} F_m(x) = \lim_{m\to\infty} \sum_{k=0}^{\infty} p_m(k)x^k = \sum_{k=0}^{\infty} \lim_{m\to\infty} p_m(x)x^k = \sum_{k=0}^{\infty} p(k)x^k,$$

which proves Euler's identity for $0 \leq x < 1$. We extend it by analytic continuation to the unit disk $|x| < 1$. □

Table 14.1 Generating functions

Generating function	The number of partitions of n into parts which are
$\prod_{m=1}^{\infty} \frac{1}{1 - x^{2m-1}}$	odd
$\prod_{m=1}^{\infty} \frac{1}{1 - x^{2m}}$	even
$\prod_{m=1}^{\infty} \frac{1}{1 - x^{m^2}}$	squares
$\prod_{p} \frac{1}{1 - x^p}$	primes
$\prod_{m=1}^{\infty} (1 + x^m)$	unequal
$\prod_{m=1}^{\infty} (1 + x^{2m-1})$	odd and unequal
$\prod_{m=1}^{\infty} (1 + x^{2m})$	even and unequal
$\prod_{m=1}^{\infty} (1 + x^{m^2})$	distinct squares
$\prod_{p} (1 + x^p)$	distinct primes

By similar arguments we can readily find the generating functions of many other partition functions. We mention a few examples in Table 14.1.

14.4 Euler's pentagonal-number theorem

We consider next the partition function generated by the product $\prod(1 - x^m)$, the reciprocal of the generating function of $p(n)$. Write

$$\prod_{m=1}^{\infty}(1 - x^m) = 1 + \sum_{n=1}^{\infty} a(n)x^n.$$

To express $a(n)$ as a partition function we note that every partition of n into unequal parts produces a term x^n on the right with a coefficient $+1$ or -1. The coefficient is $+1$ if x^n is the product of an even number of terms, and -1 otherwise. Therefore,

$$a(n) = p_e(n) - p_o(n),$$

where $p_e(n)$ is the number of partitions of n into an even number of unequal parts, and $p_o(n)$ is the number of partitions into an odd number of unequal parts. Euler proved that $p_e(n) = p_o(n)$ for all n except those belonging to a special set called pentagonal numbers.

The pentagonal numbers 1, 5, 12, 22, ... were mentioned in the Historical Introduction. They are related to the pentagons shown in Figure 14.1.

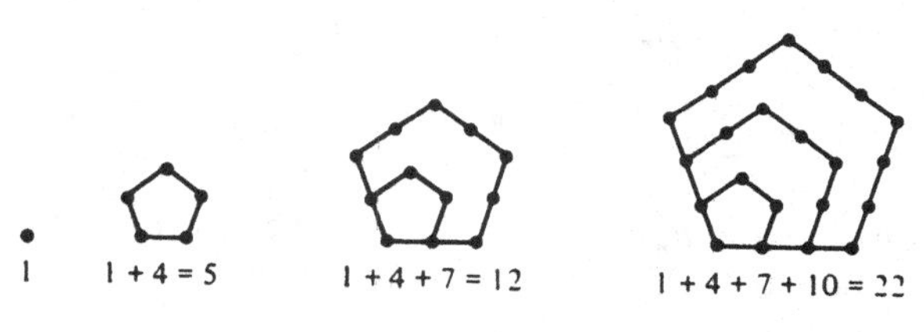

Figure 14.1

These numbers are also the partial sums of the terms in the arithmetic progression

$$1, 4, 7, 10, 13, \ldots, 3n + 1, \ldots$$

If $\omega(n)$ denotes the sum of the first n terms in this progression then

$$\omega(n) = \sum_{k=0}^{n-1}(3k + 1) = \frac{3n(n - 1)}{2} + n = \frac{3n^2 - n}{2}.$$

The numbers $\omega(n)$ and $\omega(-n) = (3n^2 + n)/2$ are called the *pentagonal numbers*.

Theorem 14.3 Euler's pentagonal-number theorem. *If* $|x| < 1$ *we have*

$$\prod_{m=1}^{\infty}(1 - x^m) = 1 - x - x^2 + x^5 + x^7 - x^{12} - x^{15} + \cdots$$

$$= 1 + \sum_{n=1}^{\infty}(-1)^n\{x^{\omega(n)} + x^{\omega(-n)}\} = \sum_{n=-\infty}^{\infty}(-1)^n x^{\omega(n)}.$$

PROOF. First we prove the result for $0 \le x < 1$ and then extend it to the disk $|x| < 1$ by analytic continuation. Define $P_0 = S_0 = 1$ and, for $n \ge 1$, let

$$P_n = \prod_{r=1}^{n}(1 - x^r) \qquad \text{and } S_n = 1 + \sum_{r=1}^{n}(-1)^r\{x^{\omega(r)} + x^{\omega(-r)}\}.$$

The infinite product $\prod(1 - x^m)$ converges so $P_n \to \prod(1 - x^m)$ as $n \to \infty$. We will prove (using a method of Shanks [63]) that

$$|S_n - P_n| \le nx^{n+1}. \tag{6}$$

Since $nx^{n+1} \to 0$ as $n \to \infty$ this will prove Euler's identity for $0 \le x < 1$.

To prove (6) we let $g(r) = r(r + 1)/2$ and introduce the sums

$$F_n = \sum_{r=0}^{n}(-1)^r \frac{P_n}{P_r} x^{rn+g(r)}.$$

We show first that F_n is a disguised form of S_n. It is easily verified that $F_1 = S_1 = 1 - x - x^2$. Therefore, if we show that

$$F_n - F_{n-1} = S_n - S_{n-1}, \qquad \text{or } F_n - S_n = F_{n-1} - S_{n-1},$$

this will prove that $F_n = S_n$ for all $n \ge 1$. Now

$$F_n - F_{n-1} = \sum_{r=0}^{n}(-1)^r \frac{P_n}{P_r} x^{rn+g(r)} - \sum_{r=0}^{n-1}(-1)^r \frac{P_{n-1}}{P_r} x^{r(n-1)+g(r)}.$$

In the first sum we write $P_n = (1 - x^n)P_{n-1}$ and separate the term with $r = n$. Then we distribute the difference $1 - x^n$ to obtain

$$F_n - F_{n-1} = (-1)^n x^{n^2+g(n)} + \sum_{r=0}^{n-1}(-1)^r \frac{P_{n-1}}{P_r} x^{rn+g(r)}$$

$$- \sum_{r=0}^{n-1}(-1)^r \frac{P_{n-1}}{P_r} x^{(r+1)n+g(r)} - \sum_{r=0}^{n-1}(-1)^r \frac{P_{n-1}}{P_r} x^{r(n-1)+g(r)}.$$

Now combine the first and third sums and note that the term with $r = 0$ cancels. In the second sum we shift the index and obtain

$$F_n - F_{n-1} = (-1)^n x^{n^2+g(n)} + \sum_{r=1}^{n-1}(-1)^r \frac{P_{n-1}}{P_r} x^{r(n-1)+g(r)}(x^r - 1)$$

$$- \sum_{r=1}^{n}(-1)^{r-1} \frac{P_{n-1}}{P_{r-1}} x^{rn+g(r-1)}.$$

But $(x^r - 1)/P_r = -1/P_{r-1}$ and $r(n-1) + g(r) = rn + g(r-1)$ so the last two sums cancel term by term except for the term with $r = n$ in the second sum. Thus we get

$$F_n - F_{n-1} = (-1)^n x^{n^2+g(n)} + (-1)^n x^{n^2+g(n-1)}.$$

But

$$n^2 + g(n) = n^2 + \frac{n(n+1)}{2} = \omega(-n) \qquad \text{and } n^2 + g(n-1) = \omega(n),$$

so

$$F_n - F_{n-1} = (-1)^n\{x^{\omega(n)} + x^{\omega(-n)}\} = S_n - S_{n-1},$$

and hence $F_n = S_n$ for all $n \geq 1$. In the sum defining F_n the first term is P_n so

$$F_n = P_n + \sum_{r=1}^{n} (-1)^r \frac{P_n}{P_r} x^{rn+g(r)}. \tag{7}$$

Note that $0 < P_n/P_r \leq 1$ since $0 \leq x < 1$. Also, each factor $x^{rn+g(r)} \leq x^{n+1}$ so the sum on the right of (7) is bounded above by nx^{n+1}. Therefore $|F_n - P_n| \leq nx^{n+1}$ and, since $F_n = S_n$, this proves (6) and completes the proof of Euler's identity. □

14.5 Combinatorial proof of Euler's pentagonal-number theorem

Euler proved his pentagonal-number theorem by induction in 1750. Later proofs were obtained by Legendre in 1830 and Jacobi in 1846. This section describes a remarkable combinatorial proof given by F. Franklin [22] in 1881.

We have already noted that

$$\prod_{m=1}^{\infty} (1 - x^m) = 1 + \sum_{n=1}^{\infty} \{p_e(n) - p_o(n)\}x^n,$$

where $p_e(n)$ is the number of partitions of n into an *even* number of unequal parts, and $p_o(n)$ is the number of partitions into an *odd* number of unequal parts. Franklin used the graphical representation of partitions by lattice points to show that there is a one-to-one correspondence between partitions of n into an odd and even number of unequal parts, so that $p_e(n) = p_o(n)$, except when n is a pentagonal number.

Consider the graph of any partition of n into unequal parts. We say the graph is in *standard form* if the parts are arranged in decreasing order, as illustrated by the example in Figure 14.2. The longest line segment connecting points in the last row is called the *base* of the graph, and the number of lattice points on the base is denoted by b. Thus, $b \geq 1$. The longest 45° line segment joining the last point in the first row with other points in the

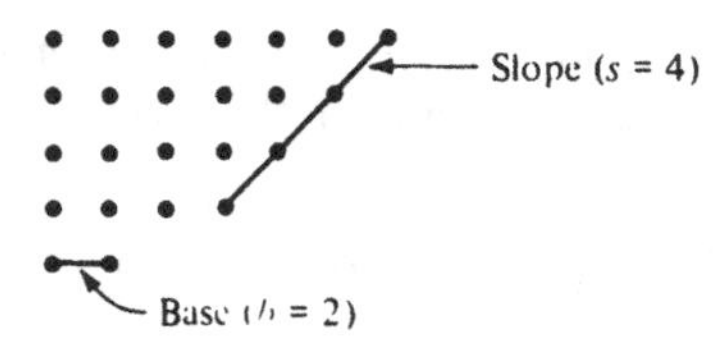

Figure 14.2

graph is called the *slope*, and the number of lattice points on the slope is denoted by s. Thus, $s \geq 1$. In Figure 14.2 we have $b = 2$ and $s = 4$.

Now we define two operations A and B on this graph. Operation A moves the points on the base so that they lie on a line parallel to the slope, as indicated in Figure 14.3(a). Operation B moves the points on the slope so that they lie on a line parallel to the base, as shown in Figure 14.3(b). We say an operation is *permissible* if it preserves the standard form of the graph, that is, if the new graph again has unequal parts arranged in descending order.

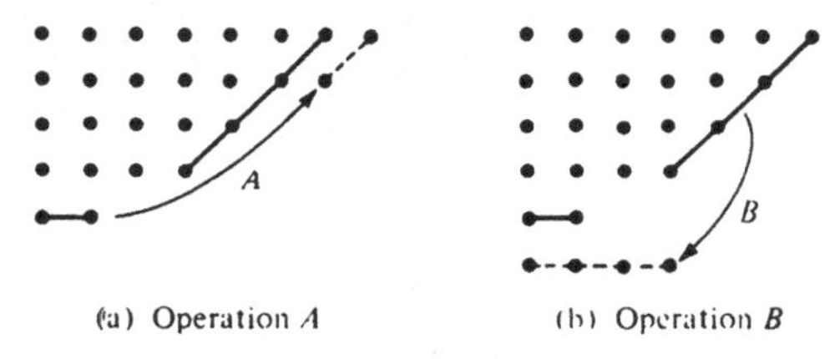

(a) Operation A (b) Operation B

Figure 14.3

If A is permissible we get a new partition of n into unequal parts, but the number of parts is one less than before. If B is permissible we get a new partition into unequal parts, but the number of parts is one greater than before. Therefore, if for every partition of n exactly one of A or B is permissible there will be a one-to-one correspondence between partitions of n into odd and even unequal parts, so $p_e(n) = p_o(n)$ for such n.

To determine whether A or B is permissible we consider three cases: (1) $b < s$; (2) $b = s$; (3) $b > s$.

Case 1: If $b < s$ then $b \leq s - 1$ so operation A is permissible but B is not since B destroys the standard form. (See Figure 14.3.)

Case 2: If $b = s$, operation B is not permissible since it results in a new graph not in standard form. Operation A is permissible except when the base and slope intersect, as shown in Figure 14.4(a), in which case the new graph is not in standard form.

Case 3: If $b > s$, operation A is not permissible, whereas B is permissible except when $b = s + 1$ and the base and slope intersect, as shown in Figure 14.4(b). In this case the new graph contains two equal parts.

Therefore, exactly one of A or B is permissible with the two exceptions noted above. Consider the first exceptional case, shown in Figure 14.4(a),

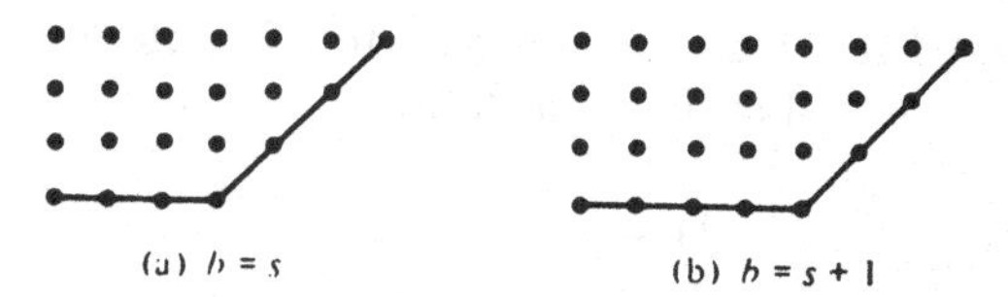

Figure 14.4 Neither A nor B is permissible.

and suppose there are k rows in the graph. Then $b = k$ also so the number n is given by

$$n = k + (k + 1) + \cdots + (2k - 1) = \frac{3k^2 - k}{2} = \omega(k).$$

For this partition of n we have an extra partition into even parts if k is even, and an extra partition into odd parts if k is odd, so

$$p_e(n) - p_o(n) = (-1)^k.$$

In the other exceptional case, shown in Figure 14.4(b), there is an additional lattice point in each row so

$$n = \frac{3k^2 - k}{2} + k = \frac{3k^2 + k}{2} = \omega(-k)$$

and again $p_e(n) - p_o(n) = (-1)^k$. This completes Franklin's proof of Euler's identity.

14.6 Euler's recursion formula for $p(n)$

Theorem 14.4 *Let $p(0) = 1$ and define $p(n)$ to be 0 if $n < 0$. Then for $n \geq 1$ we have*

(8) $$p(n) - p(n - 1) - p(n - 2) + p(n - 5) + p(n - 7) + \cdots = 0,$$

or, what amounts to the same thing,

$$p(n) = \sum_{k=1}^{\infty} (-1)^{k+1}\{p(n - \omega(k)) + p(n - \omega(-k))\}.$$

PROOF. Theorems 14.2 and 14.3 give us the identity

$$\left(1 + \sum_{k=1}^{\infty} (-1)^k\{x^{\omega(k)} + x^{\omega(-k)}\}\right)\left(\sum_{m=0}^{\infty} p(m)x^m\right) = 1.$$

If $n \geq 1$ the coefficient of x^n on the right is 0 so we immediately obtain (8) by equating coefficients. □

MacMahon used this recursion formula to compute $p(n)$ up to $n = 200$. Here are some sample values from his table.

$$\begin{aligned}
p(1) &= 1\\
p(5) &= 7\\
p(10) &= 42\\
p(15) &= 176\\
p(20) &= 627\\
p(25) &= 1{,}958\\
p(30) &= 5{,}604\\
p(40) &= 37{,}338\\
p(50) &= 204{,}226\\
p(100) &= 190{,}569{,}292\\
p(200) &= 3{,}972{,}999{,}029{,}388
\end{aligned}$$

These examples indicate that $p(n)$ grows very rapidly with n. The largest value of $p(n)$ yet computed is $p(14{,}031)$, a number with 127 digits. D. H. Lehmer [42] computed this number to verify a conjecture of Ramanujan which asserted that $p(14{,}031) \equiv 0 \pmod{11^4}$. The assertion was correct. Obviously, the recursion formula in (8) was not used to calculate this value of $p(n)$. Instead, Lehmer used an asymptotic formula of Rademacher [54] which implies

$$p(n) \sim \frac{e^{K\sqrt{n}}}{4n\sqrt{3}} \quad \text{as } n \to \infty,$$

where $K = \pi(2/3)^{1/2}$. For $n = 200$ the quantity on the right is approximately 4×10^{12} which is remarkably close to the actual value of $p(200)$ given in MacMahon's table.

In the sequel to this volume we give a derivation of Rademacher's asymptotic formula for $p(n)$. The proof requires considerable preparation from the theory of elliptic modular functions. The next section gives a crude upper bound for $p(n)$ which involves the exponential $e^{K\sqrt{n}}$ and which can be obtained with relatively little effort.

14.7 An upper bound for $p(n)$

Theorem 14.5 *If $n \geq 1$ we have $p(n) < e^{K\sqrt{n}}$, where $K = \pi(2/3)^{1/2}$.*

PROOF. Let

$$F(x) = \prod_{n=1}^{\infty}(1 - x^n)^{-1} = 1 + \sum_{k=1}^{\infty} p(k)x^k,$$

and restrict x to the interval $0 < x < 1$. Then we have $p(n)x^n < F(x)$, from which we obtain $\log p(n) + n \log x < \log F(x)$, or

$$\log p(n) < \log F(x) + n \log \frac{1}{x}. \tag{9}$$

We estimate the terms $\log F(x)$ and $n \log(1/x)$ separately. First we write

$$\log F(x) = -\log \prod_{n=1}^{\infty} (1 - x^n) = -\sum_{n=1}^{\infty} \log(1 - x^n) = \sum_{n=1}^{\infty} \sum_{m=1}^{\infty} \frac{x^{mn}}{m}$$

$$= \sum_{m=1}^{\infty} \frac{1}{m} \sum_{n=1}^{\infty} (x^m)^n = \sum_{m=1}^{\infty} \frac{1}{m} \frac{x^m}{1 - x^m}.$$

Since we have

$$\frac{1 - x^m}{1 - x} = 1 + x + x^2 + \cdots + x^{m-1},$$

and since $0 < x < 1$, we can write

$$mx^{m-1} < \frac{1 - x^m}{1 - x} < m,$$

and hence

$$\frac{m(1 - x)}{x} < \frac{1 - x^m}{x^m} < \frac{m(1 - x)}{x^m}.$$

Inverting and dividing by m we get

$$\frac{1}{m^2} \frac{x^m}{1 - x} \leq \frac{1}{m} \frac{x^m}{1 - x^m} \leq \frac{1}{m^2} \frac{x}{1 - x}.$$

Summing on m we obtain

$$\log F(x) = \sum_{m=1}^{\infty} \frac{1}{m} \frac{x^m}{1 - x^m} \leq \frac{x}{1 - x} \sum_{m=1}^{\infty} \frac{1}{m^2} = \frac{\pi^2}{6} \frac{x}{1 - x} = \frac{\pi^2}{6t},$$

where

$$t = \frac{1 - x}{x}.$$

Note that t varies from ∞ to 0 through positive values as x varies from 0 to 1.

Next we estimate the term $n \log(1/x)$. For $t > 0$ we have $\log(1 + t) < t$. But

$$1 + t = 1 + \frac{1 - x}{x} = \frac{1}{x}, \qquad \text{so } \log \frac{1}{x} < t.$$

Now

$$\text{(10)} \qquad \log p(n) < \log F(x) + n \log \frac{1}{x} < \frac{\pi^2}{6t} + nt.$$

The minimum of $(\pi^2/6t) + nt$ occurs when the two terms are equal, that is, when $\pi^2/(6t) = nt$, or $t = \pi/\sqrt{6n}$. For this value of t we have

$$\log p(n) < 2nt = 2n\pi/\sqrt{6n} = K\sqrt{n}$$

so $p(n) < e^{K\sqrt{n}}$, as asserted. □

Note. Dr. Neville Robbins points out that by refining a method of J. H. van Lint [48], with a little more effort we can obtain the improved inequality

$$\text{(11)} \qquad p(n) < \frac{\pi e^{K\sqrt{n}}}{\sqrt{6n}} \quad \text{for } n > 1.$$

Since $p(k) \geq p(n)$ if $k \geq n$, we have, for $n > 1$,

$$F(x) > \sum_{k=n}^{\infty} p(k)x^k \geq p(n) \sum_{k=n}^{\infty} x^k = \frac{p(n)x^n}{1-x}.$$

Taking logarithms we obtain, instead of (9), the inequality

$$\log p(n) < \log F(x) + n \log \frac{1}{x} + \log(1 - x).$$

Since $1 - x = tx$ we have $\log(1 - x) = \log t - \log(1/x)$, hence (10) can be replaced by

$$\text{(12)} \qquad \log p(n) < \frac{\pi^2}{6t} + (n - 1)t + \log t < f(t),$$

where $f(t) = (\pi^2/6t) + nt + \log t$. An easy calculation with derivatives shows that $f(t)$ has its absolute minimum at

$$t_0 = \frac{-1 + \sqrt{1 + 2n\pi^2/3}}{2n},$$

so $f(t_0) \leq f(t)$ for all $t > 0$. From (12) we get

$$\log p(n) < f(t_0) \leq f(\pi/\sqrt{6n}) = K\sqrt{n} + \log(\pi/\sqrt{6n})$$

which implies (11).

14.8 Jacobi's triple product identity

This section describes a famous identity of Jacobi from the theory of theta functions. Euler's pentagonal number theorem and many other partition identities occur as special cases of Jacobi's formula.

Theorem 14.6 Jacobi's triple product identity. *For complex x and z with $|x| < 1$ and $z \neq 0$ we have*

$$\prod_{n=1}^{\infty}(1 - x^{2n})(1 + x^{2n-1}z^2)(1 + x^{2n-1}z^{-2}) = \sum_{m=-\infty}^{\infty} x^{m^2}z^{2m}. \tag{13}$$

PROOF. The restriction $|x| < 1$ assures absolute convergence of each of the products $\prod(1 - x^{2n})$, $\prod(1 + x^{2n-1}z^2)$, $\prod(1 + x^{2n-1}z^{-2})$, and of the series in (13). Moreover, for each fixed x with $|x| < 1$ the series and products converge uniformly on compact subsets of the z-plane not containing $z = 0$ so each member of (13) is an analytic function of z for $z \neq 0$. For fixed $z \neq 0$ the series and products also converge uniformly for $|x| \leq r < 1$ hence represent analytic functions of x in the disk $|x| < 1$.

To prove (13) we keep x fixed and define $F(z)$ for $z \neq 0$ by the equation

$$F(z) = \prod_{n=1}^{\infty}(1 + x^{2n-1}z^2)(1 + x^{2n-1}z^{-2}). \tag{14}$$

First we show that F satisfies the functional equation

$$xz^2F(xz) = F(z). \tag{15}$$

From (14) we find

$$\begin{aligned} F(xz) &= \prod_{n=1}^{\infty}(1 + x^{2n+1}z^2)(1 + x^{2n-3}z^{-2}) \\ &= \prod_{m=2}^{\infty}(1 + x^{2m-1}z^2)\prod_{r=0}^{\infty}(1 + x^{2r-1}z^{-2}). \end{aligned}$$

Since $xz^2 = (1 + xz^2)/(1 + x^{-1}z^{-2})$, multiplication of the last equation by xz^2 gives (15).

Now let $G(z)$ denote the left member of (13) so that

$$G(z) = F(z)\prod_{n=1}^{\infty}(1 - x^{2n}). \tag{16}$$

Then $G(z)$ also satisfies the functional equation (15). Moreover, $G(z)$ is an even function of z which is analytic for all $z \neq 0$ so it has a Laurent expansion of the form

$$G(z) = \sum_{m=-\infty}^{\infty} a_m z^{2m} \tag{17}$$

where $a_{-m} = a_m$ since $G(z) = G(z^{-1})$. (The coefficients a_m depend on x.) Using the functional equation (15) in (17) we find that the coefficients satisfy the recursion formula

$$a_m = x^{2m-1}a_{m-1}$$

which, when iterated, gives

$$a_m = a_0 x^{m^2} \quad \text{for all } m \geq 0$$

since $1 + 3 + \cdots + (2m - 1) = m^2$. This also holds for $m < 0$. Hence (17) becomes

$$G_x(z) = a_0(x) \sum_{m=-\infty}^{\infty} x^{m^2} z^{2m}, \tag{18}$$

where we have written $G_x(z)$ for $G(z)$ and $a_0(x)$ for a_0 to indicate the dependence on x. Note that (18) implies $a_0(x) \to 1$ as $x \to 0$. To complete the proof we must show that $a_0(x) = 1$ for all x.

Taking $z = e^{\pi i/4}$ in (18) we find

$$\frac{G_x(e^{\pi i/4})}{a_0(x)} = \sum_{m=-\infty}^{\infty} x^{m^2} i^m = \sum_{n=-\infty}^{\infty} (-1)^n x^{(2n)^2} \tag{19}$$

since $i^m = -i^{-m}$ if m is odd. From (18) we see that the series on the right of (19) is $G_{x^4}(i)/a_0(x^4)$ so we have the identity

$$\frac{G_x(e^{\pi i/4})}{a_0(x)} = \frac{G_{x^4}(i)}{a_0(x^4)}. \tag{20}$$

We show next that $G_x(e^{\pi i/4}) = G_{x^4}(i)$. In fact, (14) and (16) gives us

$$G_x(e^{\pi i/4}) = \prod_{n=1}^{\infty} (1 - x^{2n})(1 + x^{4n-2}).$$

Since every even number is of the form $4n$ or $4n - 2$ we have

$$\prod_{n=1}^{\infty} (1 - x^{2n}) = \prod_{n=1}^{\infty} (1 - x^{4n})(1 - x^{4n-2})$$

so

$$G_x(e^{\pi i/4}) = \prod_{n=1}^{\infty} (1 - x^{4n})(1 - x^{4n-2})(1 + x^{4n-2}) = \prod_{n=1}^{\infty} (1 - x^{4n})(1 - x^{8n-4})$$

$$= \prod_{n=1}^{\infty} (1 - x^{8n})(1 - x^{8n-4})(1 - x^{8n-4}) = G_{x^4}(i).$$

Hence (20) implies $a_0(x) = a_0(x^4)$. Replacing x by $x^4, x^{4^2}, \ldots$, we find

$$a_0(x) = a_0(x^{4^k}) \quad \text{for } k = 1, 2, \ldots$$

But $x^{4^k} \to 0$ as $k \to \infty$ and $a_0(x) \to 1$ as $x \to 0$ so $a_0(x) = 1$ for all x. This completes the proof. □

14.9 Consequences of Jacobi's identity

If we replace x by x^a and z^2 by x^b in Jacobi's identity we find

$$\prod_{n=1}^{\infty}(1 - x^{2na})(1 + x^{2na-a+b})(1 + x^{2na-a-b}) = \sum_{m=-\infty}^{\infty} x^{am^2+bm}.$$

Similarly, if $z^2 = -x^b$ we find

$$\prod_{n=1}^{\infty}(1 - x^{2na})(1 - x^{2na-a+b})(1 - x^{2na-a-b}) = \sum_{m=-\infty}^{\infty} (-1)^m x^{am^2+bm}.$$

To obtain Euler's pentagonal number theorem simply take $a = 3/2$ and $b = 1/2$ in this last identity.

Jacobi's formula leads to another important formula for the cube of Euler's product.

Theorem 14.7 *If* $|x| < 1$ *we have*

$$(21)\qquad \prod_{n=1}^{\infty}(1 - x^n)^3 = \sum_{m=-\infty}^{\infty} (-1)^m m x^{(m^2+m)/2}$$

$$= \sum_{m=0}^{\infty}(-1)^m(2m + 1)x^{(m^2+m)/2}.$$

PROOF. Replacing z^2 by $-xz$ in Jacobi's identity we obtain

$$\prod_{n=1}^{\infty}(1 - x^{2n})(1 - x^{2n}z)(1 - x^{2n-2}z^{-1}) = \sum_{m=0}^{\infty}(-1)^m x^{m^2+m}(z^m - z^{-m-1}).$$

Now we rearrange terms on both sides, using the relations

$$\prod_{n=1}^{\infty}(1 - x^{2n-2}z^{-1}) = (1 - z^{-1})\prod_{n=1}^{\infty}(1 - x^{2n}z^{-1})$$

and

$$z^m - z^{-m-1} = (1 - z^{-1})(1 + z^{-1} + z^{-2} + \cdots + z^{-2m})z^m.$$

Canceling a factor $1 - z^{-1}$ we obtain

$$\prod_{n=1}^{\infty}(1 - x^{2n})(1 - x^{2n}z)(1 - x^{2n}z^{-1})$$

$$= \sum_{m=0}^{\infty}(-1)^m x^{m^2+m} z^m(1 + z^{-1} + z^{-2} + \cdots + z^{-2m}).$$

Taking $z = 1$ and replacing x by $x^{1/2}$ we obtain (21). □

14.10 Logarithmic differentiation of generating functions

Theorem 14.4 gives a recursion formula for $p(n)$. There are other types of recursion formulas for arithmetical functions that can be derived by logarithmic differentiation of generating functions. We describe the method in the following setting.

Let A be a given set of positive integers, and let $f(n)$ be a given arithmetical function. Assume that the product

$$F_A(x) = \prod_{n \in A} (1 - x^n)^{-f(n)/n}$$

and the series

$$G_A(x) = \sum_{n \in A} \frac{f(n)}{n} x^n$$

converge absolutely for $|x| < 1$ and represent analytic functions in the unit disk $|x| < 1$. The logarithm of the product is given by

$$\log F_A(x) = -\sum_{n \in A} \frac{f(n)}{n} \log(1 - x^n) = \sum_{n \in A} \frac{f(n)}{n} \sum_{m=1}^{\infty} \frac{x^{mn}}{m} = \sum_{m=1}^{\infty} \frac{1}{m} G_A(x^m).$$

Differentiating and multiplying by x we obtain

$$x \frac{F_A'(x)}{F_A(x)} = \sum_{m=1}^{\infty} G_A'(x^m)x^m = \sum_{m=1}^{\infty} \sum_{n \in A} f(n)x^{mn} = \sum_{m=1}^{\infty} \sum_{n=1}^{\infty} \chi_A(n) f(n) x^{mn},$$

where χ_A is the characteristic function of the set A,

$$\chi_A(n) = \begin{cases} 1 & \text{if } n \in A, \\ 0 & \text{if } n \notin A. \end{cases}$$

Collecting the terms with $mn = k$ we find

$$\sum_{m=1}^{\infty} \sum_{n=1}^{\infty} \chi_A(n) f(n) x^{mn} = \sum_{k=1}^{\infty} f_A(k) x^k,$$

where

$$f_A(k) = \sum_{d|k} \chi_A(d) f(d) = \sum_{\substack{d|k \\ d \in A}} f(d).$$

Therefore we have the following identity,

$$xF_A'(x) = F_A(x) \sum_{k=1}^{\infty} f_A(k) x^k. \tag{22}$$

Now write the product $F_A(x)$ as a power series,

$$F_A(x) = \sum_{n=0}^{\infty} p_{A,f}(n)x^n, \quad \text{where } p_{A,f}(0) = 1,$$

and equate coefficients of x^n in (22) to obtain the recursion formula (24) in the following theorem.

Theorem 14.8 *For a given set A and a given arithmetical function f, the numbers $p_{A,f}(n)$ defined by the equation*

$$\prod_{n \in A}(1 - x^n)^{-f(n)/n} = 1 + \sum_{n=1}^{\infty} p_{A,f}(n)x^n \tag{23}$$

satisfy the recursion formula

$$np_{A,f}(n) = \sum_{k=1}^{n} f_A(k)p_{A,f}(n-k), \tag{24}$$

where $p_{A,f}(0) = 1$ and

$$f_A(k) = \sum_{\substack{d|k \\ d \in A}} f(d).$$

EXAMPLE 1 Let A be the set of all positive integers. If $f(n) = n$, then $p_{A,f}(n) = p(n)$, the unrestricted partition function, and $f_A(k) = \sigma(k)$, the sum of the divisors of k. Equation (24) becomes

$$np(n) = \sum_{k=1}^{n} \sigma(k)p(n-k),$$

a remarkable relation connecting a function of multiplicative number theory with one of additive number theory.

EXAMPLE 2 Take A as in Example 1, but let $f(n) = -n$. Then the coefficients in (23) are determined by Euler's pentagonal-number theorem and the recursion formula (24) becomes

$$np_{A,f}(n) = -\sum_{k=1}^{n} \sigma(k)p_{A,f}(n-k) = -\sigma(n) - \sum_{k=1}^{n-1} p_{A,f}(k)\sigma(n-k), \tag{25}$$

where

$$p_{A,f}(n) = \begin{cases} (-1)^m & \text{if } n \text{ is a pentagonal number } \omega(m) \text{ or } \omega(-m) \\ 0 & \text{if } n \text{ is not a pentagonal number.} \end{cases}$$

Equation (25) can also be written as follows:

$$\sigma(n) - \sigma(n-1) - \sigma(n-2) + \sigma(n-5) + \sigma(n-7) - \cdots$$

$$= \begin{cases} (-1)^{m-1}\omega(m) & \text{if } n = \omega(m), \\ (-1)^{m-1}\omega(-m) & \text{if } n = \omega(-m), \\ 0 & \text{otherwise.} \end{cases}$$

The sum on the left terminates when the term $\sigma(k)$ has $k \leq 1$. To illustrate, when $n = 6$ and $n = 7$ this gives the relations

$$\sigma(6) = \sigma(5) + \sigma(4) - \sigma(1),$$
$$\sigma(7) = \sigma(6) + \sigma(5) - \sigma(2) - 7.$$

14.11 The partition identities of Ramanujan

By examining MacMahon's table of the partition function, Ramanujan was led to the discovery of some striking divisibility properties of $p(n)$. For example, he proved that

(26) $$p(5m + 4) \equiv 0 \pmod{5},$$

(27) $$p(7m + 5) \equiv 0 \pmod{7},$$

(28) $$p(11m + 6) \equiv 0 \pmod{11}.$$

In connection with these discoveries he also stated without proof two remarkable identities,

(29) $$\sum_{m=0}^{\infty} p(5m + 4)x^m = 5\frac{\varphi(x^5)^5}{\varphi(x)^6},$$

and

(30) $$\sum_{m=0}^{\infty} p(7m + 5)x^m = 7\frac{\varphi(x^7)^3}{\varphi(x)^4} + 49x\frac{\varphi(x^7)^7}{\varphi(x)^8},$$

where

$$\varphi(x) = \prod_{n=1}^{\infty}(1 - x^n).$$

Since the functions on the right of (29) and (30) have power series expansions with integer coefficients, Ramanujan's identities immediately imply the congruences (26) and (27).

Proofs of (29) and (30), based on the theory of modular functions, were found by Darling, Mordell, Rademacher, Zuckerman, and others. Further proofs, independent of the theory of modular functions, were given by Kruyswijk [36] and later by Kolberg. Kolberg's method gives not only the Ramanujan identities but many new ones. Kruyswijk's proof of (29) is outlined in Exercises 11–15.

Exercises for Chapter 14

1. Let A denote a nonempty set of positive integers.

   (a) Prove that the product

$$\prod_{m \in A} (1 - x^m)^{-1}$$

   is the generating function of the number of partitions of n into parts belonging to the set A.

   (b) Describe the partition function generated by the product

$$\prod_{m \in A} (1 + x^m).$$

   In particular, describe the partition function generated by the finite product $\prod_{m=1}^{k} (1 + x^m)$.

2. If $|x| < 1$ prove that

$$\prod_{m=1}^{\infty} (1 + x^m) = \prod_{m=1}^{\infty} (1 - x^{2m-1})^{-1},$$

   and deduce that the number of partitions of n into unequal parts is equal to the number of partitions of n into odd parts.

3. For complex x and z with $|x| < 1$, let

$$f(x, z) = \prod_{m=1}^{\infty} (1 - x^m z).$$

   (a) Prove that for each fixed z the product is an analytic function of x in the disk $|x| < 1$, and that for each fixed x with $|x| < 1$ the product is an entire function of z.

   (b) Define the numbers $a_n(x)$ by the equation

$$f(x, z) = \sum_{n=0}^{\infty} a_n(x) z^n.$$

   Show that $f(x, z) = (1 - xz)f(x, zx)$ and use this to prove that the coefficients satisfy the recursion formula

$$a_n(x) = a_n(x)x^n - a_{n-1}(x)x^n.$$

   (c) From part (b) deduce that $a_n(x) = (-1)^n x^{n(n+1)/2}/P_n(x)$, where

$$P_n(x) = \prod_{r=1}^{n} (1 - x^r).$$

   This proves the following identity for $|x| < 1$ and arbitrary z:

$$\prod_{m=1}^{\infty} (1 - x^m z) = \sum_{n=0}^{\infty} \frac{(-1)^n}{P_n(x)} x^{n(n+1)/2} z^n.$$

4. Use a method analogous to that of Exercise 3 to prove that if $|x| < 1$ and $|z| < 1$ we have

$$\prod_{m=1}^{\infty} (1 - x^m z)^{-1} = \sum_{n=0}^{\infty} \frac{z^n}{P_n(x)}$$

where $P_n(x) = \prod_{r=1}^{n} (1 - x^r)$.

5. If $x \neq 1$ let $Q_0(x) = 1$ and for $n \geq 1$ define

$$Q_n(x) = \prod_{r=1}^{n} \frac{1 - x^{2r}}{1 - x^{2r-1}}.$$

(a) Derive the following finite identities of Shanks:

$$\sum_{m=1}^{2n} x^{m(m-1)/2} = \sum_{s=0}^{n-1} \frac{Q_n(x)}{Q_s(x)} x^{s(2n+1)},$$

$$\sum_{m=1}^{2n+1} x^{m(m-1)/2} = \sum_{s=0}^{n} \frac{Q_n(x)}{Q_s(x)} x^{s(2n+1)}.$$

(b) Use Shanks' identities to deduce Gauss' triangular-number theorem:

$$\sum_{m=1}^{\infty} x^{m(m-1)/2} = \prod_{n=1}^{\infty} \frac{1 - x^{2n}}{1 - x^{2n-1}} \quad \text{for } |x| < 1.$$

6. The following identity is valid for $|x| < 1$:

$$\sum_{m=-\infty}^{\infty} x^{m(m+1)/2} = \prod_{n=1}^{\infty} (1 + x^{n-1})(1 - x^{2n}).$$

(a) Derive this from the identities in Exercises 2 and 5(b).
(b) Derive this from Jacobi's triple product identity.

7. Prove that the following identities, valid for $|x| < 1$, are consequences of Jacobi's triple product identity:

(a) $$\prod_{n=1}^{\infty} (1 - x^{5n})(1 - x^{5n-1})(1 - x^{5n-4}) = \sum_{m=-\infty}^{\infty} (-1)^m x^{m(5m+3)/2}.$$

(b) $$\prod_{n=1}^{\infty} (1 - x^{5n})(1 - x^{5n-2})(1 - x^{5n-3}) = \sum_{m=-\infty}^{\infty} (-1)^m x^{m(5m+1)/2}.$$

8. Prove that the recursion formula

$$np(n) = \sum_{k=1}^{n} \sigma(k) p(n-k),$$

obtained in Section 14.10, can be put in the form

$$np(n) = \sum_{m=1}^{n} \sum_{k \leq n/m} mp(n - km).$$

9. Suppose that each positive integer k is written in $g(k)$ different colors, where $g(k)$ is a positive integer. Let $p_g(n)$ denote the number of partitions of n in which each part k appears in at most $g(k)$ different colors. When $g(k) = 1$ for all k this is the unrestricted partition function $p(n)$. Find an infinite product which generates $p_g(n)$ and prove that there is an arithmetical function f (depending on g) such that

$$np_g(n) = \sum_{k=1}^{n} f(k)p_g(n-k).$$

10. Refer to Section 14.10 for notation. By solving the first-order differential equation in (22) prove that if $|x| < 1$ we have

$$\prod_{n \in A} (1 - x^n)^{-f(n)/n} = \exp\left\{\int_0^x \frac{H(t)}{t}\, dt\right\},$$

where

$$H(x) = \sum_{k=1}^{\infty} f_A(k)x^k \quad \text{and} \quad f_A(k) = \sum_{\substack{d \mid k \\ d \in A}} f(d).$$

Deduce that

$$\prod_{n=1}^{\infty} (1 - x^n)^{\mu(n)/n} = e^{-x} \quad \text{for } |x| < 1,$$

where $\mu(n)$ is the Möbius function.

The following exercises outline a proof of Ramanujan's partition identity

$$\sum_{m=0}^{\infty} p(5m + 4)x^m = 5\,\frac{\varphi(x^5)^5}{\varphi(x)^6}, \quad \text{where } \varphi(x) = \prod_{n=1}^{\infty} (1 - x^n),$$

by a method of Kruyswijk not requiring the theory of modular functions.

11. (a) Let $\varepsilon = e^{2\pi i/k}$ where $k \geq 1$ and show that for all x we have

$$\prod_{h=1}^{k} (1 - x\varepsilon^h) = 1 - x^k.$$

(b) More generally, if $(n, k) = d$ prove that

$$\prod_{h=1}^{k} (1 - x\varepsilon^{nh}) = (1 - x^{k/d})^d,$$

and deduce that

$$\prod_{h=1}^{k} (1 - x^n e^{2\pi i n h/k}) = \begin{cases} 1 - x^{nk} & \text{if } (n, k) = 1, \\ (1 - x^n)^k & \text{if } k \mid n. \end{cases}$$

12. (a) Use Exercise 11(b) to prove that for prime q and $|x| < 1$ we have

$$\prod_{n=1}^{\infty} \prod_{h=1}^{q} (1 - x^n e^{2\pi i n h/q}) = \frac{\varphi(x^q)^{q+1}}{\varphi(x^{q^2})}.$$

(b) Deduce the identity

$$\sum_{m=0}^{\infty} p(m)x^m = \frac{\varphi(x^{25})}{\varphi(x^5)^6} \prod_{h=1}^{4} \prod_{n=1}^{\infty} (1 - x^n e^{2\pi i n h/5}).$$

13. If q is prime and if $0 \leq r < q$, a power series of the form

$$\sum_{n=0}^{\infty} a(n)x^{qn+r}$$

is said to be of *type r* mod q.

(a) Use Euler's pentagonal number theorem to show that $\varphi(x)$ is a sum of three power series,

$$\varphi(x) = \prod_{n=1}^{\infty}(1 - x^n) = I_0 + I_1 + I_2,$$

where I_k denotes a power series of type k mod 5.

(b) Let $\alpha = e^{2\pi i/5}$ and show that

$$\prod_{h=1}^{4}\prod_{n=1}^{\infty}(1 - x^n\alpha^{nh}) = \prod_{h=1}^{4}(I_0 + I_1\alpha^h + I_2\alpha^{2h}).$$

(c) Use Exercise 12(b) to show that

$$\sum_{m=0}^{\infty} p(5m + 4)x^{5m+4} = V_4\frac{\varphi(x^{25})}{\varphi(x^5)^6},$$

where V_4 is the power series of type 4 mod 5 obtained from the product in part (b).

14. (a) Use Theorem 14.7 to show that the cube of Euler's product is the sum of three power series,

$$\varphi(x)^3 = W_0 + W_1 + W_3,$$

where W_k denotes a power series of type k mod 5.

(b) Use the identity $W_0 + W_1 + W_3 = (I_0 + I_1 + I_2)^3$ to show that the power series in Exercise 13(a) satisfy the relation

$$I_0 I_2 = -I_1{}^2.$$

(c) Prove that $I_1 = -x\varphi(x^{25})$.

15. Observe that the product $\prod_{h=1}^{4}(I_0 + I_1\alpha^h + I_2\alpha^{2h})$ is a homogeneous polynomial in I_0, I_1, I_2 of degree 4, so the terms contributing to series of type 4 mod 5 come from the terms $I_1{}^4$, $I_0 I_1{}^2 I_2$ and $I_0{}^2\, I_2{}^2$.

(a) Use Exercise 14(c) to show that there exists a constant c such that

$$V_4 = cI_1{}^4,$$

where V_4 is the power series in Exercise 13(c), and deduce that

$$\sum_{m=0}^{\infty} p(5m + 4)x^{5m+4} = cx^4\frac{\varphi(x^{25})^5}{\varphi(x^5)^6}.$$

(b) Prove that $c = 5$ and deduce Ramanujan's identity

$$\sum_{m=0}^{\infty} p(5m + 4)x^m = 5\frac{\varphi(x^5)^5}{\varphi(x)^6}.$$

Bibliography

MR denotes reference to *Mathematical Reviews.*

1. Apostol, Tom M. (1970) Euler's φ-function and separable Gauss sums. *Proc. Amer. Math. Soc.*, *24*: 482–485; *MR* 41, #1661.
2. Apostol, Tom M. (1974) *Mathematical Analysis*, 2nd ed. Reading, Mass.: Addison-Wesley Publishing Co.
3. Ayoub, Raymond G. (1963) *An Introduction to the Analytic Theory of Numbers.* Mathematical Surveys, No. 10. Providence, R. I.: American Mathematical Society.
4. Bell, E. T. (1915) An arithmetical theory of certain numerical functions. *University of Washington Publ. in Math. and Phys. Sci., No. 1, Vol. 1*: 1–44.
5. Borozdkin, K. G. (1956) K voprosu o postoyanni I. M. Vinogradova. *Trudy tretego vsesoľuznogo matematičeskogo siezda, Vol. I,* Moskva [Russian].
6. Buhštab, A. A. (1965) New results in the investigation of the Goldbach–Euler problem and the problem of prime pairs. [Russian]. *Dokl. Akad. Nauk SSSR*, *162*: 735–738; *MR* 31, #2226. [English translation: (1965) *Soviet Math. Dokl. 6*: 729–732.]
7. Chandrasekharan, Komaravolu (1968) *Introduction to Analytic Number Theory.* Die Grundlehren der Mathematischen Wissenschaften, Band 148. New York: Springer-Verlag.
8. Chandrasekharan, Komaravolu (1970) *Arithmetical Functions.* Die Grundlehren der Mathematischen Wissenschaften, Band 167. New York. Springer-Verlag.
9. Chebyshev, P. L. Sur la fonction qui détermine la totalité des nombres premiers inferieurs à une limite donée. (a) (1851) *Mem. Ac. Sc. St. Pétersbourg*, *6*: 141–157. (b) (1852) *Jour. de Math* (1) *17*: 341–365. [*Oeuvres*, *1*: 27–48.]
10. Chen, Jing-run (1966) On the representation of a large even integer as the sum of a prime and the product of at most two primes. *Kexue Tongbao* (Foreign Lang. Ed.), *17*: 385–386; *MR* 34, #7483.

11. Clarkson, James A. (1966) On the series of prime reciprocals. *Proc. Amer. Math. Soc.*, *17*: 541; *MR* 32, #5573.

12. Davenport, Harold (1967) *Multiplicative Number Theory*. Lectures in Advanced Mathematics, No. 1. Chicago: Markham Publishing Co.

13. Dickson, Leonard Eugene (1919) *History of the Theory of Numbers*. (3 volumes). Washington, D. C.: Carnegie Institution of Washington. Reprinted by Chelsea Publishing Co., New York, 1966.

14. Dickson, Leonard Eugene (1930) *Studies in the Theory of Numbers*. Chicago: The University of Chicago Press.

15. Dirichlet, P. G. Lejeune (1837) Beweis des Satzes, dass jede unbegrenzte arithmetische Progression, deren erstes Glied und Differenz ganze Zahlen ohne gemeinschlaftichen Factor sind, unendliche viele Primzahlen enthält. *Abhand. Ak. Wiss. Berlin*: 45-81. [*Werke*, *1*: 315-342.]

16. Dirichlet, P. G. Lejeune (1840) Ueber eine Eigenschaft der quadratischen Formen. *Bericht Ak. Wiss. Berlin*: 49-52. [*Werke*, *1*: 497-502.]

17. Edwards, H. M. (1974) *Riemann's Zeta Function*. New York and London: Academic Press.

18. Ellison, W. J. (1971) Waring's problem. *Amer. Math. Monthly*, *78*: 10–36.

19. Erdös, Paul (1949) On a new method in elementary number theory which leads to an elementary proof of the prime number theorem. *Proc. Nat. Acad. Sci. U.S.A.*, *35*: 374-384: *MR* 10, 595.

20. Euler, Leonhard (1737) Variae observationes circa series infinitas. *Commentarii Academiae Scientiarum Imperialis Petropolitanae*, *9*: 160–188. [*Opera Omnia* (1), *14*; 216–244.]

21. Euler, Leonhard (1748) *Introductio in Analysin Infinitorum*, Vol. 1. Lausanne: Bousquet. [*Opera Omnia* (1), *8*.]

22. Franklin, F. (1881) Sur le développement du produit infini $(1 - x)(1 - x^2)(1 - x^3)(1 - x^4) \cdots$. *Comptes Rendus Acad. Sci. (Paris)*, *92*: 448–450.

23. Gauss, C. F. (1801) *Disquisitiones Arithmeticae*. Lipsiae. [English translation: Arthur A. Clarke (1966) New Haven: Yale University Press.

24. Gauss, C. F. (1849) Letter to Encke, dated 24 December. [*Werke*, Vol. II, 444–447.]

25. Gerstenhaber, Murray (1963) The 152nd proof of the law of quadratic reciprocity. *Amer. Math. Monthly*, *70*: 397–398; *MR* 27, #100.

26. Goldbach, C. (1742) Letter to Euler, dated 7 June.

27. Grosswald, Emil (1966) *Topics from the Theory of Numbers*. New York: The Macmillan Co.

28. Hadamard, J. (1896) Sur la distribution des zéros de la fonction $\zeta(s)$ et ses conséquences arithmétiques. *Bull. Soc. Math. France*, *24*: 199–220.

29. Hagis, Peter, Jr. (1973) A lower bound for the set of odd perfect numbers. *Math. Comp.*, *27*: 951–953; *MR* 48, #3854.

30. Hardy, G. H. (1940) *Ramanujan. Twelve Lectures on Subjects Suggested by His Life and Work*. Cambridge: The University Press.

31. Hardy, G. H. and Wright, E. M. (1960) *An Introduction to the Theory of Numbers*, 4th ed. Oxford: Clarendon Press.

32. Hemer, Ove (1954) Notes on the Diophantine equation $y^2 - k = x^3$. *Ark. Mat.*, 3: 67–77; *MR* 15, 776.

33. Ingham, A. E. (1932) *The Distribution of Prime Numbers*. Cambridge Tracts in Mathematics and Mathematical Physics, No. 30. Cambridge: The University Press.

34. Jacobi, C. G. J. (1829) *Fundamenta Nova Theoriae Functionum Ellipticarum.* [*Gesammelte Werke*, Band I, 49–239.]

35. Kolesnik, G. A. (1969) An improvement of the remainder term in the divisor problem. (Russian). *Mat. Zametki*, *6*: 545–554; *MR* 41, #1659. [English translation: (1969), *Math. Notes*, 6: 784–791.]

36. Kruyswijk, D. (1950) On some well-known properties of the partition function $p(n)$ and Euler's infinite product. *Nieuw Arch. Wisk.*, (2) *23*: 97–107; *MR* 11, 715.

37. Landau, E. (1909) *Handbuch der Lehre von der Verteilung der Primzahlen.* Leipzig: Teubner. Reprinted by Chelsea, 1953.

38. Landau, E. (1927) *Vorlesungen über Zahlentheorie* (3 volumes). Leipzig: Hirzel. Reprinted by Chelsea, 1947.

39. Leech, J. (1957) Note on the distribution of prime numbers. *J. London Math. Soc.*, *32*: 56–58; *MR* 18, 642.

40. Legendre, A. M. (1798) *Essai sur la Theorie des Nombres.* Paris: Duprat.

41. Lehmer, D. H. (1959) On the exact number of primes less than a given limit. *Illinois J. Math.*, *3*: 381–388; *MR* 21, #5613.

42. Lehmer, D. H. (1936) On a conjecture of Ramanujan. *J. London Math. Soc.*, *11*: 114–118.

43. Lehmer, D. N. (1914) List of prime numbers from 1 to 10, 006, 721. Washington, D.C.: Carnegie Institution of Washington, Publ. No. 165.

44. LeVeque, W. J. (1956) *Topics in Number Theory* (2 volumes). Reading, Mass.: Addison–Wesley Publishing Co.

45. LeVeque, W. J. (1974) *Reviews in Number Theory* (6 volumes). Providence, RI: American Mathematical Society.

46. Levinson, N. (1969) A motivated account of an elementary proof of the prime number theorem. *Amer. Math. Monthly*, *76*: 225–245; *MR* 39, #2712.

47. Levinson, Norman (1974) More than one third of zeros of Riemann's zeta-function are on $\sigma = 1/2$. *Advances Math.*, *13*: 383–436.

48. van Lint, Jacobus Hendricus (1974) *Combinatorial Theory Seminar* (Eindhoven University of Technology), Lecture Notes in Mathematics 382. Springer–Verlag, Chapter 4.

49. Littlewood, J. E. (1914) Sur la distribution des nombres premiers. *Comptes Rendus Acad. Sci.* (*Paris*), *158*: 1869–1872.

50. Mills, W. H. (1947) A prime-representing function. *Bull. Amer. Math. Soc.*, *53*: 604; *MR* 8, 567.

51. Nevanlinna, V. (1962) Über den elementaren Beweis des Primzahlsatzes. *Soc. Sci. Fenn. Comment. Phys.-Math,. 27 No. 3*, 8 pp.; *MR* 26, #2416.

52. Niven, I. and Zuckerman, H. S. (1972) *An Introduction to the Theory of Numbers*, 3rd ed. New York: John Wiley and Sons, Inc.

53. Prachar, Karl (1957) *Primzahlverteilung.* Die Grundlehren der Mathematischen Wissenschaften, Band 91. Berlin–Göttingen–Heidelberg: Springer–Verlag.

54. Rademacher, Hans (1937) On the partition function $p(n)$. *Proc. London Math. Soc.*, *43*: 241–254.

55. Rademacher, Hans (1964) *Lectures on Elementary Number Theory.* New York: Blaisdell Publishing Co.

56. Rademacher, Hans (1973) *Topics in Analytic Number Theory.* Die Grundlehren der Mathematischen Wissenschaften, Band 169. New York–Heidelberg–Berlin: Springer–Verlag.

57. Rényi, A. (1948) On the representation of an even number as the sum of a single prime and a single almost-prime number. (Russian). *Izv. Akad. Nauk SSSR Ser. Mat.*, *12*: 57–78; *MR* 9, 413. [English translation: (1962) *Amer. Math. Soc. Transl.* *19* (2): 299–321.]

58. Riemann, B. (1859) Über die Anzahl der Primzahlen unter einer gegebener Grösse. *Monatsber. Akad. Berlin*, 671–680.

59. Robinson, R. M. (1958) A report on primes of the form $k \cdot 2^n + 1$ and on factors of Fermat numbers. *Proc. Amer. Math. Soc.*, *9*: 673–681; *MR* 20, #3097.

60. Rosser, J. Barkley, and Schoenfeld, Lowell (1962) Approximate formulas for some functions of prime number theory. *Illinois J. Math.*, *6*: 69–94; *MR* 25, #1139.

61. Schnirelmann, L. (1930) On additive properties of numbers. (Russian). *Izv. Donskowo Politechn. Inst. (Nowotscherkask)*, 14 (2–3): 3–28.

62. Selberg, Atle (1949) An elementary proof of the prime number theorem. *Ann. of Math.*, *50*: 305–313; *MR* 10, 595.

63. Shanks, Daniel (1951) A short proof of an identity of Euler. *Proc. Amer. Math. Soc.*, *2*: 747–749; *MR* 13, 321.

64. Shapiro, Harold N. (1950) On the number of primes less than or equal x. *Proc. Amer. Math. Soc.*, *1*: 346–348; *MR* 12, 80.

65. Shapiro, Harold N. (1952) On primes in arithmetic progression II. *Ann. of Math.* *52*: 231–243: *MR* 12, 81.

66. Shen, Mok-Kong (1964) On checking the Goldbach conjecture. *Nordisk Tidskr. Informations-Behandling*, *4*: 243–245; *MR* 30, #3051.

67. Sierpiński, Waclaw (1964) *Elementary Theory of Numbers*. Translated from Polish by A. Hulanicki. Monografie Matematyczne, Tom 42. Warsaw: Państwowe Wzdawnictwo Naukowe.

68. Tatuzawa, Tikao, and Iseki Kaneshiro (1951) On Selberg's elementary proof of the prime number theorem. *Proc. Japan Acad.*, *27*: 340–342; *MR* 13, 725.

69. Titchmarsh, E. C. (1951) *The Theory of the Riemann Zeta Function*. Oxford: Clarendon Press.

70. Uspensky, J. V., and Heaslett, M. A. (1939) *Elementary Number Theory*. New York: McGraw-Hill Book Co.

71. Vallée Poussin, Ch. de la (1896) Recherches analytiques sur la théorie des nombres premiers. *Ann. Soc. Sci. Bruxelles*, 20_2: 183–256, 281–297.

72. Vinogradov, A. I. (1965) The density hypothesis for Dirichlet L-series. (Russian). *Izv. Akad. Nauk SSSR, Ser. Math.* *29*: 903–934; *MR* 33, #5579. [Correction: (1966) *ibid.*, *30*: 719–720; *MR* 33, #2607.]

73. Vinogradov, I. M. (1937) The representation of an odd number as the sum of three primes. (Russian.) *Dokl. Akad. Nauk SSSR*, 16: 139–142.

74. Vinogradov, I. M. (1954) *Elements of Number Theory*. Translated by S. Kravetz. New York: Dover Publications.

75. Walfisz, A. (1963) *Weylsche Exponentialsummen in der neueren Zahlentheorie*. Mathematische Forschungsberichte, XV, V E B Deutscher Verlag der Wissenschaften, Berlin.

76. Williams, H. C., and Zarnke, C. R. (1972) Some prime numbers of the form $2A3^n + 1$ and $2A3^n - 1$. *Math. Comp.* *26*: 995–998; *MR* 47, #3299.

77. Wrathall, Claude P. (1964) New factors of Fermat numbers. *Math. Comp.*, *18*: 324–325; *MR* 29, #1167.

78. Yin, Wen-lin (1956) Note on the representation of large integers as sums of primes. *Bull. Acad. Polon. Sci. Cl. III*, *4*: 793–795; *MR* *19*, 16.

Index of Special Symbols

Index

(continued from page ii)

Frazier: An Introduction to Wavelets Through Linear Algebra
Gamelin: Complex Analysis.
Gordon: Discrete Probability.
Hairer/Wanner: Analysis by Its History. *Readings in Mathematics.*
Halmos: Finite-Dimensional Vector Spaces. Second edition.
Halmos: Naive Set Theory.
Hämmerlin/Hoffmann: Numerical Mathematics. *Readings in Mathematics.*
Harris/Hirst/Mossinghoff: Combinatorics and Graph Theory.
Hartshorne: Geometry: Euclid and Beyond.
Hijab: Introduction to Calculus and Classical Analysis.
Hilton/Holton/Pedersen: Mathematical Reflections: In a Room with Many Mirrors.
Hilton/Holton/Pedersen: Mathematical Vistas: From a Room with Many Windows.
Iooss/Joseph: Elementary Stability and Bifurcation Theory. Second edition.
Irving: Integers, Polynomials, and Rings: A Course in Algebra
Isaac: The Pleasures of Probability. *Readings in Mathematics.*
James: Topological and Uniform Spaces.
Jänich: Linear Algebra.
Jänich: Topology.
Jänich: Vector Analysis.
Kemeny/Snell: Finite Markov Chains.
Kinsey: Topology of Surfaces.
Klambauer: Aspects of Calculus.
Lang: A First Course in Calculus. Fifth edition.
Lang: Calculus of Several Variables. Third edition.
Lang: Introduction to Linear Algebra. Second edition.
Lang: Linear Algebra. Third edition.
Lang: Short Calculus: The Original Edition of "A First Course in Calculus."
Lang: Undergraduate Algebra. Second edition.
Lang: Undergraduate Algebra. Third edition
Lang: Undergraduate Analysis.
Laubenbacher/Pengelley: Mathematical Expeditions.
Lax/Burstein/Lax: Calculus with Applications and Computing. Volume 1.
LeCuyer: College Mathematics with APL.
Lidl/Pilz: Applied Abstract Algebra. Second edition.
Logan: Applied Partial Differential Equations, Second edition.
Lovász/Pelikán/Vesztergombi: Discrete Mathematics.
Macki-Strauss: Introduction to Optimal Control Theory.
Malitz: Introduction to Mathematical Logic.
Marsden/Weinstein: Calculus I, II, III. Second edition.
Martin: Counting: The Art of Enumerative Combinatorics.
Martin: The Foundations of Geometry and the Non-Euclidean Plane.
Martin: Geometric Constructions.
Martin: Transformation Geometry: An Introduction to Symmetry.
Millman/Parker: Geometry: A Metric Approach with Models. Second edition.
Moschovakis: Notes on Set Theory.
Owen: A First Course in the Mathematical Foundations of Thermodynamics.
Palka: An Introduction to Complex Function Theory.
Pedrick: A First Course in Analysis.
Peressini/Sullivan/Uhl: The Mathematics of Nonlinear Programming.

Prenowitz/Jantosciak: Join Geometries.
Priestley: Calculus: A Liberal Art. Second edition.
Protter/Morrey: A First Course in Real Analysis. Second edition.
Protter/Morrey: Intermediate Calculus. Second edition.
Pugh: Real Mathematical Analysis.
Roman: An Introduction to Coding and Information Theory.
Roman: Introduction to the Mathematics of Finance: From Risk Management to Options Pricing.
Ross: Differential Equations: An Introduction with Mathematica®. Second edition.
Ross: Elementary Analysis: The Theory of Calculus.
Samuel: Projective Geometry. *Readings in Mathematics.*
Saxe: Beginning Functional Analysis
Scharlau/Opolka: From Fermat to Minkowski.
Schiff: The Laplace Transform: Theory and Applications.
Sethuraman: Rings, Fields, and Vector Spaces: An Approach to Geometric Constructability.
Sigler: Algebra.
Silverman/Tate: Rational Points on Elliptic Curves.
Simmonds: A Brief on Tensor Analysis. Second edition.
Singer: Geometry: Plane and Fancy.
Singer/Thorpe: Lecture Notes on Elementary Topology and Geometry.
Smith: Linear Algebra. Third edition.
Smith: Primer of Modern Analysis. Second edition.
Stanton/White: Constructive Combinatorics.
Stillwell: Elements of Algebra: Geometry, Numbers, Equations.
Stillwell: Elements of Number Theory.
Stillwell: Mathematics and Its History. Second edition.
Stillwell: Numbers and Geometry. *Readings in Mathematics.*
Strayer: Linear Programming and Its Applications.
Toth: Glimpses of Algebra and Geometry. Second Edition. *Readings in Mathematics.*
Troutman: Variational Calculus and Optimal Control. Second edition.
Valenza: Linear Algebra: An Introduction to Abstract Mathematics.
Whyburn/Duda: Dynamic Topology.
Wilson: Much Ado About Calculus.

Printed in the United States
By Bookmasters